T0188695

Tectonically Active Landscapes

Tectonically Active Landscapes

William B. Bull

A John Wiley & Sons, Ltd., Publication

Blackwell Publishing was acquired by John Wiley & Sons in February 2007. Blackwell's publishing program has been merged with Wiley's global Scientific, Technical and Medical business to form Wiley-Blackwell.

Registered office: John Wiley & Sons Ltd, The Atrium, Southern Gate, Chichester, West Sussex, PO19 8SQ, UK

Editorial offices: 9600 Garsington Road, Oxford, OX4 2DQ, UK
                        The Atrium, Southern Gate, Chichester, West Sussex, PO19 8SQ, UK
                        111 River Street, Hoboken, NJ 07030-5774, USA

For details of our global editorial offices, for customer services and for information about how to apply for permission to reuse the copyright material in this book please see our website at www.wiley.com/wiley-blackwell

Library of Congress Cataloguing-in-Publication Data
Bull, William B., 1930-
  Tectonically active landscapes / William B. Bull.
      p. cm.
  Includes bibliographical references and index.
  ISBN 978-1-4051-9012-1 (hardcover : alk. paper)
  1. Morphotectonics. 2. Landscape changes. I. Title.

  QE511.44.B853 2009
  551.41--dc22
                              2008052159
ISBN: 978-1-4051-9012-1

A catalogue record for this book is available from the British Library.
Set in 11.8/12pt AGaramond
by William B. Bull
Printed and bound in Malaysia by Vivar Printing Sdn Bhd
1[Mspace]2009

Contents

4 Sediment Yield and Landslides

5 A Debate About Steady State

Preface

This book is about tectonic controls on landscape shapes and processes, with streams as the essential connecting link between upstream and downstream reaches of fluvial systems. Uplift on a range-bounding fault increases relief, changes rates of geomorphic processes, and modifies the shapes of hills and streams. Streams respond to rock uplift by deepening their valley floors, thus changing the base levels for adjacent hillslopes. We seek to learn more about landscape responses to rock uplift as a function of:

1) Rate, style, location of tectonic deformation,
2) Rock mass strength,
3) Climatic controls on erosion and deposition.

Intriguing diverse challenges include:

1) To determine of how quickly nearby and distant parts of hilly landscapes change in consecutive reaches upstream from an active structure.

2) Recognition of how cumulative rock uplift over Late Quaternary time spans creates landscape assemblages with distinctive topographic signatures. Tectonically active and inactive landscapes are very different.

3) Assessment of whether or not a landform, or landscape, achieves a time-independent configuration after tectonic and climatic perturbations change geomorphic processes on hills and along streams.

4) How does the arrival of a propagating tip of a normal, thrust, or strike-slip fault change geomorphic processes?

5) How do fluvial systems of mountain ranges respond to regional extension?

6) How quickly do streams of small watersheds react to a pulse of uplift created by rising hot asthenosphere associated with a migrating triple junction?

7) What have been the geomorphic consequences of 30 million years of uplift and intrusion of India into the Tibetan Plateau on the rivers of southern Asia? Is areal variation in summer monsoon rainfall important?

8) How are hills and streams affected by coastal erosion during times of sea-level highstands?

9) How do rising ground-water tables cause landslides?

10) Do fluvial systems have sufficient continuity to fully adjust to rock uplift?

These topics are indicative of the diversity of subject matter in "Tectonically Active Landscapes". Geomorphic processes are discussed for time spans ranging from 1,000 to 80,000,000 years. The sizes of landforms studied here ranges from a tectonic microplate to a hillside gully.

My intent is to develop basic principles of landscape evolution by examining hills, streams, and piedmonts of selected mountain ranges. These were chosen to include arid to extremely humid climates, inactive to highly active tectonic settings, and fractured, soft to massive, tough rock types. Structural styles include normal, thrust, strike-slip, and listric faults. I explore and test concepts so you can better understand the landscape evolution of your favorite mountains.

Tectonic settings of the specific study areas are summarized in Chapter 1. Most are in the broad San Andreas transform boundary between the North America and Pacific Plates, so we return to this tectonic setting in every subsequent chapter. This large, diverse region is arid to humid, includes the normal and strike-slip faults of the Basin and Range Province and the thrust faults of the Coast Ranges. It includes the landscapes of the Sierra Nevada microplate and the Mendocino triple junction. The Southern Alps of New Zealand are an essential study area because of their rapid uplift, high erosion rates, and extremely humid climate. The Himalayas provide a setting of maximum relief and a very long time span of tectonic deformation and landscape changes.

Chapters 2–4 about how hillslopes adjust to the static or changing valley floors to which they are graded. Small narrow drainage basins that characterize youthful scarps become larger and more circular over the long time spans needed to create a mountain range (Chapter 2).

Stream capture events are important. Chapter 3 defines topographic signatures in hillslopes from the standpoint of rates of base-level change. Chapter 4 examines mass movements in tectonically active regions in the context of how rain changes ground-water levels, seepage forces, and hillslope stability.

Important issues in tectonics and topography are the focus of the concluding four chapters. The common presumption of the past 50 years that entire landscapes achieve unchanging (steady state) configurations is tested in Chapter 5. Chapter 6 examines a much newer subject. How does erosion change crustal loading and influence styles and rates of fracturing, faulting, and exhumation of deep crustal materials? How streams respond to propagation of the tips of active faults is explored in Chapter 7. Concluding Chapter 8 has a plate-tectonics emphasis. It discusses how hills and streams of a granitic microplate and a migrating triple junction respond to temporary uplift caused by upwelling of hot asthenosphere.

These essays are written for scientists and students with an interest in geomorphic processes, landscape evolution, and tectonics of plate boundaries. This book is appropriate for upper division and graduate-level courses in active tectonics, tectonic geomorphology, physical geography and geomorphology, plate-tectonics geology, engineering geology, and environmental studies.

This project began in 1975 when Luna Leopold encouraged me to write books as a career development tool. Fieldwork was done in many countries. "Geomorphic Responses to Climatic Change" (Bull, 1991) reveals pervasive Late Quaternary impacts on geomorphic processes of arid and humid regions. Tectonic Geomorphology of Mountains – a New Approach to Paleoseismology (Bull, 2007) has an applied theme. Tectonically Active Landscapes could just as well be entitled Landscape Evolution.

I appreciate the formal reviews of the entire book manuscript by Ed Keller and an anonymous reviewer. Their suggestions improved clarity of writing and completeness of thought. I received valuable suggestions about hydrogeologic discussions in Chapter 4 from Leo Leonhart of Hargis & Associates, Gray Wilson, and Erik Nelson of Engineering Analytics. Lewis Owen reviewed an earlier version of Chapter 8.

Correspondence with colleagues provided essential insights:
Bodo Bookhagen, Himalayan monsoons,
John Bradshaw, New Zealand greywackes,
Bill Dickinson, volcanism, triple junctions,
Kurt Frankel, Fish Lake fault zone,
Roger Hooke, Panamint Range tilt,
Oliver Korup, Southern Alps landslides,
Kurt Frankel, Fish Lake fault zone,
Peter Koons, Southern Alps tectonics,
Les McFadden, soils and landscapes,
Jim McKean, hillslope processes,
Dorothy Merritts, triple-junction landscapes,
Jarg Pettinga, Southern Alps tectonics,
Fred Phillips, Owens Valley–Sierra faulting,
Tom Sawyer, Sierra, Diablo Range tectonics,
Greg Stock, Sierra Nevada rivers incisement,
Kirk Vincent, triple-junction landscapes,
John Wakabayashi, Sierra Nevada tectonics.

Generous Earth scientists provided essential technical illustrations:
Bodo Bookhagen: Figures 6.20–6.25,
Malcolm Clark: Figures 8.3, 8.6,
Dykstra Eusden: Figure 4.1,
Ryan Elley: Figure 2.18,
Tom Farr: Figure 8.4,
Kurt Frankel: Figures 7.21–7.23,
Bernard Hallet: Figures 6.16–6.18,
Oliver Korup: Figures 5.9–5.11,
Scott Miller: Figures 7.15A, 7.16,
Fred Phillips: Figure 8.17,
Gerald Roberts: Figures 7.1, 7.3–7.5,
Tom Sawyer, Jeff Unruh: Figures 2.13–2.15,
Jamie Schulmeister: Figure 4.3,
Greg Stock: Figure 8.21.

Photographs illustrate many landscape complexities allowing readers to make their own interpretations.

Tim Davies: Figure 6.9,
Peter Haeussler: Figure 6.2,
Ed Keller: Chapter 7 banner photo,
Peter Kresan: Figures 2.3, 3.15, 3.20,
Jim McCalpin: Figure 6.3,
Karl Mueller: Figure 7.20,
Bill Tucker: Figures 3.13A–D, 3.14,
Alistair Wright: Chapter 2 banner photo, Figure 4.19A.

Images are important for landscape analysis and portrayal. Tom Farr helped locate NASA-Jet Propulsion Laboratory images used here; Figures 1.8, 1.9, 1.13, 3.20, 4.14, 6.15, 8.4. Environment Canterbury, New Zealand made the map of alluvial-fan flooding used as Figure 2.18. Malcolm Clark, U. S. Geological Survey supplied the high altitude photos in Figures 8.3 and 8.6.

I thank you all for adding depth to discussions and for illustrations that added much diversity to this book.

It was a pleasure to work with the production staff at Blackwell Publishing including Ian Francis, Delia Sandford, and especially Rosie Hayden.

Essential financial and logistical support for this work was supplied by the U.S. National Science Foundation, National Earthquake Hazards Reduction Program of the U.S. Geological Survey, National Geographic Society, University of Canterbury in New Zealand, Hebrew University of Jerusalem, Royal Swedish Academy of Sciences, and Cambridge University in the United Kingdom.

Chapter 1

Tectonic Settings of the Study Regions

This tectonically active landscape is rising out of the sea on the transpressional boundary between the Pacific and Australian plates. The landforms reflect three basic controls on landscape evolution – lithology, climate, and rock uplift.

The long pair of parallel ridges of the Eocene Amuri Formation is an erosionally truncated syncline in the Puhi Puhi valley of New Zealand. The limestone rock mass strength is higher than the fractured sandstone in the middleground, so differential erosion has made the syncline higher in the landscape.

Folding and thrust faulting here have resulted in ~10 km of local crustal shortening (Crampton et al., 2003). Uplift rates increase abruptly from 1 to 6 m/ky across the Jordan thrust fault in the foreground. They also appear to increase towards the left side of the view; note the parallism of the limestone ridges and the skylined ridgecrest underlain by fractured sandstone.

Steep braided rivers at the lower-right edge of this view reflect rapid erosion of the fractured greywacke sandstone of the Torlesse Formation underlying the 2,500 m high Seaward Kaikoura Range. Flat grassy ridgecrests in the broad valley are stream terraces that record tectonic and climatic controls on this rapidly changing landscape. Late Quaternary climatic changes caused pulses of aggradation that briefly raised valley floors. Long-term tectonically induced downcutting has preserved previous valley floors as the steps and risers in flights of stream terraces.

Tectonically Active Landscapes, 1st edition. By W. B. Bull. Published 2009 by Blackwell Publishing, ISBN 978-1-4051-9012-1

The format of this book is to explore wide-ranging, important, topics in tectonic geomorphology. I mainly use the exceptionally diverse landscapes and tectonic settings of southwestern North America. Lengthy discussions presented in Chapters 2, 6, and 7 use sites in tectonically young New Zealand and Greece, and in the Himalayas where time spans of tectonic deformation are much longer. United States mountainous landscapes investigated in Chapters 2 through 8 are influenced by regional tectonic shifts of the adjacent Pacific and North American plates. So, this diverse tectonic setting is summarized first.

Chapter 1 summarizes regional tectonic settings. This format underscores the interlinked nature of recent tectonic activity of southwestern North America in one place. Tectonic settings of this region include the Basin and Range Province, Walker Lane–Eastern California shear zone, Sierra Nevada microplate, Diablo Range adjacent to the San Andreas fault, and the migrating Mendocino triple junction. Chapters 2–4 study hillslope processes and responses to tectonic base-level fall of adjacent streams. Chapters 5–8 are a series of essays that explore steady-state premises, consider how erosion influences fracturing and faulting, examines drainage-net responses to propagation of fault tips, and discuss how upwelling asthenosphere affects geomorphic processes.

1.1 Introduction

Rising mountains are different than tectonically inactive landscapes. Just being high and lofty may not tell us if mountain-building forces are still active. Instead, we should scrutinize landforms and the geomorphic processes that create them. Streams deepen their valleys in response to increases of watershed relief, and this changes the length, slope, and curvature of the adjacent hills.

Uplift may be regional and isostatic, but often is concentrated on range-bounding faults and folds. Such local tectonic deformation may be regarded as a perturbation (change in a variable of fluvial systems) that steepens stream gradients in the mountain-front reach. The effects of such tectonic displacements emanate from the range-bounding fault. Long time spans may pass before the consequences of renewed uplift arrive at distant upstream

reaches of the drainage basin. Streams are the essential connecting link between different sections of a drainage basin. Tectonically induced downcutting along a trunk stream channel steepens the footslopes of the adjacent hillsides and increases hillslope area too. Notable results include increases of stream power and hillslope sediment yield. Concurrent changes in climate also alter types and rates of geomorphic processes.

Changes in climate and rock uplift operate separately – they are independent variables – to alter geomorphic processes and landscape characteristics. Both variables originate outside of the drainage basins that comprise mountainous landscapes.

Response times for Late Quaternary climate-change perturbations are much shorter than those for mountain-front uplift. Climate change impacts all of a watershed with minimal time lag. The resulting changes in the size and amount of stream-channel bedload influence rates of tectonically induced downcutting. Climate-change induced impacts on geomorphic processes, such as changing valley-floor deposition to erosion, are pervasive. An aggradation event temporarily overwhelms the influence of concurrent tectonic displacements for a reach of a stream tending to maintain steady state valley-floor erosion (unchanging longitudinal profile underlain by a strath).

A good example is the Charwell River basin in the South Island of New Zealand. Repeated surface ruptures on a range-bounding fault created a 40 m high sub-alluvial fault scarp during a climate-change-induced aggradation event. Aggradation disrupted the continuity of an erosional fluvial system and prevented the tectonic perturbation from migrating upstream. The mountain range was raised 42 m between 26 and 9 ka[1], but climatic controls delayed the upstream transmission of the tectonic perturbation for 17 ky. Sediment flux modeling (Gasparini et al., 2007) illustrates the strength of climatic controls on bedrock incision rates. "Geomorphic Responses to Climatic Change" (Bull, 1991) elucidates the effects of Late Quaternary climatic and tectonic perturbations in mountain ranges whose present climates range from extremely arid to extremely humid.

[1]   1 ky = 1000 years; 1 ka = 1 ky before present. 1 My = 1 million years; 1 Ma = 1 My before present.

We have much to learn about the possible impacts associated with the present acceleration of human-induced global climate changes. Geomorphic impacts may resemble the consequences of the Late Quaternary climate changes, but the types, rates, and magnitudes of change in geomorphic processes could be quite different.

Both the Pleistocene and Holocene styles of climate change are now history, having been replaced by the Anthropocene (Crutzen and Stoermer, 2000). Intensification of human activities with the onset of the industrial revolution now influences the climate of planet Earth. Amounts of temperature change seem modest, but rates of change are more than an order of magnitude faster than during the Quaternary. Models based on slowly changing Quaternary climates may not be appropriate now.

We could be entering significantly new territory with local demise of permafrost (Cheng and Wu, 2007; Gruber and Haeberli, 2007), and eradication of so much tropical rain forest (Aitken and Leigh, 1992). But human impacts on forests are not a recent development (Bjorse and Bradshaw, 2000; Ruddiman, 2003).

Clearing of forests for agriculture began at 8 ka causing a modest increase in atmospheric carbon dioxide, CO_2 (Ruddiman, 2003 and the polar ice cores (Ferretti et al., 2005) record the increase of methane, CH_4, at the time of major increase of Asian rice paddies at 5 ka. Such human impacts in the early Anthropocene, with major acceleration since the onset of the industrial revolution, have more than offset the Holocene decrease of solar radiation at 65° N that is a function of Milankovitch orbital parameters (Ruddiman, 2003).

A key aspect will be crossings of the geomorphic threshold separating erosional and depositional modes of stream processes. Local uplift that changes slope of stream channels may put a given geomorphic process closer to, or further from, a threshold separating contrasting styles of landscape behavior.

This book emphasizes hills and their relation to streams. Streams catch our attention with visually impressive flows of water and debris, the power of boulder-transport processes, and as being sources of water, electric power, and fertile land. Tectonically induced changes in the altitude of a reach of a stream affect the base level to which the adjacent hillslopes are graded. Hills yield water and sediment to valley floors. Mountain ranges are nearly all hills, but their shape is dependent on the behavior of streams.

The landforms of tectonically active landscapes have definitive characteristics that reflect local rates of rock uplift, climate, and rock mass strength. This book is about the nature of these signatures and how the hills and streams of fluvial systems change with the passage of time. The three blocks of Figure 1.1 diagram the flow of topics presented in the three parts of the book. A discourse that focuses mainly on streams is worthy of a separate book, perhaps "Fluvial Tectonic Geomorphology".

We use tectonic signatures in landforms to explore several topics in tectonics and topography. I present data and analyses from study areas in quite different tectonic, climatic, and lithologic settings. Insight gained from study sites in the southwestern United States, Greece, New Zealand, and the Himalayas should aid you in resolving similar problems in your favorite geographical settings. Study site locations of this book are shown in Figures 1.2, 1.9, and 1.13. The captions for these three figures contain links to chapter section numbers. A fluvial emphasis here excludes permafrost and glaciers, sand seas, lacustrine and tidal settings, and active volcanoes – but includes marine terraces.

I use a variety of geomorphic concepts, and presume that you know these key principles. A broad base of essential tools lets us explore diverse approaches in tectonic geomorphology. They include a sensitive erosional–depositional threshold, time lags of response to perturbations, type 1 and type 2 dynamic equilibrium, local and ultimate base levels, impediments to the continuity of fluvial systems, and the process of tectonically induced downcutting to the base level of erosion. See the Glossary for basic definitions. Chapter 2 in a companion book is devoted to defining and discussing these essential tenets; see "Tectonic Geomorphology of Mountains" (Bull, 2007). Chapter 1 of that book assesses the nuances of scrunch and stretch bedrock uplift.

Quaternary temporal terms (Table 1.1) have been assigned conventional ages. The 12-ka age assignment for the beginning of the Holocene

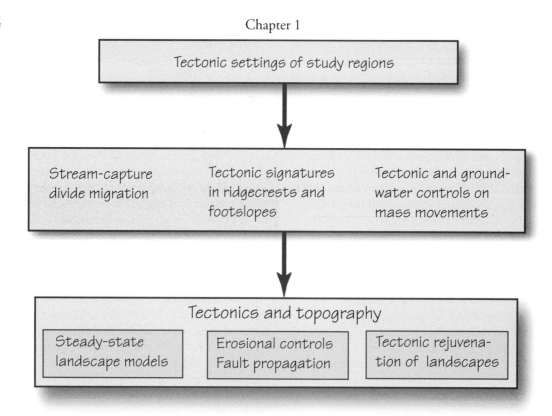

Figure 1.1 Flow chart of major topics of this book by chapter.

is arbitrary and is preceded by the transition be-tween full-glacial and interglacial climatic condi-tions. Unless specifically noted, radiocarbon ages are conventional (using the old 5,568 year half-life allows comparison with dates in the older literature) and have been corrected for isotope fractionation. The term "radiocarbon age" means that the correct 5,730 year half-life is used and that variations in at-mospheric ^{14}C have been accounted for, using the techniques of Stuiver et al. (1998). Calibration of radiocarbon ages (Bard et al., 1990) shows that the peak of full-glacial conditions may be 22 ka instead of the conventional radiocarbon age estimate of 18 ka. The 125- and 790-ka ages are radiometric and paleomagnetic ages that have been fine-tuned using the astronomical clock (Johnson, 1982; Edwards et al., 1987a, b). The 1,650-ka age is near the top of the Olduvai reversed polarity event (Berggren et al., 1995).

1.2 North America–Pacific Plate Boundary

The San Andreas transform boundary between the North America and Pacific plates in the southwestern United States is a 200–800 km wide transition zone extending from Pacific Ocean coastal fault zones far into the Basin and Range Province. Two primary components control many secondary tectonic fea-tures of the transition zone. The following synop-sis emphasizes coincidences of timing of important tectonic events in the boundary between the Pacific and North American plates. The San Andreas fault is the primary plate-boundary fault zone at the pres-ent time. This right-lateral continental transform fault slices through batholithic complexes to create the Peninsular and Transverse Ranges of southern California. It then continues through the central and northern Coast Ranges, and turns west at Cape Mendocino to join the Mendocino fracture zone (Fig. 1.3).

The Sierra Nevada microplate is an equally important component. This 650 km long mountain range was created by batholithic intrusions in the Mesozoic. The mountain range is long but the microplate is immense because it also includes the adjacent Central Valley of California (Fig. 1.3). This tectonic block has minimal internal deformation but the eastern side of the Sierra Nevada was raised abruptly at about 4 ka (Jones et al., 2004; Saleeby and Foster, 2004). An impressive escarpment now rises 1,000 m in the north and 2,000 m in the south (Fig. 1.4).

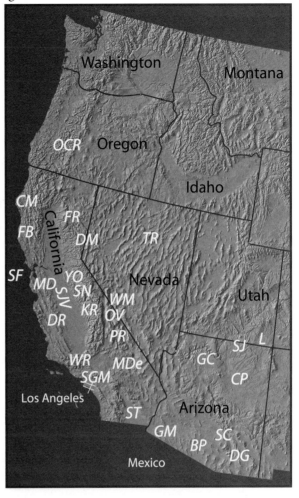

Age	ka
Holocene	
Late	0–4
Middle	4–8
Early	8–12
Pleistocene	
Latest	12–22
Late	12–125
Middle	125–790
Early	790–1,650

Table 1.1 Assigned ages of Quaternary temporal terms, in thousands of years before present (ka).

Figure 1.2 Locations and study sites in the western United States and their book section numbers [3.5]. BP, Baboquivari Peak [3.6.1]; CM, Cape Mendocino [1.2.4, 4.2.2, 8.3]; CP, Colorado Plateau [3.5]; DG, Downpour Gulch [3.4.2]; DM, Diamond and Fort Sage Mountains [8.1]; DR, Diablo Range sites including the Ciervo Hills, Dormant Hollow, Kettleman Hills, Laguna Seca Hills, Panoche Hills, Tumey Hills, and Vigorous Vale [1.2.3, 2.3.2.2, 3.2, 3.2, 3.7, 5.1, 5.4]; FB, Fort Bragg [8.3]; GC, Grand Canyon [3.5]; GM, Gila Mountains [3.6]; KR, Kings River [8.2]; MDe, Mojave Desert [1.2.1, 8.1]; FR, Feather River and Mount Lassen and the southern end of the subduction related Cascade volcanoes [8.2.2]; L, The Loop abandoned meander [3.5.1]; MD, Mount Diablo [2.3.2.2]; OCR, Oregon Coast Ranges; [3.4, 3.7]; OV, Owens Valley[1.2.1, 8.1, 8.2]; PR, Panamint Range and Death Valley, [1.2.1.1, 2.2.1, 8.1.1]; SC, Santa Catalina Mountains [2.3.1]; SF, San Francisco Bay region [2.3.2.2, 4.2.1.2.2]; SGM, San Gabriel Mountains [3.6.2]; SJ, San Juan River [3.5.1]; SJV, San Joaquin Valley [1.2.3, 3.4.1, 7.5.2]; SN, Sierra Nevada microplate [1.2.2, 5.3.2, 8.2]; ST, Salton Trough [1.2.1]; TR, Tobin Range [7.1.1]; WM, White Mountains [1.2.1, 7.3]; WR, Wheeler Ridge [7.2.2]; YO, Yosemite National Park [6.1]. Digital topography courtesy of Richard J. Pike, US Geological Survey.

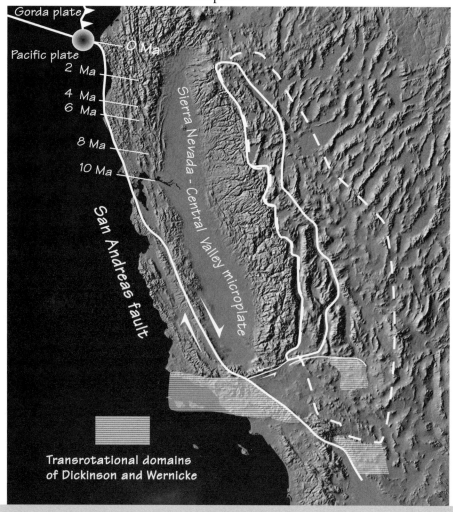

Figure 1.3 Tectonic setting of southwestern United States. Numbered lines show the northward migration of the south edge of the Gorda plate (from Atwater and Stock, 1998) to its present position at the Mendocino triple junction. Mendocino fracture zone extends northwest, and the Cascadia subduction zone north, from the triple junction. The Sierra Nevada microplate was wider at 5 Ma because it extended west almost to the San Andreas fault, and extended east into the Basin and Range Province. Area within the solid line outlines the area of accelerated extensional faulting since 5 Ma (Jones et al., 2004, figure 1). This area and the area inside the dashed line approximate the province of the Walker Lane–Eastern California shear zone, that has accommodated >10% of the plate-boundary dextral shear since 5 Ma. Areas of sinestral shear act as transrotational domains that facilitate tectonic linking within the plate boundary (Dickinson and Wernicke, 1997).

A major tectonic event – detachment of the Sierra Nevada batholithic root – (called delamination) resulted in a pulse of uplift of the microplate, and changed important tectonic processes elsewhere in the San Andreas transform boundary. Crustal extension accelerated into the eastern margin of the Sierra

Nevada, thus making the Basin and Range Province ever broader. See Figure 1.3. The San Andreas fault dextral shear zone split to create the seemingly diffuse Walker Lane–Eastern California shear zone in eastern California and western Nevada. Parts of the central Coast Ranges were created since the delamination event. The following summaries are made in the context of how this batholithic-root detachment affected the rates and styles in each distinct tectonic province.

Geomorphic features of hills and streams in of each study area tell us more about present and past tectonic activity. The rates and styles of tectonic activity, briefly noted in this introductory chapter, will be deciphered and discussed further when the tectonic geomorphology is presented for each study area.

1.2.1 Walker Lane–Eastern California Shear Zone

About 10 to 20% of the dextral shear between the Pacific and North American plates split off from the San Andreas fault at 3 to 5 Ma. Unlike the San Andreas fault, movements along this much younger shear zone seem dispersed. Diffuse, and appearing spatially intermittent, this is called the "Eastern California shear zone" in the south and the "Walker Lane shear zone" in the north. Two prominent dextral active faults in the south are separated by 100 km – the 310 km long Death Valley–Fish Lake and the 110 km long Owens Valley fault zones (Reheis and Dixon, 1996). Two active dextral fault zones in the north are separated by 60 km – Honey Lake and Mohawk Valley (Wills and Borchardt, 1993). Field studies and seismic analyses reveal additional widespread dextral faulting.

The importance of the San Andreas as a plate boundary fault is obvious because of its 300 km of cumulative displacement since 30 Ma (Dickinson, 1996). Magnitude M_w >7 earthquakes occurred on this fault in A.D. 1812, 1838, 1857, and 1906.

In marked contrast the Walker Lane–Eastern California shear zone initially appeared so intermittent and diffuse that it has taken a century to recognize its importance. Nevertheless, historic earthquakes attest to its tectonic significance. It too has

had four M_w >7 dextral earthquakes since A.D. 1800 including the magnitude M_w 7.6 Owens Valley earthquake of 1872, the M_w 7.2 Cedar Mountain event of 1932 in western Nevada, the M_w 7.3 Landers earthquake in 1992 and M_w 7.1 Hector Mine earthquake in 1999; both recent events were in the central Mojave Desert.

Geodetic measurements indicate right-lateral shear between the Pacific and North American plates of about 10 to 14 m/ky (Bennett et al., 1999; Dixon et al., 2000; Argus and Gordon, 2001). The Sierra Nevada microplate has been rotated counterclockwise in response to deformation of adjacent tectonic provinces.

Extension of the distinctive Basin and Range Province began at 35 Ma in the north and had propagated to the south by 20 Ma (Dilles and Gans, 1995). Such normal faulting migrated westward during the late Cenozoic by encroaching into the Sierra Nevada microplate. The microplate was much wider just 5 My ago (Fig 1.3).

Uplift and eastward tilting of the lofty White Mountains as a separate block (Guth, 1997) began at 12 Ma (Stockli et al., 2003), which is the same time as encroachment near Reno created the Carson Range (Henry and Perkins, 2001). This early episode of synchronous normal faulting occurred along nearly 300 km of the Sierra Nevada–Basin and Range boundary. The Panamint Range also may have been created at this time.

Range-bounding normal faulting after ~3.5 Ma created the eastern escarpment of the Sierra Nevada (Fig. 1.4) and increased relief of the White and Inyo Mountains on the other side of Owens Valley. Neither synchronous episode of tectonic rejuvenation was related to the continued migration of the Mendocino triple junction (Unruh, 1991).

This encroachment has reduced the width of the Sierra Nevada microplate. Detached batholithic rocks are now Basin and Range Province mountain ranges. Examples include the plutonic rocks of the Panamint and Argus Ranges in the south and the Diamond and Fort Sage Mountains in the north (Dilles and Gans, 1995; Wakabayashi and Sawyer, 2001; Surpless et al., 2002). Encroachment began before onset of the Walker Lane–Eastern California shear zone dextral shearing, and is continuing.

8   Chapter 1

Encroachment of Walker Lane–Eastern California shear zone near Reno, Nevada was associated with pulses of volcanic activity at 12 and 3 Ma (Henry and Perkins, 2001). Putirka and Busby (2007) suggest that high K$_2$O volcanism in the central Sierra Nevada was synchronous with the onset of Walker Lane–Eastern California shear zone transtensional tectonic deformation. In an interesting conclusion they state "We speculate that high-K$_2$O lavas in the southern Sierra are similarly related to the onset of transtensional stresses, not delamination".

Diverse recent studies provide a clearer tectonics picture for the Coast Ranges and the Sierra Nevada, but tectonic deformation of the western Basin and Range Province is not as well understood. Local changes from oblique-slip extension to nearly pure strike-slip faulting at Honey Lake and Owens Valley (Monastero et al., 2005) underscore changes in the importance of dextral faulting. Rates of strike-slip displacement on individual faults of 5 to 9 m/ky (Klinger, 1999) are not visually obvious if mountain fronts are not created.

The change from regional extension to regional transtension east of the Sierra Nevada microplate suggests early stages of a new plate-boundary shear zone. This interesting possibility is based on seismic (Wallace, 1984), geologic (Wesnousky, 1986, Dokka and Travis, 1990 a; b; Putirka and Busby, 2007), and geodetic studies (Sauber et al., 1986, 1994; Savage et al., 1993). Wallace speculated that future earthquakes would fill gaps in historical seismicity. The subsequent north-trending dextral

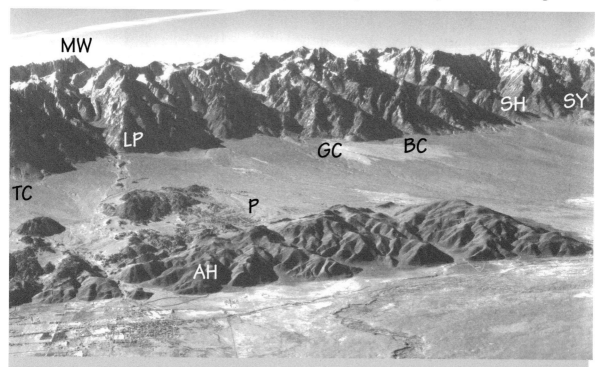

Figure 1.4 Eastern escarpment of the Sierra Nevada microplate. Aerial view west across Owens Valley. Oblique right-lateral movement on the Owens Valley fault zone ruptured through the town of Lone Pine in the left foreground in 1872 (Mw 7.6). The eastern escarpment of the Sierra Nevada rises from the range-bounding fault zone to summit of Mount Whitney, MW (altitude 4,418 m). TC, Tuttle Creek; LP, Lone Pine Creek; GC, George Creek; BC, Bairs Creek; SH Shepherd Creek; SY, Symmes Creek. A pediment surface, P, in the Alabama Hills, AH, extends under the adjacent piedmont alluvium. 1955 Photograph GS–OAl–5–13, courtesy of U. S. Geological Survey.

surface ruptures of the Landers earthquake of 1992 and Hector Mine earthquake of 1999 corroborate his 1984 prediction. Some workers include these surface-rupture events in a model of a new plate-boundary fault zone (Nur et al., 1993; Sauber et al., 1994; Du and Aydin, 1996).

This introduction reveals the subtle, but important, nature of this part of a changing plate boundary. The tectonics of the Walker Lane–Eastern California shear zone, as augmented by tectonic geomorphology input, is worthy of a more far-reaching discourse presented in Section 8.1.

1.2.1.1 Panamint Range

The raised, tilted Panamint Range block was chosen as a study area within the Walker Lane–Eastern California shear zone because of its diverse tectonic landforms. The Range rises 3,400 m above adjacent Death Valley whose low point has sunk to an altitude of − 89 m. Granitic and gneissic rocks, quartzite, dolomite, and argillite are common lithologies. Granitic rocks of the Panamint Range are part of the former Sierran batholithic arc; several plutons were emplaced at about 70 to 100 Ma (Hodges et al., 1990; Mahood et al., 1996). From west to east, present exposure of granitic rocks decreases from 99% in the Sierra Nevada, to >80% in the Argus Range, to <40% in the Panamint Range. This trend suggests an eastward increase in thickness of sedimentary rock cover. Since 5 Ma, Basin and Range Province extensional faulting has topographically separated the Panamint and Argus Ranges from the Sierra Nevada.

Large, exhumed fault surfaces in the Black Mountains flanking the east side of Death Valley are referred to as "turtlebacks". Realizing that these antiformal-shaped fault planes are the result of late Cenozoic normal faulting led to further work (Wright et al., 1974) including articles about the magnitude, timing, and style of extension (Burchfiel and Davis, 1981; Stewart, 1983; Wernicke et al., 1988; and Burchfiel et al., 1995). Cichanski's study (2000) shows that similar planar to curviplanar low-angle faults were a common and recent style of tectonic extension in many mountain ranges of the region. Accordant ridgecrests in a highly rilled bedrock surface that is dipping 15° to 35° characterize exposures of these faults, but locally these low-angle faults dip as much as 35° to 40°.

Changes in style of range-front faulting in Panamint Valley and nearby ranges (Loomis and Burbank, 1988) occurred during the Late Cenozoic. Changing plate-boundary tectonics includes initiation of more than 64 km of left-lateral displacement on the Garlock fault since about 10 Ma (Burbank and Whistler, 1987; Dawson et al., 2000, 2003). It marks the southwestern extent of the Basin and Range Province (Davis and Burchfiel, 1973). Panamint and Saline Valleys appear to be genetically linked rhombochasms that began to open at the same time (Zellmer, 1980; Burchfiel et al., 1987). Such wrench-fault tectonics has been important (Densmore and Anderson, 1997) and may have set the stage for low-angle range-front faulting.

Cichanski describes a recent change from low-angle to vertical range-bounding faulting on the west side of the Panamint Range. Zhang et al. (1990) suggest that such faulting began in the Holocene, but soil profiles on the older faulted fan remnants indicate a Late Pleistocene age to me. A key conclusion by Cichanski is that prolonged low-angle normal faulting, not the presently active oblique-dextral Panamint Valley fault zone, is responsible for the present range-front topography on the west side of the Panamint Range. Dextral faulting at ~0.5 m/ky is occurring on the Panamint Valley fault, which is on the Argus Range side of Panamint Valley (Densmore and Anderson, 1997).

Varying rates of range-front uplift have resulted in tectonic blocks being raised and tilted. The Panamint Range block is being tilted eastward (Maxson, 1950; Green, in Hunt and Mabey, 1966; Hooke, 1972) to help create the topographic depression of Death Valley, which was formed by a complex mix of E–W crustal extension and N–S transtension. Pleistocene tilting about a N–S axis west of the valley trough has resulted in a 20 to 30 m downward displacement of ~120 ka to ~180 ka lacustrine shorelines of Late Quaternary age at the toe of the Panamint Range piedmont relative to patches of shoreline tufas clinging to the escarpment of the Black Mountains on the east side of Death Valley (Hooke, 1972).

1.2.2 Sierra Nevada

The Sierra Nevada microplate is a major tectonic element of western North America. The 650 km long Sierra Nevada was created by enormous batholithic intrusions during the Mesozoic. The majestic landscapes of these mountains are the focus of conceptual models in landscape evolution in Chapters 5 and 8.

Mesozoic subduction beneath the North American plate produced granitoid melts that rose above their heavier denser ultramafic residues, which were held in place by the subducting slab. Eventual failure of the slab allowed the ultramafic root to detach from the granitic microplate and sink into the mantle. This major tectonic event – detachment of the Sierra Nevada batholithic root, delamination – affected tectonic elements in much of the broad transform boundary. Ongoing synchronous subduction removed most of the oceanic Farallon plate, creating the Mendocino triple junction; both processes influenced microplate and transform-boundary tectonics. This introduction summarizes apparent coincidences of timing of several changes of tectonic style between the Pacific and North American tectonic plates.

Late Cenozoic rock uplift rates of the Sierra Nevada were slow to moderate. Unruh's 1991 long-term estimate is based on initiation of late Cenozoic uplift at roughly 5 Ma and maximum possible uplift of the range crest of 2500 m. This conservative estimate, <0.5 m/ky, is fairly slow because of the assumption that uplift was uniformly distributed throughout the 5 My time span. A nonuniform uplift model, presented later in Section 8.2.2, suggests that short-term uplift rates were faster, being comparable to the present uplift of the Southern Alps of New Zealand, Taiwan, and parts of the Himalayas.

Diverse mechanisms have been proposed for late Cenozoic renewed uplift of the Sierra Nevada. These include flexural isostasy (Chase and Wallace, 1986), support by a buoyant asthenosphere (Liu and Shen, 1998), isostatic erosional unloading in the mountains combined with depositional loading in the San Joaquin Valley (Small and Anderson, 1995), and extension of the adjacent Basin and Range Province (Wernicke et al., 1996; McQuarrie and Wernicke, 2005). Tectonic reconstruction for southwestern North America shows that the Sierra Nevada microplate has moved ~235 km N78°W with respect to the Colorado Plateau since ~ 16 Ma (Geological Society of America Penrose Conference, 21–26 April 2005, "Kinematics and Geodynamics of Intraplate Dextral Shear in Eastern California and Western Nevada"). Manley et al. (2000) suggest that foundering of the eclogitic root occurred as recently as 3.5 Ma. To these potential uplift mechanisms I would add a flexural uplift component associated with accelerated deposition on the western edge of the microplate as a result of major increases in the sediment yield from the Coast Ranges.

Distinctive early Sierran landscapes were eroded during times of tectonic quiescence. Using the chronology of Wakabayashi and Sawyer (2001), uplift related to batholithic intrusion was complete by 99 Ma. Erosion had reduced the northern Sierra Nevada to a tectonically inactive landscape by 52 Ma with remnants of mountains rising as much as 800 m above adjacent pediment base levels. Virtually no uplift occurred between 57 Ma and 5 Ma. Renewed volcanism during this interval covered all but a few scattered basement highs with the Mehrten Formation. This volcanic activity ceased at about 5 Ma as the Mendocino triple junction migrated ever farther to the north. Eocene-to-Pliocene tectonically induced downcutting by Sierra Nevada rivers was miniscule. Wakabayashi and Sawyer, using altitudes of dated stream gravels, estimated valley floor degradation to be less than 0.007 m/1000 years. The southern Sierra Nevada, although greatly eroded, still had local relief of 2,000 m that was inherited from uplift that occurred between 99 Ma and 57 Ma.

Tectonic quiescence ended at about 3.5 Ma with initiation of uplift along range bounding normal faults on the eastern flank of the Sierra Nevada. Encroachment by normal faults continued, perhaps at an accelerated pace in the northern Sierra Nevada. Part of the impressive relief of the Sierra Nevada eastern escarpment (Fig. 1.4) may be the result of dropping of adjacent the Owens Valley as extensional processes promoted collapse of formerly higher terrain in much same way as the valleys between the transition ranges in Arizona fell away from the Colorado Plateau (Lucchitta, 1979; Mayer, 1979). The Yuba and Stanislaus Rivers were beheaded as a new range crest developed. Similar beheading of the San Joaquin River dropped at least 40 km of its ancestral headwaters into what is now Owens Valley (Huber, 1981). Valley floor lava flows of the past 10 Ma (Huber, 1990) were tilted and incised to become meandering flat-topped ridgecrests. These non-steady-state landscapes document recent bedrock uplift along the east-side range-bounding fault.

The effect of tilting caused by uplift of only ~0.1m/ky was enhanced incision by glaciers and large rivers that now flow through spectacular canyons in plutonic rocks. Wakabayashi and Sawyer point out that late Cenozoic tilting raised the entire length of the Sierra Nevada, with uplift ranging from 1,710 to 1,930 m. Rock uplift and surface uplift of the Sierra Nevada crest were about the same because Late Quaternary erosion rates are very low (Small et al., 1997; House et al., 2001; Stock et al., 2005). We need to further examine the causes of this recent uplift.

Uplift at >1m/ky may have been a consequence of the crustal delamination. Detachment of the crustal root of the Sierra Nevada batholith followed by upwelling of hot asthenosphere would initiate a sudden pulse of rock uplift that would fade away as adjustments between the new set of crustal mountain-building forces stabilized. I find the diverse evidence for a delamination event fairly convincing.

Data from a geophysical transect extending from the San Joaquin Valley across the Sierra Nevada and Panamint Ranges constrains tectonic models (Wernicke et al., 1996; Liu and Shen, 1998; Fliedner et al., 2000). The Cretaceous batholith had a thick residual root, but a crustal root is no longer present to support the Sierra Nevada.

Xenolith (pieces of older rock engulfed by rising magma) composition changes (Ducea and Saleeby, 1996, 1998) indicate that a former dense root beneath the Sierra Nevada crest was convectively removed before 3 Ma. Saleeby and Foster (2004) say "the sharp contrast in peridotite mantle facies fields between the mid-Miocene and Pliocene–Quaternary xenolith suites, and a Pliocene change in the composition of lavas erupted in the region to more primitive compositions are interpreted to indicate the removal of the sub-batholith mantle lithosphere and its replacement by asthenosphere. Removal of dense (3.5 g/cm^3) eclogite and replacement with lower density peridotite (3.3 g/cm^3) may have increased buoyancy sufficiently to account for as much as 2 km of range-crest uplift (Liu and Shen, 1998). Similar delamination-magmatic pulses may have occurred beneath the Puna Plateau of Argentina and the Tibetan Plateau (Glazner, 2003).

Mantle lithosphere is now abnormally thin beneath the Sierra Nevada and Panamint Ranges (Jones et al., 1994; Wernicke et al., 1996; Ruppert et al., 1998). They seem to be largely supported by a buoyant upwelling of hot asthenosphere. Modeling by Liu and Shen (1998) indicates that mantle upwelling beneath the Basin and Range extensional province caused ductile material within the lithosphere to flow southwestward. Zandt (2003) refers to this flow as "mantle wind" that shifted this part of the detached Sierra Nevada batholithic root – a mantle drip – to the southwest.

Xenolith studies suggest recent removal of garnet-bearing rocks triggered a brief (3.5 ± 0.25 Ma) pulse of potassium-rich basaltic vulcanism (Fig. 1.5) in a 200 km diameter circular area in the central Sierra Nevada (Manley et al., 2000; Farmer et al., 2002; Jones et al., 2004). This event indicates the most likely timing and location of the main delamination event. Zandt (2003) points out that the highest peaks of the Sierra Nevada occur in the area where the delamination event began – maximum strength of the geophysical perturbation coincides with maximum uplift. I like the interpretation that recent uplift of the Sierra Nevada also could be this young, and that volcanism and uplift resulted from a common delamination-triggering mechanism. Section 8.2.2 discusses the landscape-evolution consequences of the "Post-4-Ma uplift event", focusing mainly on this same part of the Sierra Nevada, which includes the Kings River, Owens Valley, and the highest part of the east-side escarpment.

The interval between 5 and 3 Ma fits a model for regional synchronous changes in many plate-boundary tectonic processes, and of landscapes that record changes in tectonic base-level controls. Geophysical detection of the lack of a crustal root beneath the lofty Sierra Nevada has led to much creative thinking about tectonics of the region.

Chapter 1

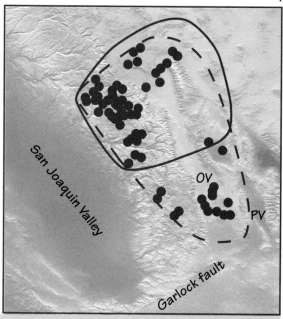

Figure 1.5 Times of volcanism north of the Garlock fault. Dashed line outlines area of 4 to 3 Ma volcanism and solid line outlines area of highly potassic lavas. OV is Owens Valley and PV is Panamint Valley. From Manley et al., 2000.

A brilliant synthesis by Jones et al. (2004) recognizes the late Pliocene foundering of the Sierra Nevada crustal root as a perturbation that changed regional plate tectonics. They reason that this crustal delamination event increased the total gravitational potential energy of the lithosphere (Jones et al., 1996), thus increasing both extensional strain rates in the western Basin and Range Province and the altitude of the eastern Sierra Nevada. They also conclude "an increase in extensional displacement rates must be accommodated by a decrease in rates of extension or an increase in rates of shortening somewhere in the vicinity of the Sierra Nevada" (Jones et al., 2004, p. 1411). The timing of recent thrust faulting that created the Coast Ranges bordering the Central Valley coincides nicely (Wentworth and Zoback, 1989; Namson et al., 1990; Lettis and Hanson, 1991; Wakabayashi and Smith, 1994). And, "Lithospheric removal may also be responsible for shifting of the distribution of transform slip from the San Andreas

Fault system to the Eastern California shear zone" (Jones et al., 2004, p. 1408). Details and confirmation of their model of synchronous regional changes will be tested by future work. Initial geomorphic tests of the model presented in Section 8.2 support uplift of the Sierra Nevada at about 3.5 Ma, concurrent with changes in tectonic style of the Coast Ranges and formation of the Walker Lane–Eastern California shear zone.

Cenozoic tectonism of western North America is related to the Mendocino plate boundary triple junction, which migrated northward creating the San Andreas zone. This is a continental right-lateral transform fault system (Atwater, 1970; Atwater and Stock, 1998) that presently includes the Maacama and Bartlett Springs faults. They conclude that no discernible change in rates of motion of the Pacific–North American plate boundary has occurred since 8 Ma. This is a key assumption of the Jones et al. model where horizontal velocities across the Sierra before and after the delamination event match the boundary condition of Pacific–North American plate motion. The assumption is supported by the work of Argus and Gordon (2001). They say "the Sierran microplate changed motion relative to North America at the same time (~ 8 to 6 Ma) as did the Pacific plate. Local acceleration of extensional encroachment in a belt east of the Sierra Nevada may match increased rates of Coast Ranges shortening to the west. The Sierra Nevada microplate has become narrower as a result of both tectonic processes.

Interactions between plate-boundary kinematics and locally derived forces generated by the foundering of Sierra Nevada eclogitic lithosphere appear pervasive. Although not as nicely constrained as the 3.5 ±0.25 Ma brief pulse of potassic vulcanism associated with recent Sierra Nevada uplift, key plate tectonics and geomorphic events seem to be regional consequences of the delamination event. The degree of synchroneity is a bit fuzzy because 1) there is minimal information about the time span needed for the delamination process to occur in different parts of the microplate, and 2) establishment of times of tectonic events elsewhere in the plate boundary may be rather crude where dating uncertainties are large.

Vertical and horizontal displacements commonly do not occur on the same fault. Instead they may be partitioned between adjacent faults. Partitioning refers to the common situation where vertical and horizontal components of displacement occur on separate faults within a fault zone. Partitioning styles are defined later in this chapter (Fig. 1.10). A parallel style of partitioning is common in the Walker Lane–Eastern California shear zone. Examples include dextral fault zones in Owens and Death Valleys and adjacent range-bounding normal faults. Parallel partitioning was the result of dextral faulting being introduced into an area of pre-existing normal faulting.

How significant were the processes of extensional faulting, dextral faulting, and crustal-root delamination in the formation of Sierra Nevada and nearby mountain ranges? What is the present rate of uplift on the Sierra Nevada range-bounding fault? How important is Late Quaternary strike-slip faulting relative to normal faulting? How did the large rivers draining the western flank of the Sierra Nevada respond to pulsatory uplift? How long are the ridgecrest response times to stream-channel downcutting? These are some of the questions addressed in Sections 5.3.2 and 8.2.

1.2.3 Diablo Range

The Coast Ranges border the west side of the Sierra Nevada microplate, the only gap being where rivers leave the central valley and flow into San Francisco bay (the estuary by the 10 Ma notation in Figure 1.3). The south half of the Central Valley of California is the San Joaquin Valley, and this book discusses sites on the west side from Mt. Diablo at the north end (Section 2.3.2.2) to central districts (Sections 3.2, 3.4, 4.2.1.2.2., 4.2.2, 5.4). Wheeler Ridge (Fig. 1.2 and Section 7.2.2) at the south end of the San Joaquin Valley is another key study site.

The western edge of the central San Joaquin Valley has deceptive, but fascinating, mountain fronts. I say deceptive because I now realize that some of these low, largely barren hills are so tectonically active that they have only recently risen to modest altitudes of 300 to 900 m. With the advantage of plate-tectonics hindsight we now realize that, although low, the hills flanking the San Joaquin Valley are both rising and seismogenic.

Plate-boundary tectonics near the right-lateral oblique-slip San Andreas fault (Fig. 1.3) has created a fold-and-thrust belt (Wentworth, and Zoback, 1990; Yerkes, 1990) that is encroaching into the adjacent San Joaquin Valley. West-dipping thrust faults border about 600 km of the eastern margin of the Coast Ranges, in both the Sacramento and San Joaquin Valleys (Wentworth and Zoback, 1990; Wakabayashi and Smith, 1994). Both strike-slip and thrust faults are present locally (LaForge and Lee, 1982: Lettis, 1985). Lettis and Hanson (1991) use a regional strain-partitioning model that relates the kinematics of the fold-and-thrust belt to the San Andreas fault.

Historical earthquakes reveal how active the young foothill belt is. The first of a northwest-to-southeast sequence of earthquakes on the same 100-km long hidden thrust fault (Stein and Yeats, 1989; Stein and Ekstrom, 1992; Lin and Stein, 2006) was the Mw magnitude 5.4 New Idria earthquake of 1982 on an internal fault. The Mw magnitude 6.5 Coalinga earthquake in 1983 (Hill, 1984; Rymer and Ellsworth, 1990) emanated from beneath Anticline Ridge, and the Mw magnitude 6.1 earthquake of 1985 from beneath the north end of the Kettleman Hills (Wentworth et al., 1983). Both are range-bounding structures. The crest of Anticline Ridge rose abruptly about 500 mm during the 1983 earthquake (King and Stein, 1983). In hindsight, I now realize that comparison of first-order level-line surveys across Anticline Ridge in March, 1962 and March, 1963 recorded a precursor event to the destructive Coalinga earthquake. Benchmarks rose as much as 24 mm in a pattern that mimicked the topographic profile of Anticline Ridge (Bull, 1975b, fig. 4). These "hidden earthquakes" (Stein and Yeats, 1989) underscore the continuing active nature of thrust faults in the cores of these folds. The landscape evolution this reflects is rapid, continuing uplift. But how does this thrust faulting fit into the Jones et al. (2004) model of regional tectonic activity as influenced by the delamination of the Sierra Nevada batholithic crustal root?

Sedimentary basins can contain a wealth of information about the geomorphic consequences of tectonic events in their source mountain ranges (Dickinson, 1974; Angevine et al., 1990;

Frostick, and Steel, 1994; Emery and Myers, 1996; Cloetingh et al., 1998; Busby and Ingersoll, (1995). Geomorphologists look at the consequences of cumulative erosion on landscapes. Sedimentologists have access to sequences of outputs from fluvial systems. In the example used here, the distributions of thick Quaternary basin fill in the Central Valley of California may be regarded as a competition for a finite amount of space for newly arriving deposits from Sierra Nevada and Coast Ranges source areas. A tectonic story is revealed by the depositional contact between basin-fill derived from these two opposing source areas.

Geologic logs of many water wells in San Joaquin Valley provide a wealth of information about this contest for space. The granitic rocks of the Sierra Nevada to the east supply distinctive micaceous arkosic sands to the basin. Sierran watersheds and rivers are large. Terrains and rock types along the west side of the San Joaquin Valley are much different. These mountains consist of a main range and a younger foothill belt. Water-laid sediments derived from the main Diablo Range consist primarily of silty to clayey sand transported by fairly large streams, such as Panoche Creek, through gaps in the foothill belt. Clay contents are high in the poorly sorted water-laid and debris flow alluvial-fan deposits derived from the foothill belt watersheds. These small drainage basins are underlain by mudstone, diatomaceous shale, and soft sandstone.

Rates of deposition in the San Joaquin Valley have accelerated during the Quaternary. Sierra Nevada sediment yield was increased by tectonically induced downcutting of deep canyons in response to 1 to 2 km of post-delamination uplift of the range crest and by glacial erosion. Rapid, continuing uplift of the Diablo Range foothill belt created new sources of sediment. Mountain-front encroachment reduced the area of the large San Joaquin Valley depositional basin. All three factors accelerated rates of accumulation of valley fill, but did either source area win the contest for depositional space?

Competition between Diablan and Sierran alluvial fans for space in the San Joaquin Valley can be described by cross sections such as Figure 1.6 (also see Lettis, 1985). Alluvial fans occupy space that is proportional to the amount of sediment deposited

on each fan (Hooke, 1968). Contacts between components of basin fill are vertical where aggradation rates are similar, and sloping where rates are dissimilar. Amounts of sediment derived from Diablan and Sierran sources varied with Late Quaternary climatic change and with uplift of the ranges. I assume that the overall Sierra Nevada sediment yield has not changed during the past 600 ky. Of course there were temporary and large increases during times of glacier advances and decreases during interglacial times (Lettis and Unruh, 1991). Glacial induced sediment yield increases may be responsible for temporary advances of Sierran alluvium toward the Diablo Range (Fig. 1.6). It is further assumed that the proportion of sediment flushed out to the Pacific Ocean has not changed.

The gently sloping contact between Diablan and Sierran alluvium (Fig. 1.6) records major changes in relative source-area sediment yields. The area of the Diablo Range piedmont has doubled. Progradation of Diablan alluvial-fan deposits over Sierran alluvium during the past 620 ky since deposition of the Corcoran Lake Clay indicates that alluvium derived from the Diablo Range was deposited at progressively faster rates, relative to rates of accumulation of arkosic alluvium derived from the Sierra Nevada. Transgression of Diablo Range basin fill over Sierran fill continued despite the depression of the basin that resulted from tectonic loading of the rising Coast Ranges. Much of the Corcoran lake clay is now far below sea level.

Mean rates of aggradation of basin fill above the Corcoran and above the much younger "A" lake clay increased from about 0.30 m/ky for the time span since 620 ka to about 0.55 m/ky for the time span since 27 ka (Croft, 1968, 1972). These increases in area and rate of deposition suggest a three- to six-fold increase in sediment yield from the rising Diablo Range. Diablo Range denudation rates may have increased from roughly 0.1 to 0.5 m/ky. Rates of sedimentation were sufficiently rapid to exceed tectonic subsidence (maximum rates are along the western margin of the valley) and shift the axis of deposition eastward.

The consequences of the delamination event for the crustal root of the Sierra Nevada batholith caused rapid uplift of the Sierra Nevada

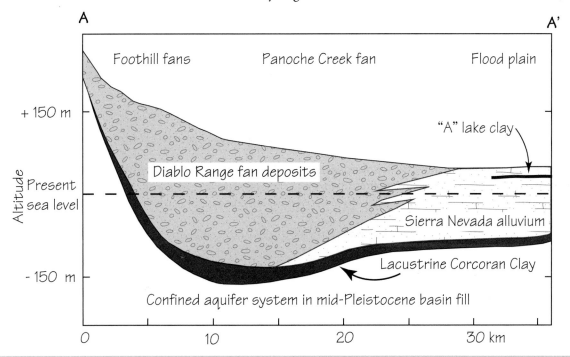

Figure 1.6 Cross section of valley fill deposited since the 620 ka Corcoran Clay Member of the Tulare Formation. From Magleby and Klein, 1965, plate 4, and Croft, 1968, 1972). Vertical exaggeration is 40:1

that occurred before 2 Ma. But, using the model of Jones et al. (2004), the regional consequences were perhaps even more profound for the California Coast Ranges. Compression near the San Andreas fault created a fold-and-thrust belt that continues to encroach northeastward into the microplate. The present eastern edge of the Coast Ranges was not formed by a synchronous pulse of uplift, as was the eastern front of the Sierra Nevada. Uplift progressed from southeast to northwest. This migration of thrust-faulted mountain fronts profoundly influenced landscape evolution and has continued to the present.

1.2.4 Mendocino Triple Junction

Innovative plate-tectonic analyses by Wilson (1965), McKenzie and Morgan (1969), and Atwater (1970) deciphered the consequences of a spreading ridge between the Farallon and Pacific plates moving into the

North American–Farallon subduction trench at –30 Ma. These plate movements were relative. North America might have progressively overridden parts of a Pacific plate oceanic spreading center, and the ratio of movement between the two plates probably changed with time. This arrival of the Pacific plate split the Farallon plate into the now distant Cocos and Juan de Fuca plates (Fig. 1.7B, C). Right-lateral continental transform faults formed – represented at present by the San Andreas fault. The southern part of the Juan de Fuca plate is named the Gorda plate.

The oceanic ridge of upwelling new crust met the North American plate at an oblique angle, but continued to spread out to both sides. The compressional side of the ridge moved off to the southeast as a triple junction between the underthrusting side of the ridge, the subduction trench in front of it and a newly created right-lateral and ever-lengthening transform fault extending behind. This was the birth of a system of transform faults. The San

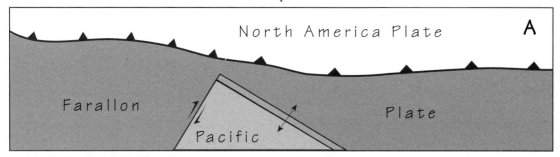

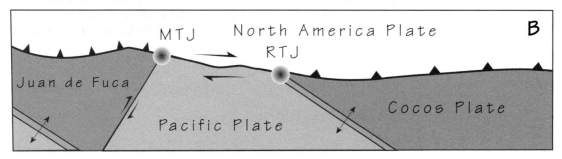

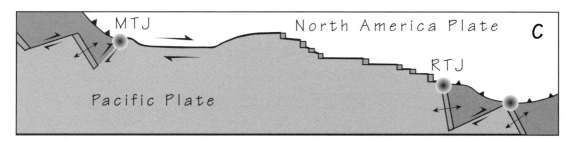

Figure 1.7 Plate-tectonic evolution of the San Andreas continental transform boundary. Paired arrows indicate dextral relative motion of transform faults. Double-headed arrows are ocean ridge spreading centers. Teeth are on the upper plate of subduction thrusts. Triple-junction locations are shown by the filled circles. From Dickinson, 1981.
A. Pacific plate approaches the North America plate at about 30 Ma.
B. At about 20 Ma. Rivera triple junction, RTJ, is moving southeast. Mendocino triple junction, MTJ, is moving northwest.
C. Plate boundary at the present time. San Andreas fault ends at string of rhombochasms at the head of the Gulf of California, which did not exist before 5 Ma. Rivera and Cocos plates are at far right. Gorda–Juan de Fuca plate is at far left.

Andreas fault is one of the more recent faults. This ridge–transform–trench Rivera triple junction now is off the coast of Mexico.

Migration to the northwest also formed part of the ever-lengthening right-lateral San Andreas transform fault. Being a transform–transform–trench tectonic system, the types of plate boundary interactions were different for this triple junction. New lithosphere generated on the north flank of the spreading ridge moved northward along the North American plate margin (Furlong and Govers, 1999; Furlong and Schwartz, 2004). They conclude that the Pacific–North America relative plate motion was subparallel to the North America plate margin. Subduction did not play a role. Instead, northwest migration of the Mendocino triple junction created a crustal void – a slab window – into which hot asthenosphere flowed. These processes thickened the crust ahead of the triple junction and thinned the crust in the wake of the migrating crustal welt. Rapid uplift rates decreased as the Mendocino triple junction continued its northwest migration, allowing the new crustal materials created by asthenospheric inflow to become cooler.

Uplift was not the result of convergence or underthrusting processes for the transform–transform–trench system. In part it resulted from delayed upwelling of hot asthenosphere into the crustal abyss created by the northwest movement of the Gorda plate beneath a presumably rigid North American plate (Dickinson and Snyder, 1978; Lachenbruch and Sass, 1980; Zandt and Furlong, 1982; Furlong and Schwartz, 2004). Concurrent lengthening of the San Andreas continental transform has continued to the present.

The northern tip of the San Andreas fault is not a single fault that joins the Mendocino fracture zone. This is a 30 km long restraining bend whose characteristics in relation to the enclosed King Range terrain are discussed thoroughly by Merritts (1996). The right-lateral characteristics change where the trace of the San Andreas fault comes ashore at Point Delgada. The obvious 1906 surface rupture quickly becomes obscure as tectonic displacements are dispersed along several oblique thrust faults. Merritts has one active trace of the divided San Andreas fault continuing along the east side of the King Range terrain, making a key left bend, and then joining the east–west Mendocino fracture zone. The zone is a mix of right-lateral and up-to-the-south motion (McLaughlin et al., 1994) in the near-shore terminus of the restraining bend. Having the San Andreas fault onshore, rather than at the previously thought offshore position, makes sense in terms of the 3–4 m/ky Late Quaternary uplift rates in the Punta Gorda region. The restraining-bend model for the northern tip of the San Andreas fault accounts for a long-term component of uplift associated with a migrating Mendocino triple junction. Each increment of fault slip should advance the restraining bend further north producing rock uplift, and converting more of the trailing edge to predominately dextral displacements.

Flights of marine terraces document a rising coastline north of the advancing triple junction (Carver and Burke, 1992). Furlong and Schwartz (2004) describe why crustal thickening followed by crustal thinning causes this tectonic-uplift perturbation. They also note that some accretion occurred as hot asthenosphere flowed into the slab window to cause local thickening of the overlying North American crust. Landscape evolution of the Mendocino coast reflects formation of an ephemeral crustal welt resulting from transpression at the fault tip, upwelling of hot asthenosphere, and local accretion. The associated waxing and waning of a regional pulse of uplift shifted north at ~5.6 km/My.

All of these processes take time. Uplift does not change abruptly with the approach or passage of the Mendocino triple junction. Deciphering the tectonic-geomorphology history of northwestern California is an exercise in understanding reaction times and response times to the passage of the Mendocino triple junction. Changes in rock uplift rate have profound consequences on geomorphic processes and landscapes, from headwaters to basin mouths of fluvial systems, that have a substantial range of reaction times and response times. In Section 8.3 we seek to understand the timing of perturbations and subsequent adjustments in both tectonic and geomorphic systems.

1.3 Australia–Pacific Plate Boundary

I like the challenges posed by diverse landscapes associated with continental transform faults. New Zealand is truly an ideal terrain for geologists interested in tectonics and topography. The 700 km long Southern Alps of the South Island are rising at an average rate of more than 5 m/ky. Crumbly fractured greywacke sandstone and weathered schist have a low rock mass strength that results in spectacular erosion rates in the humid to extremely humid mesic climate. It is the tectonically induced fractured nature of such rocks (Molnar et al., 2007) that contributes to watershed sediment yields in excess of 4,000 tonnes/km^2/yr.

Landscape evolution is rapid. The valley floor of the Waiho River downstream from the Franz Josef glacier experienced an aggradation event of 30 to 40 m, which was followed by creation of a flight of stream terraces during the subsequent degradation. It took only 10 years to create this flight of terraces, whereas in western North America the time span needed for the same scale of landscape change usually would be more than 10 ky. The landscapes are so sensitive that individual Alpine fault earthquakes cause brief aggradation (landslide sediment-yield increase) followed by net tectonically induced downcutting (Bull, 2007, figure 6.18). Such rapid rates of landscape change improve the likelihood of dating young geomorphic events. These advantages complement the drier North American study areas.

The Alpine Fault is a continent–continent collision zone between the Chatham Rise–Campbell Plateau to the east and the Challenger Plateau to the west (Suggate, 1963). It connects the Hikurangi and Puysegur subduction zones, in the north and south respectively, which have opposite convergence directions (Berryman et al., 1992; Norris and Cooper, 1997, 2000).

The Alpine fault (Figs. 1.8, 1.9) together with the Marlborough fault system of northeast-trending oblique strike-slip faults is a transpressional transform boundary between the Australian and Pacific plates in the South Island of New Zealand. A single nearly vertical fault zone south of Haast is characterized by right-lateral displacements. The Marlborough faults link west-dipping Hikurangi trough subduction with the east-dipping Alpine fault (Pettinga et al., 1998; Nicol and Van Dissen, 2002).

The importance of the Alpine fault was first recognized by Wellman and Willett (1942) and Wellman (1953). It became a plate boundary fault zone in the mid-Cenozoic (Lu et al., 2005). Rapid convergence of the tectonic plate has occurred since ~5 Ma with a shifting pole of plate rotation. Total dextral displacement is now ~450 km and rock uplift of the Haast Schist is 25 km (Sutherland, 1999; Cooper, 1980).

Present-day plate motion vectors on the central section of the Alpine fault (from Chase, 1978; Walcott, 1978) and DeMets et al. (1990) indicate rates of 39 to 45 m/ky that cause 37 to 40 m/ky of parallel slip between the Australian and Pacific plates, and 11 to 22 m/ky of contraction perpendicular to the fault. The Alpine fault accommodates ~75% of the fault-parallel interplate motion (Norris and Cooper, 1995, 2000; Sutherland, 1994). Milford Sound–Hokitika strike-slip displacement is 27 ±5 m/ky. Regional oblique thrust faults accommodate the remaining ~25% (Norris et al., 1990; Litchfield and Norris, 2000; Litchfield, 2001).

Figure 1.8 The Southern Alps in the South Island of New Zealand are being raised rapidly along the amazingly linear Alpine fault (between the two arrows). The width of this view from space is 490 km. This is a grayscale version of Shuttle Radar Topography Mission image PIA06661 furnished courtesy of NASA and JPL-Caltech.

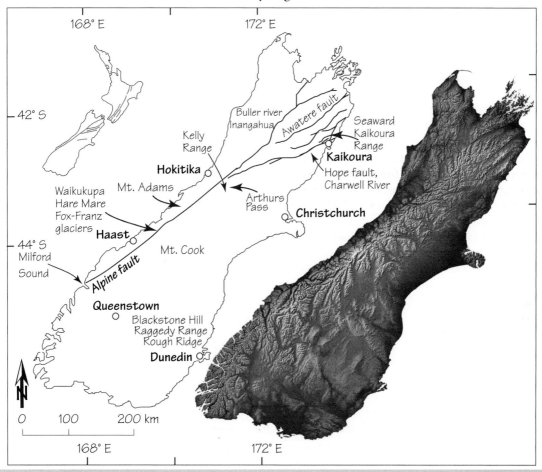

Figure 1.9 Locations of place names and study sites in the South Island of New Zealand and their book Section numbers [5.5]. Alpine fault [1.3, 5.3.1, 6.2, 6.3]; Arthurs Pass [1.3, 5.3.1]; Awatere fault [4.3.1]; Blackstone Hill [7.2.1]; Charwell River [1.1]; Fox and Franz Josef glaciers [6.3]; Hare Mare [6.3]; Hope fault [2.4, 6.2]; Kelly Range [6.2]; Milford Sound [3.6.2]; Mt. Adams [6.2]; Raggedy Range [7.2.1]; Rough Ridge [7.2.1]; Seaward Kaikoura Range [1.3, 2.4, 5.3.1, 6.2]; Waikukupa [6.3]. This is a grayscale version of Shuttle Radar Topography Mission image PIA06662 furnished courtesy of NASA and JPL-Caltech.

The 500–600 km long Southern Alps are rising fast with the highest uplift rates being in the central Westland District north of the New Zealand's highest peak (Mt. Cook). Wellman's uplift-rate map (1979) for the entire South Island shows maximum uplift of a bit more than 10 m/ky, but Adams (1980) increased that estimate to >20 m/ky. A geomorphic study of tectonically sensitive landforms tested these estimates.

Broad level benches in the Southern Alps near Haast (Fig. 1.8) have widely scattered rounded quartz pebbles. These ridgecrest notches were thought to be uplifted marine terraces (Cooper and Bishop, 1979). This seemed preposterous to geologists looking up at alpine ridgecrests 1,500 m above present sea level. It was simpler to have the quartz pebbles as being of fluvial origin or as crop stones from the now extinct large flightless bird – the moa.

Moa crop stones studied at the Canterbury Museum, Christchurch have a much different surface texture than pebbles on high-energy beaches. Slow churning of rocks in a moa's crop with siliceous vegetation imparts a delicate patina, like that of pebbles tumbled in a commercial rock polisher.

Bull and Cooper (1986) noted that such round quartz pebbles on benches notched into hillslopes at Arthurs Pass (Fig. 1.9)on the main divide could not be of fluvial origin because streams do not occur on ridgecrests.

Lacking isotopic means of dating the flights of notched benches, Bull and Cooper compared the spacing of these New Zealand uplifted marine terraces with the spacings on another South Pacific island – New Guinea – where coral provides a means of dating global marine terraces. Each uplift rate has a unique spacing of marine terraces. This approach to estimating ages of marine-terrace remnants has its detractors (Ward, 1988; Pillans, 1990), but has been used successfully by other workers (Merritts, 1987; Bishop, 1985, 1991; Carver and Burke, 1992; Grant et al., 1999; MacInness, 2004). Potential pitfalls and promises of the method are discussed in Section 8.3.1.

Bull and Cooper concluded that uplift rates on the Alpine Fault attained maximum values of 8 m/ky in central Westland, decreasing to ~5 m/ky towards the northeast and southwest. Their results were the same as the previous estimate by Wellman (1979), and the 9 m/ky exhumation rate since 0.9 My based on fission-track analyses (Tippett and Kamp, 1993). Uplift rates as estimated by Bull and Cooper (1986, 1988) are compatible with all rate estimates determined subsequently.

The 220–km long Hope fault is the most active splay of the transpressional transform plate boundary in the Marlborough section of the Alpine fault system (Knuepfer, 1992; Pettinga and Wise, 1994; Cowan et al., 1996). The Hope segment of the Hope fault extends at least 25 km west from the Hanmer pull-apart basin and is the only segment of the Alpine fault system to have ruptured since colonists arrived in substantial numbers in A.D. 1840. This was the North Canterbury Ms magnitude ~7.1 earthquake of 1888 (Cowan, 1991). The 58 km long Conway segment (Kahutara segment of Van Dissen

and Yeats, 1991) extends northeast from the 15 km long Hanmer pull-part basin to a 5 km right step at Mt. Fyffe near the town of Kaikoura (Fig. 1.9).

The Hope fault (Fig. 1.9) is as active as the San Andreas fault of California (Sieh and Jahns, 1984). Estimated slip rates are rapid (Bull, 1991, section 5.1.4); latest Holocene horizontal slip is estimated to be 33 (+2–4) m/ky. The Amuri and Seaward Kaikoura Ranges on the hanging-wall block have been rising during the Late Pleistocene at about 2.5 to 3.8 ± 0.4 m/ky. Hilly terrain on the footwall block has been rising at 0.5 to 1.3 m/ky. The fault then splits into several oblique thrust faults and uplift rates increase to about 4.5 to 6 m/ky. Erosion rates are estimated to be 5,000 tonnes/km²/yr (O'Loughlin and Pearce, 1982; Mackay, 1984) in that part of the Seaward Kaikoura Range discussed in Section 5.3.1.

A transpressive setting, and alternating thrust and dextral faulting along the Alpine fault is used in Figure 1.10A to illustrate serial and parallel partitioning styles of the range-bounding faulting. Serial partitioning is common along transpressive plate boundaries because the basal dip of a thrust-fault complex becomes less as a fold-and-thrust complex expands out from a transpressional mountain front. Serial partitioning also develops where rapid erosion of deep valleys prevents further expansion of a low-angle thrust complex (Norris and Cooper, 1995, 1997). Parallel partitioning is where displacement styles occur on two adjacent fault zones that are approximately parallel. The range-bounding fault may be the principal location of either lateral or vertical displacement.

Conway segment strain southwest of Mt. Fyffe has a parallel style of fault partitioning between the range-bounding dextral Hope fault that dips 75° to 85°, and an internal thrust fault with large amounts of vertical displacement. The internal Kowhai fault is clearly apparent in the topography 16 km west-southwest of Mt. Fyffe (Fig. 1.11). Further southwest, surficial expression of this internal fault zone is progressively less obvious but it may continue as a blind thrust.

The South Island of New Zealand also has an interesting fold-and-thrust fault belt in the Otago district northwest between Dunedin and the

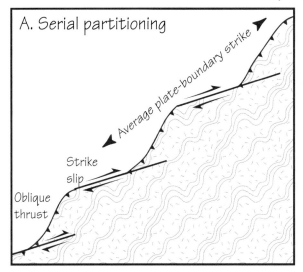

A. Serial partitioning

Average plate-boundary strike

Strike
slip

Oblique
thrust

Kowhai fault

Hope fault →

Figure 1.11 Parallel partitioning of the Conway segment of the Hope fault, New Zealand. The range-bounding fault has a dextral slip rate of 30 ± 3 m/ky) and an internal fault has a vertical slip rate of 2.5 ± 0.5 m/ky.

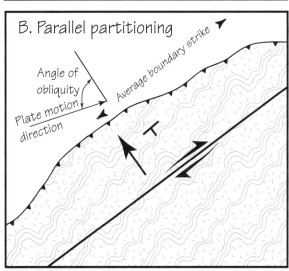

B. Parallel partitioning

Angle of
obliquity
Plate motion
direction

Average boundary strike

Figure 1.10 Styles of partitioning of surface fault traces on the Alpine fault, New Zealand transpressive plate boundary. Figure 19 of Norris and Cooper (1997).
A. Serial partitioning that consists of alternating oblique thrust and strike-slip fault segments.
B. Parallel partitioning that consists of thrust and strike-slip faulting in two separate fault zones. Direction of thrust plate movement shown by the large arrow.

Southern Alps (Fig. 1.9). Tectonic geomorphology studies in this area are based on deformation of a Late Cretaceous to mid-Tertiary erosion surface beveled in basement schist. The erosion surface is called the Waipounamu erosion surface (Landis et al., 2008). It is nearly flat, except where Late Cenozoic folds and thrust faults have created a set of 15 to 30 km long parallel ridges such as Rough Ridge and the Raggedy Range. These asymmetric anticlines have formed above buried reverse faults. The set of tectonically active structures accommodates much less than 10% of the oblique convergence between the Pacific and Australian plates. However, the frequency of magnitude Mw 6.5 to 7 thrust-fault earthquake events is sufficient to make this a good area in which to better understand how fault propagation affects evolution of stream networks (Section 7.2.1; Jackson et al., 1996; Markley and Norris, 1999).

1.4 India–Asia Plate Collision

Inclusion of Himalayan stream studies opens an active tectonics window to a much longer time span than the Plio-Pleistocene events in the southwestern United States and of New Zealand. Rapid tectonic

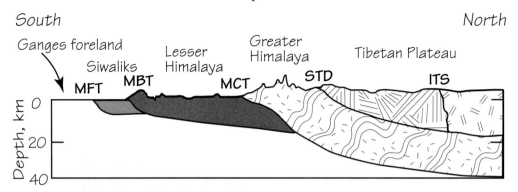

Figure 1.12 Generalized geologic cross section of the Himalaya from Hodges et al. (2001). The Greater Himalaya has peaks higher than 7 km. MFT, Main frontal thrust fault. MBT, Main boundary thrust fault. MCT, Main central thrust fault. STD, Southern Tibetan detachment normal fault. ITS, Indus–Tsangpo suture.

deformation of the southern Asia continent has continued an order of magnitude longer.

The initial impact of the rapidly moving Indian plate with Asian continental crust at approximately 55 mm/yr did little more than start to close the Tethys Sea, terminate marine sedimentation by 25 Ma, and fold sediments. The extent and magnitude of uplift caused by continental collision since 40 Ma was sufficiently large to change Earth's atmospheric circulation (Raymo and Ruddiman, 1992), and affect rock metamorphism (Zeitler et al., 2001). About 2,000 km of northward movement of the Indian plate has compressed the crust into twice its normal thickness, allowing the Tibetan Plateau and adjacent mountains (Fig. 1.12) to rise several kilometers by means of crustal buoyancy. This requires a comparable magnitude of right-lateral displacement of the indented continental mass towards China.

The tectonically driven indentor block – the Indian plate – has corners at its leading-edge, so some crustal materials are shoved aside (Fig. 1.13). Both ends of the Himalayas are areas where orogenic forces turn sharply about a vertical axis. These two tectonic syntaxes developed fairly late in the collisional history and are areas of exceptionally rapid erosion, rock uplift, and metamorphism. In a general sense they can be regarded as large antiformal structures that attest to the large magnitude

and continuing high rates of continental shortening. They are named after lofty peaks in their cores – Nanga Parbat in the west and Namche Barwa in the east (Wadia, 1931; Gansser, 1991). Both syntaxes are areas where high-grade metamorphism has overprinted basement rocks just since the Miocene Epoch. Zeitler et al. (2001) believe they "owe their origin to rapid exhumation by great orogen-scale rivers". Rapid degradation of the Himalayas since 4 Ma has influenced creation of syntaxial massifs that are developed atop weak crust that is relatively hot, dry, and thin.

Data obtained from many global positioning satellite monitoring stations provide key information regarding present-day tectonics of the Tibetan Plateau (Jade et al., 2004). These data allow the assessment of whether the Tibetan Plateau deforms as rigid plates and blocks that float on the lower crust, or by continuous deformation of the entire lithosphere. They conclude that Tibet behaves more like a fluid than like a plate. Crustal material is moving eastward (Lavé et al., 1996) and flowing around the eastern end of the Himalayas. Of course some deformation of the brittle upper crust occurs along fault zones between rigid blocks. Zhang et al. (2004) conclude that "deformation of a continuous medium best describes the present-day tectonics of the Tibetan plateau; . . . crustal thickening dominates

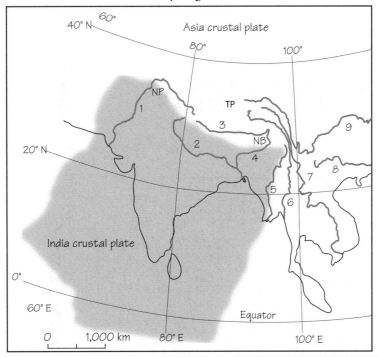

Figure 1.13 The Himalayan–Tibetan region discussed in Section 6.4. Metamorphic massifs (NP) and (NB) are near corners of the India crustal plate, which has been moving north-ward into the Asia crustal plate since ~40 Ma. (India crustal plate image provided cour-tesy of NASA). Study-site locations are Tibetan Plateau [TP], Nanga Parbat [NP], Nam-che Barwa [NB]. Rivers shown are the Indus [1], Ganges [2], Tsangpo [3], Brahmaputra [4], Irrawaddy [5], Salween [6], Mekong [7], Xi Jang [8], Chang Jiang [9].

deformation on its eastern margin except near the eastern syntaxis, where rapid clockwise flow around the syntaxis, not rigid body movement occurs". . . . , "material within the plateau interior moves roughly eastward with speeds that increase toward the east, and then flows southward around the eastern end of the Himalaya".

1.5 Aegean Transtension

The Mediterranean Sea region is tectonically ac-tive. The Aegean Sea and adjacent coastal Greece and Turkey are undergoing rapid crustal extension. Much of land has dropped below the sea (McKenzie 1972, 1978; Le Pichon and Angelier 1981; Jackson and McKenzie, 1988). The tectonic setting summa-rized here is mainly from Armijo et al. (1996, 1999).

In Section 7.1.2 we discuss propagation of normal faults and landscape evolution of the Corinth Gulf rift of Greece. Geodetic surveys across the Gulf of Corinth over the past century suggest that the present rate of north-south extension is 7–10 m/ky (Billiris et al., 1991; Armijo et al., 1999).

Local rifting results from larger-scale tec-tonic deformation between the Arabia, Africa, and Eurasia plates (Fig. 1.14). This again illustrates the importance of wrench-fault tectonics. The Arabia–Eurasia collision zone closed in the Miocene and con-tinued plate movements created one of the worlds most active continental transform faults (McKenzie, 1972). The right-lateral North Anatolian fault forms the boundary between Eurasia and Anatolia. Transtensional deformation north of the Hellenic

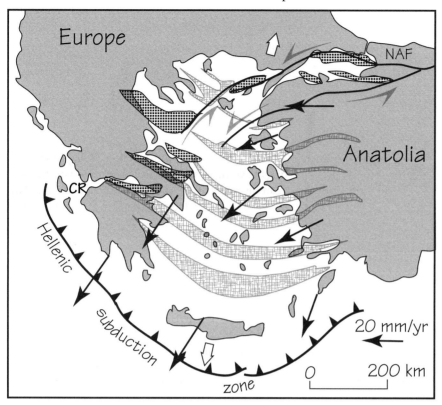

Figure 1.14 Tectonic setting of southwestward continental extrusion where the Aegean block is being forced westward along the North Anatolian fault zone (NAF) as a result of the collision of the Arabian and Eurasian tectonic plates far to the east. Two deformation regimes are widespread, slow extension starting before 5 Ma (light gray pattern) followed by faster transtension associated with propagation of the North Anatolian fault (dark gray pattern). The amount of extension decreases toward the western and eastern margins creating the curvature of the arcs of graben extension. Propagation of the southern branch of the North Anatolian fault zone across the Aegean progressively reactivated grabens such as the Corinth Rift (CR). Black arrows represent current motion with respect to the European plate based on space geodetic measurements. Present rapid Corinth Rift extension absorbs 20–30% of the total right-lateral motion on the North Anatolian fault zone. Slow extension persists in the rest of the Aegean and western Turkey (areas of light gray pattern and the stubby white arrows). Figure 25D of Armijo et al. (1996).

subduction zone is related to the westward propagation of the North Anatolian fault. This implies extrusion of the Anatolia–Aegean block as the North Anatolian fault propagated 2,000 km to the west of the Karliova triple junction; the Eurasia–Anatolia–Arabia plate junction. The fault originated at ~10 Ma, entered the Aegean Sea at 5 Ma, and reached central Greece at 1 Ma (Armijo et al., 1996).

Aegean transtension accelerated as the North Anatolian fault propagated ever closer to Greece. Armijo et al. (1996) see this as a two-stage process. Widespread, slow extension initiated the process (light gray patterns in Figure 1.14), and was followed by more local, faster transtension associated with later, further westward propagation of the North Anatolian fault (darker gray diverging patterns in Figure 1.14). Note how nicely the basic wrench-fault model (presented later as Figure 2.12) agrees with the direction of Aegean transtension. The post-1-Ma phase of rapid Corinth rifting (dark gray patterns in Figure 1.14) appears to be the result of southwestward propagation of the southern branch of the North Anatolian fault.

The seismically active 130 km long Corinth Rift (Ambraseys and Jackson, 1990) is an asymmetric half-graben with an uplifted southern footwall and downward flexed northern hanging wall, with minor antithetic faulting. The resulting coastlines are linear in the south and sinuous in the north. The minimum thickness of sediment in the Corinth Rift gulf indicates a fast slip rate of 6 m/ky, the more likely slip rate on the main rift-bounding fault since 350 ka is an amazing ~11 m/ky. The footwall blocks of the active faults have risen as the hanging-wall blocks plunged into the Corinth Gulf. Regionally, this footwall-block uplift does not appear to exceed a slow 0.3 m/ky. But this was sufficient to preserve flights of marine terraces and profoundly alter patterns of streamflow (Section 7.1.2).

1.6 Summary

The study areas used in this book are representative of common geologic structures and tectonic landforms on planet Earth. Pure strike-slip faulting is not emphasized because it does not change the base level of streams. Admittedly, this chapter summarizing local tectonics is brief, almost casual. That is because I prefer to emphasize geomorphic discussions of tectonics. I may have omitted your favorite mountain range by making sure that this book includes a broad range of tectonic, lithologic, and climatic settings. My apologies for not using the Andes of South America, the Alps of Europe, Mongolia, Japan, Indonesia, Alaska, and the tectonically inactive expanses of Australia. I invite you to apply the ideas presented here to these equally fascinating landscapes.

Even an essay as long as this book is merely a thought-provoking introduction. Geographic settings have been selected to maximize diversity. The crumbly rocks of the extremely humid Southern Alps of New Zealand have impressive rates of erosion as the result of Pleistocene rock uplift. The arid to semiarid Panamint Range of southeastern California has such little annual stream power that tectonic landforms persist for millions of years. The young Coast Ranges of central California provide key study areas of soil-mantled hillslopes that provide essential comparisons of tectonically active and inactive fluvial systems underlain by soft rocks. The Himalayas and Sierra Nevada were created 40 to 60 million years ago, long before the Southern Alps. They provide comparisons with the topography of young mountain ranges.

The landscapes discussed here have immense variability that we sense remotely, fly over, and hike across, and photograph. Mountainous topography contains signatures that proclaim the importance of rock uplift on landscape evolution, as well as climate and rock type.

Let us proceed with discussions of how hills and streams respond to tectonic perturbations that include thrust faulting, tilt, and upwelling of hot asthenosphere. The main purpose of the next three chapters is to better understand how rock uplift, and rates of stream-channel downcutting, influence hillslope processes and shapes.

Drainage Basins

Mountains consist of hillslopes except for the few percent of area in valley floors. Hillslopes change gradually because most hillslope geomorphic processes are slow compared to those in stream channels. Changes in hillslope shape involve removal of large masses of rock, much of which is first weathered to colluvium, saprolite, and soil. So the consequences of continuing or renewed rock uplift on hillslopes are gradually integrated into mountainous landscapes over long time spans. Such landscape changes may require more than a million years in hot, dry deserts with resistant lithologies, and < 10 ky in steep humid mountains composed of crumbly rock types. Delving into such complex landscape evolution can be approached with hydrologic, engineering, or land-use goals in mind, but the emphasis here is different.

The fairly ambitious purpose of Chapters 2–4 is to explore how tectonic base-level change affects common hillslope processes and shapes. Complex hillslope topography is the product of many controlling variables, but contrasts between tectonically active and inactive soil-mantled terrains are obvious and profound to me. Every landscape element of hillslopes including planimetric watershed shape, topographic hollows and noses, and longitudinal profiles of valleys is affected by past regional uplift and present local tectonic activity. Let us give you some insight so that such comparisons are easy and styles of landscape behavior are obvious.

The scope of the next three chapters is diverse. Chapter 2 emphasizes the larger scale and the longest time span – the planimetric aspects of

A 1.7 km wide aerial view of contrasting geomorphic processes of two drainage basins in the Seaward Kaikoura Range, New Zealand that are rising ~4.5 m/ky. Both watersheds are underlain by fractured greywacke sandstone, but higher altitude and greater frost wedging favor dendritic talus slopes in one basin as compared to forested slopes and stream channels in the other. Image furnished courtesy of Alistair Wright of Environment Canterbury, New Zealand.

Tectonically Active Landscapes, 1st edition. By W. B. Bull. Published 2009 by Blackwell Publishing, ISBN 978-1-4051-9012-1

watershed shape. Ridgecrest convexity, footslope concavity, and cross-valley morphologies are the subjects of Chapter 3 with an emphasis on landscape signatures of tectonically active and inactive settings. Smaller-scale aspects are examined in Chapter 4, with the emphasis on sediment yield and mass movements. Each hillslope process has a typical response time to perturbations caused by changes in tectonic activity and other independent variables of a fluvial system. Climatic changes affect entire drainage basins, commonly with short response times of less than 1 ky (Fernandes and Dietrich, 1997). Consequences of local uplift along faults and folds are superimposed on those of regional uplift. Reaction times are less than 1 ky, but relaxation times (Bull, 2007, Section 2.5) typically exceed 1 my and may be much longer.

Tectonic perturbations commonly emanate from fault zones. In marked contrast to climatic perturbations, watershed responses to pulses of uplift have spatial as well as temporal lags of response to uplift events. Tectonically induced increases of relief along a fault or fold accelerate channel downcutting and steepen stream gradients.

Downcutting or backfilling of the valley floor changes local base levels and eventually the shapes of adjacent hillslopes (Davis, 1898, 1899). Base-level fall at a faulted mountain front is transmitted throughout the drainage net and up the adjacent hillslopes. Footslopes are both extended and steepened as each consecutive reach of a valley floor is deepened and steepened. In this manner, the rates and types of processes operating on hillslopes are changed by uplift along nearby or distant faults. The stream is the connecting link between an active fault or fold and the headwaters of a drainage basin.

Fluvial landscapes consist of drainage basins – watersheds whose divides direct flow of water down hillslopes into stream channels. Chapter 2 is a story of how drainage basins compete for available space in mountain ranges. Watershed–divide shift can be exceedingly gradual or abrupt.

The abrupt nature of stream capture involves situations of abnormal concentrations of stream power in unusual locations within a watershed. This chapter considers some fundamental processes that change planimetric shapes of watersheds. We examine the effects of differing uplift regimes on basin shape for opposite sides of the tilted Panamint Range in southeastern California. Tectonic controls of a Diablo Range fold-and-thrust belt and how they promote stream capture is the next topic. First, let us define locations within drainage basins with an emphasis on power available to do geomorphic work.

2.1 Hydraulic Coordinates

A good way to relate tectonics to topography is with models that compare landforms and geomorphic processes at different spatial or altitudinal positions in a fluvial system. I do this with dimensionless hydraulic coordinates that quantify spatial and altitudinal positions within a fluvial system. Two different classes for positions are used; coordinates for hillslope points and coordinates for locations along the trunk stream channel of a drainage net. Tectonic geomorphologists use hydraulic coordinates to relate the position of an active fault or fold to nearby or distant parts of a fluvial system, and to compare the morphologies of hills and stream channels in different tectonic, lithologic, and climatic settings.

Two planimetric coordinates are described in terms of ratios of horizontal distances of flow direction down hills and streams. The *hillslope-position coordinate*, H_{pc}, is the planimetric length from a ridgecrest divide to a point on the hillslope, L_h, divided by the total length from the divide to the edge of the valley floor, L_t.

$$H_{pc} = \frac{L_h}{L_t} \qquad (2.1)$$

Hillslope-position coordinates range from 0.00 at the ridgecrest to 1.00 at the base of the footslope.

The *basin-position coordinate*, B_{pc}, is the planimetric length, L_v, from the headwaters divide of a drainage basin along the trunk valley to a streambed point divided by the total length of the valley, L_{vt}, from the headwaters divide to the mouth of the drainage basin. The point in the valley floor should be directly down the flow line from a hillslope point of interest, which commonly is directly opposite the stream-channel point.

$$B_{pc} = \frac{L_v}{L_{vt}} \qquad 2.2$$

Total length is not measured along a sinuous stream channel, because this is a landform that changes too quickly for our longer-term perspective, nor along a chord between two endpoints. Instead, the basin-position coordinate describes distance along the trend of a valley. Interpretation of the numerical values is fairly straightforward. For example, 0.50R/0.67 describes a point half way down the right side hillslope for a point whose flux of sediment and water is two-thirds of the distance through a drainage basin. Right (R) and left (L) sides are when looking downstream.

The aerial photos used in Figure 2.1 illustrate descriptions of hillslope- and basin-position coordinates. Locations of points along trunk streams are easy to define in low-order basins (Fig. 2.1A); flow proceeds from points 1 to 2 to 3 whose hydraulic coordinates are listed in Table 2.1. Point 2 marks the location where fluxes of water and sediment from a small tributary valley enter the trunk stream. Point 4 is a third of the way down the right-hand side hillslope opposite a basin-position coordinate of 0.48. Note that I have made a subjective decision to regard rills as part of this hillslope instead of as lower order tributary valleys. This type of subjective decision is always present and is largely a matter of scale and mode of depicting drainage nets, and the purposes of your investigation. The rills, which would not be apparent on conventional topographic maps, are useful. In this case, they alert us to measure distance along flow lines that are not at right angles to the trunk valley floor in much of this watershed. The left-side position of point 5 is denoted with an L, and this slope drains directly into the trunk channel.

Basin-position coordinates can also be used to describe points on hillslopes within multiple orders of nested tributary valleys. The first step in defining hydraulic coordinates in more complex watersheds is to identify the trunk channel of the longest subbasin, from the headwater divide to the mouth of the drainage basin. We then identify the junctions at which water and sediment from tributaries join this trunk stream. By working our way upstream we go from the mouth of the master stream to the sources of sediment and water.

Two large subbasins are shown in Figure 2.1B. The longest flow path is through the southern subbasin, which in this case has a larger watershed area than the northern subbasin. Flow begins at point 1 and leaves the watershed at point 6. Points 2, 3, 4, and 5 are the locations of several stream junctions whose tributary valleys contain the hillslope points of Table 2.2. Each stream junction marks an abrupt increase of discharge from tributary watersheds that here are progressively larger in the downstream direction.

Hillslope-position coordinates are the same as for the simple watershed of Figure 2.1A, but basin-position coordinates should include the location of the tributary in the nested hierarchy of stream orders. For example, points 7 through 12 are all located in the tributary that joins the trunk channel at a basin-position coordinate of 0.85, so (0.85) is placed

Hydraulic coordinate number	Basin-position coordinate	Hillslope-position coordinate
1	0.00	0.00
2	0.27	1.00
3	1.00	1.00
4	0.48	0.30R
5	0.70	0.72L

Table 2.1 Basin-position and hillslope-position coordinates for points in the low-order drainage basin shown in Figure 2.1A.

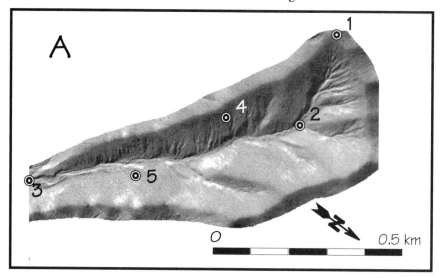

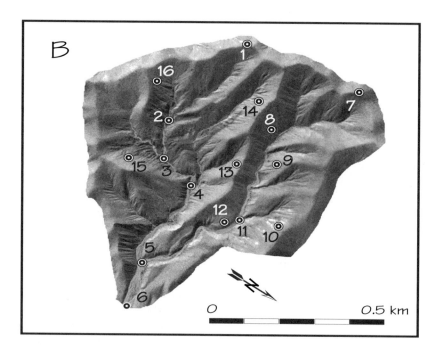

Figure 2.1 Aerial photographs of drainage basins eroded into soft rocks of the eastern Rodman Mountains, central Mojave Desert, southern California. Hydraulic coordinates for each numbered point are listed in Tables 2.1 and 2.2.
A. Low-order watershed with points to illustrate basic concepts of hydraulic coordinates.
B. Complex watershed with points to illustrate numbering for nested basin-position coordinates.

Hydraulic coordinate number	Basin-position coordinate	Hillslope-position coordinate
1	0.00	0.00
2	0.33	1.00
3	0.46	1.00
4	0.57	1.00
5	0.85	1.00
6	1.00	1.00
7	(0.85) 0.00	0.00
8	(0.85) 0.31	0.67R
9	(0.85) 0.44	0.51L
10	(0.85, 0.68) 0.70	0.27L
11	(0.85) 0.68	1.00
12	(0.85) 0.70	0.81R
13	(0.57, 0.90) 0.53	0.40L
14	(0.57) 0.28	0.52L
15	(0.46) 0.87	0.39R
16	(0.33) 0.45	0.36R

Table 2.2 Basin-position and hillslope-position coordinates for points in the more complex drainage basin shown in Figure 2.1B.

before the basin-position coordinate in the tributary. Point 10 is on a left-side hillslope of a small basin that has a third-order relation to the trunk valley floor. Its location would be described by [(0.85, 0.68) 0.70; 0.27L]. The number 0.85 tells us where flow from the large second-order tributary joins the third-order trunk stream. The number 0.68 refers to the fact that this first-order stream joins a second-order stream at a position that is 68% of the distance from the headwater divide of the second-order basin. Flow passing through hillslope-position coordinate 0.27L passes directly to first-order channel at the local basin-position coordinate of 0.70. Similarly, point 13 is 40% down a left side hillslope and 53% down a first-order basin that drains to a second-order basin at 90% of its length, which in turn drains to the third-order stream at 57% of its length: [(0.57, 0.90) 0.53; 0.40L].

Dimensionless numbers can also be used to describe relative relief positions. The hillslope-fall ratio, H_f, is the decrease in altitude from a ridgecrest divide to a point on the hillslope, H, divided by the total decrease in altitude from the divide to the base of the footslope, H_{th}.

$$H_f = \frac{H}{H_{th}} \qquad (2.3)$$

The basin-fall ratio, B_f, is the decrease in altitude from the headwaters divide along the trunk valley to a point in the valley floor, H_v, divided by the total watershed relief (decrease in altitude from the headwater divide to the mouth of the drainage basin) R.

$$B_f = \frac{H_v}{R} \qquad (2.4)$$

Hillslope and basin-fall coordinates for a sequence of nested drainage basins can be described in much the same way as for hillslope- and basin-position coordinates by using percentages of total relief in each subbasin draining to the trunk stream.

Dimensionless hydraulic coordinates and ratios minimize the factor of size in comparisons of fluvial landscapes. The basin-position coordinate can describe knickpoint migration in adjacent watersheds of different sizes and stream powers.

I use hydraulic coordinates to compare tectonic signatures in hillslope morphology. Hillslope-position coordinates are used to describe the relative locations of change from convex to concave slopes, landslide head scarps, and where rills start to incise with respect to their source ridgecrests. Use of hydraulic coordinates standardizes survey procedures. I like to use the basin-fall ratio for dimensionless analyses of stream-gradient index (Hack, 1973a) and for semi-logarithmic longitudinal profiles of streams.

2.2 Basin Shapes

Drainage basins change their planimetric shapes and areas when unequal denudation rates in adjacent watersheds cause drainage divides to migrate. Tectonically induced downcutting generated by displacement along a range-bounding fault is a base-level fall that migrates up the drainage net and eventually arrives in headwaters reaches to accelerate degradation of hillslopes and watershed divides. Highly elongate basins characterize actively rising escarpments (Davis, 1909; Shelton, 1966), but become more circular through the processes of divide migration and stream capture (Horton, 1932, 1945).

Basin shape influences conveyance of sediment load through a fluvial system. The drainage basins of rapidly rising mountainous escarpments range from short rills to long deep canyons. Strahler (1964) points out that such tectonically active (elongate) basins have lower order trunk stream channels compared to tectonically inactive (circular) basins. Philips and Schumm (1987) describe increases in basin circularity and drainage-net characteristics that occur with decreasing overall watershed slope. Cannon (1976) postulates that the preponderance of lower-order streams that empty directly into the main channel of an elongate basin represents a concentration of energy for efficient downcutting and headward erosion along the main channel. Conversely, a circular basins may be the most efficient configuration of drainage net for conveyance of sediment loads in watersheds with stationary drainage divides.

Planimetric shapes of drainage basins may be described by *elongation ratios* such as the one used by Cannon (1976). His R_e is a dimensionless index of circularity.

$$R_e = \frac{R_c}{L} \qquad (2.5)$$

R_c is the diameter of a circle with the same area as the basin and L is the map length between the two most distant points in the basin. Basin elongation increases with departure below the 1.0 value for a circle.

2.2.1 Panamint Range Watersheds

Erosional power is concentrated along valley floors where streams gather progressively more force as they flow downvalley. Hillslope erosion is infinitely slower as comparatively weak geomorphic processes gradually remove huge volumes of rock. Watershed divides have the longest reaction times to mountain-front tectonic perturbations.

The west flank of the arid to semiarid Panamint Range of southeastern California is a dramatic tectonic setting. Table 2.3 uses the tectonic signatures in such landscapes to define five *classes of relative tectonic activity*. Landforms indicative of class 1 tectonic activity include an unentrenched alluvial fan being deposited next to a straight mountain front that coincides with a fault zone, range front triangular facets, and exceptionally narrow valley floors relative to the heights of the adjacent divides.

A rugged mountainous escarpment has a conspicuous break-in-slope (Fig. 2.2A) at approximately 900–1000 m above the range base. The ancestral landscape consisted of hills and broad valley floors. Uplift of this relict landscape records an impressive 1 km increase in relief during tectonic extension of the southwestern part of the Basin and Range Province.

Different landforms have different response times to uplift along the range-bounding fault. The trunk stream channels of the six largest west-flank watersheds should have the shortest response times. Their larger streams erode downward fastest, becoming linear sites of maximal relief. Adjacent steep hillslopes have ridgecrests that slowly migrate into parallel watersheds of smaller streams. These smaller watersheds have yet to fully respond to the range-front tectonic base-level fall. The result is lesser relief and relatively slower rates of ridgecrest erosion. Thus the adjacent larger basins will gradually

Class of Relative Tectonic Activity	Relative Uplift Rate	Typical Landforms	
		Piedmont*	Mountain**
Active			
Class 1A – maximal	$\Delta u/\Delta t \geq \Delta cd/\Delta t + \Delta pa/\Delta t$	Unentrenched alluvial fan [0.6–0.9]	V-shaped valley profile in hard rock [1.1–1.4]
Class 1B – maximal	$\Delta u/\Delta t \geq \Delta cd/\Delta t + \Delta pa/\Delta t$	Unentrenched alluvial fan [0.6–0.9]	U-shaped profile in soft rock [1.0–1.2]
Class 2 – rapid	$\Delta u/\Delta t < \Delta cd/\Delta t > \Delta pd/\Delta t$	Entrenched alluvial fan [1.0–1.1]	V-shaped valley [1.1–1.3]
Class 3 – slow	$\Delta u/\Delta t < \Delta cd/\Delta t > \Delta pd/\Delta t$	Entrenched alluvial fan [1.1]	U-shaped valley [1.0–1.1]
Class 4 – minimal	$\Delta u/\Delta t < \Delta cd/\Delta t > \Delta pd/\Delta t$	Entrenched alluvial fan [1.1]	Embayed front [1.0–1.1]
Inactive			
Class 5A	$\Delta u/\Delta t << \Delta cd/\Delta t > \Delta pd/\Delta t$	Dissected pediment [1.1]	Dissected pediment embayment [1.0–1.1]
Class 5B	$\Delta u/\Delta t << \Delta cd/\Delta t = \Delta pd/\Delta t$	Undissected pediment [1.0]	Dissected pediment embayment [1.0]
Class 5C	$\Delta u/\Delta t << \Delta cd/\Delta t < \Delta pd/\Delta t$	Undissected pediment [1.1]	May be like class 1 landscapes

* Unentrenched – entire fanhead deposited recently, or only Holocene fan surfaces are entrenched. Entrenched alluvial fanhead surfaces with Pleistocene soils have incised stream channels.
** Stream power / Resisting power ratios in [ ] suggest departure from equilibrium value of 1.0.

Table 2.3 Geomorphic classification of Quaternary relative tectonic activity of mountain fronts as defined by the relative rates of uplift u; stream-channel downcutting, cd; piedmont aggradation, pa; piedmont degradation, pd; and time, t. Table 4.1 of Bull, 2007.

encroach into the small basins, changing basin circularity for both.

The east side of the range is rugged and lofty, but mountain-front landscapes are much different. Landforms indicative of Class 3 tectonic activity include permanent stream-channel incisement of the alluvial fan that has shifted the apex of fan deposition downstream, poorly defined triangular facets above a sinuous mountain–piedmont junction, and broad straths beneath the valley floors in the mountains.

The Panamint Range block continues to be tilted to the east, but lack of faulting at the mouths of east-side drainage basins during the late Cenozoic

has favored gradual widening of the drainage basins. Figure 2.2B shows a broad valley and old, inset alluvial fan surfaces – both indicate minimal recent uplift. The Figure 2.2B front is rated as being characterized by very slow uplift Class 3. The Figure 2.2A view of a west-side mountain front has the maximal active rating of Class 1.

Contrasting drainage-basin shapes of the Panamint Range reflect different tectonic settings and responses to drainage-basin uplift. Basin shapes may have been similar before ~4–7 Ma when these plutonic mountains were part of the Sierra Nevada microplate. Relief, represented by the relict uplands, was doubled as this fault block was tilted

Figure 2.2 Contrasting mountain fronts of
the west and east sides of the Panamint
Range, southeastern California.
A. Dissected low-relief upland on the west
flank. View shows 1,200 m high escarpment
at mouth of Redlands Canyon which is 10
km south of Pleasant Canyon (Fig. 2.3).
Snow-capped summit hills at altitudes of
2000 to 2200 m are part of the inherited
pre-4-Ma hilly terrain. Alluvial-fan surfaces
adjacent to the faulted mountain front are
of Holocene age.
        The large fan is representative of
a fluvial system that extends to the main
divide. The small fans are derived from adjacent watersheds that have not undergone
as much stream-channel downcutting.

by displacements on the west side normal faults.
V-shaped canyons characterize these west-side fluvial systems.

This is a typical response to large, rapid
base-level fall generated by active range bounding
fault zones. Active normal faulting on dipping fault
planes appears to have doubled the lengths of the
west-side drainage basins.

Drainage-basin widths become much narrower near the tectonically active west-side front.
Mean basin widths are 1.0 km for the west side basins and 3.2 km for the east-side basins at basin-position coordinates of 0.9. The downstream thirds of
the west-side watersheds are the youngest additions
to the fluvial systems. They generally narrow quickly.
The upstream thirds of the watersheds are remnants

Figure 2.2 Contrasting mountain fronts of
the west and east sides of the Panamint
Range, southeastern California.
B. Hanuapah Canyon mountain front on the
east side of the Panamint Range. Poorly preserved triangular facets mark approximate
location of range-bounding normal fault. Alluvial fan surfaces are Middle Pleistocene to
Holocene in age. Death Valley playa is in the
foreground.

of pre-4-Ma landscapes. As such they have planimetric shapes similar to the east-side watersheds. The middle thirds of the watersheds have been present long enough for basin widening to occur and they do not narrow downstream. Future widening will reduce the areas of the triangular facets between the six drainage basins that extend to the main divide of the Panamint Range.

The east-side drainage basins have been tilted. A Late Quaternary tilt estimate is 0.0036 $\pm 0.0007°$/ky (Roger Hooke, email of 7 July 2008). Drainage-basin lengths may have become shorter if deposition of Death Valley basin fill encroached onto the footslopes of the Panamint Range. Broad valley floors with straths characterize the east-side drainage basins. The east-side basin circularity, as compared to the west-side basin shapes, indicates much slower rates of tectonically induced downcutting.

Contrasting tectonic styles between the east and west sides of the Panamint Range suggests that planimetric drainage-basin shapes also should be different. Indeed they are. Different basin shapes and areas for the east- and west-side basins are obvious in Figure 2.3. Rock types and climate are similar. Six

narrow west-side drainage basins are crowded into the same segment of the main divide as four east-side basins. The total areas of the four east-side basins and six west-side basins are virtually the same.

The degree of watershed divide circularity varies greatly from nearly circular for Hanuapah Canyon to excessively long and narrow for Pleasant Canyon. East-side basins have elongation ratios (equation 2.5) that range from 0.66 to 0.72 (mean is 0.68) whereas west-side basins range from 0.40 to 0.63 (mean is 0.53). Larger east-side R_e values, and the extensive broad straths beneath the stream channels, indicate that > 1 My has passed since significant base-level fall occurred along the east-side mountain front. These markedly different drainage basin shapes are an introductory glimpse into the tectonic controls of long-term landscape evolution of the Panamint Range – the subject of Section 8.1.1.

2.3 Divide Migration and Stream Capture

A stream that cuts down faster than the streams of adjacent watersheds may become the dominant watershed. Adjacent drainage basins change their areas and shapes when parts of their common watershed

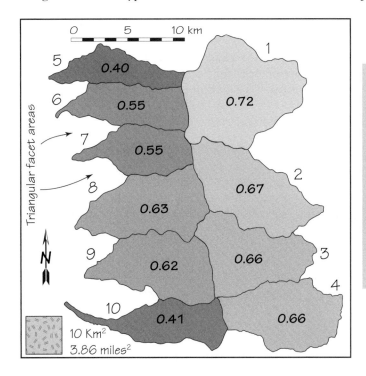

Figure 2.3 Tectonically controlled contrasts in shapes and areas of drainage basins on the east and west sides of the Panamint Range that head in the main divide. Elongation ratios of drainage basin shape describe departures from circular (1.00). East side watersheds are: 1, Hanuapah; 2, Starvation; 3, Johnson; and 4, Six Spring. West side watersheds are: 5, Tuber; 6, Jail; 7, Hall; 8, Surprise; 9, Happy; 10, Pleasant.

divide migrates. The ***dominant watershed*** is the one that is slowly expanding its area at the expense of adjacent ***subordinate watersheds***. Dominant drainage basins become larger at the expense of their neighbors. Its steeper hillslopes have ridgecrests that migrate into the more passive fluvial system. The subordinate stream is beheaded when some of its streamflow is diverted into the dominant stream. Drainage-basin shapes change very slowly as watershed divides migrate in response to differential erosion rates of opposing hillslopes (Horton, 1945). Capture of a headwaters drainage net by the stream of an adjacent watershed represents a sudden expansion by the dominant drainage basin to an ever more circular shape.

2.3.1 Stream Capture and Changing Geomorphic Processes

Stream capture is especially common in tectonically active mountain ranges because rapid uplift allows soft rocks to become ridgecrests between adjacent watersheds. Soft rocks favor faster ridgecrest denudation that may divert streamflow into the dominant watershed.

Locally resistant rocks may inhibit upstream migration of knickpoints created by tectonic

deformation of downstream reaches. This lack of fluvial continuity means that both stream channels and hillslopes in the reach upstream from the hard rocks may degrade slower than in an adjacent watershed.

Capture is a fascinating story about a base-level fall that occurs at a most unlikely location – a watershed divide. Abrupt diversion of flow occurs when a local ridgecrest altitude becomes less than stream-channel altitude. Adjacent streams may respond differently to vertical movements on a range-bounding fault that cause a similar base-level fall. Variable rock mass strength controls rates of tectonically induced downcutting and partly determines how fast a base-level fall migrates upstream.

Stream capture may occur in settings with minimal tectonic activity such as the Appalachian Mountains of the eastern United States (Judson and Kauffman, 1990). The leadoff example used here is from semiarid southern Arizona. We introduce the subject of stream capture with an example from a tectonically inactive landscape in order to better identify non-tectonic factors influencing stream-capture processes. Climate of the Santa Catalina Mountains ranges from strongly seasonal semiarid and strongly seasonal thermic to moderately seasonal subhumid and moderately seasonal mesic (Appendix A).

The range-bounding faults of the Santa Catalina Mountains have been inactive since the Basin and Range orogeny of 12 to 8 Ma. Erosion, without the influence of new tectonic perturbations, has been the agent of landscape change since then. Erosion has created deep valleys even in upstream

Figure 2.4 Aerial view of the south side of the Catalina mountains of southern Arizona. The largest drainage basin, Sabino Canyon, **S**, has eroded deeply into gneissic rocks during the past 8 My. Much of Bear Canyon, **B**, flows at a higher level because stream-channel downcutting is locally impeded by hard rock at Seven Falls, **7**. Deeper erosion of Sabino Canyon has created a situation of imminent stream capture of Sycamore Canyon, **Sy**, in the headwaters of Bear Canyon. Photograph by Peter L. Kresan ©.

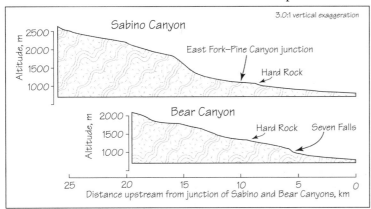

Figure 2.5 Longitudinal pro-files of the Sabino Canyon and Bear Canyon watersheds. Santa Catalina Mountains of southern Arizona, USA. Neither stream has eroded a smooth concave profile along the trunk stream channel. Impediments, such as Seven Falls, are local base levels. Even small steps (Hard Rock) may create situations of poten-tial stream capture.

reaches. The scale of long-term denudation and the time span of tectonic quiescence is revealed by 0.5–1 km of retreat of the mountain front to create a pedi-ment. This magnitude of pedimentation implies that the Santa Catalina Mountains also have been lowered by at least 500 m.

The largest watershed in the 2,970 m high mountains is 91 km² Sabino Canyon (Fig. 2.4), which has a fairly circular drainage basin (basin elongation ratio of 0.61). Large floods (historical flood discharg-es of >400 m³/second) and favorable rock mass strength have allowed the downstream reaches of Sabino Creek to gradually cut down to its base level of erosion. The result is a moderately wide valley floor flanked by spectacular cliffy hillsides. The 10 km long headwaters reach cannot respond to chan-nel deepening near the mountain front because up-stream migration of knickpoints has created a 500 m

giant step in the longitudinal profile (Fig. 2.5). Such occurrences of hard rock are local base levels that disrupt the continuity of a fluvial system.

The adjacent 43 km² watershed of Bear Canyon (Figs. 2.4, 2.5) also has a stream that cross-es patches of massive gneiss with high rock mass strength. The result is steep reaches and waterfalls. The reach at Seven Falls has a gradient of 0.269 m/m compared to 0.071 and 0.066 m/m for the ad-jacent upstream and downstream reaches. The re-sulting lack of fluvial system continuity created by such knickpoints means that the upstream reaches have yet to fully respond to base-level changes at the mountain front. This inability to fully respond to downstream base-level falls is more pronounced for Bear than for Sabino Canyon because the Seven Falls reach of Bear Canyon has not degraded to depths similar to those of Sabino Canyon. The resulting

Figure 2.6 View to the west across the junction of Bear and Sycamore Can-yons at the Hard Rock site noted on Figure 2.5. The 160 m of downcutting here is impressive, but lithologic control of local base level is still present. The band of resistant rock continues westward as a prominent ridge for >20 km (right edge of this view), crossing Sabino Canyon at the Hard Rock local-ity noted in Figure 2.5.

Drainage Basins

Figure 2.7 Future site of Sycamore Canyon stream capture, Santa Catalina Mountains, Arizona. A. The deep Pine Canyon (left side) has rugged hillslopes compared to Sycamore Canyon (right side). A splash of sunlight fortuitously highlights the location of the stream capture, SC, about to occur. Width of this view is 1.4 km. Trail at lower right for scale.

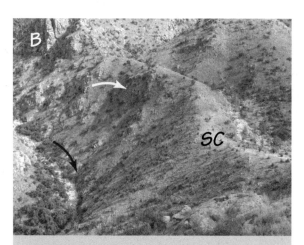

Figure 2.7B Future site of Sycamore Canyon stream capture, SC. The ever-deeper Pine Canyon cuts into the adjacent footslope (black arrow), creating a steep hillslope with outcrops of bedrock (white arrow). The ridgecrest is migrating towards Sycamore Canyon because it is steeper on the Pine Canyon side.

Figure 2.7C Future site of Sycamore Canyon stream capture, SC. The Sycamore Canyon side of the divide consists of gentle hillslopes next to the braided stream channel of Sycamore Canyon. The white arrow points to a fill-terrace exposure of recent aggradation event gravels. The black arrow points to scattered round boulders that are only 1 m below the stream-capture spillover point on the ridgecrest.

deeper Sabino Canyon valleys have steep hillslopes with ridgecrests that migrate into adjacent terrain with less relief.

I have noted seemingly minor reaches of "Hard Rock" on both longitudinal profiles. The longitudinal profile steepens downstream from both examples, but we need to note the amount of stream channel downcutting that has been accomplished. A 160 m high cliff rising above a slot-like gorge in Figure 2.6 shows that most, but not all, of the lithologic control has been removed.

The nice example used here is a large tributary of Bear Canyon – Sycamore Canyon – which is about to be captured by Pine Canyon, a small tributary to the East Fork of Sabino Canyon. Sycamore Canyon is perched far above Pine Canyon. Only a narrow low ridgecrest remains to be eroded before streamflow is diverted into the Pine–Sabino drainage basin (Fig. 2.7). This event will increase the area of the Sabino watershed by 12%, and reduce the watershed area of Bear Canyon (Fig. 2.8) by 26%. The migrating drainage-divide now is < 100 m from the trunk stream channel of Sycamore Canyon.

Stream channel and hillslope characteristics reveal interesting aspects of this situation. Pine Canyon has a catchment area that is only half that of the adjacent Sycamore Canyon, but its deep valley and steep hillsides make Sycamore landforms look relatively complacent (Fig. 2.7A). A closer view of the hillslope characteristics is shown in Figure 2.7B. Base-level fall creates a steep slope on the Pine Canyon side of the divide. Minimal base-level fall by Sycamore Canyon favors gentler footslopes. This contrast results in faster hillslope erosion on the Pine Creek side, which gradually lowers and shifts the ridgecrest towards Sycamore Canyon.

The timing of a stream capture may not be determined by erosion. Climatic fluctuations modulate sediment yields that raise and lower streambeds at a frequency that is fast compared to ridgecrest lowering. Return of summer monsoon rains caused an early Holocene aggradation event here. Aggradation raised the braided channel of Sycamore Canyon to within 1 m of the ridgecrest altitude (Fig 2.7C).

Resistant rock crosses the Bear Canyon watershed at the mouth of Sycamore Canyon (Fig. 2.6 and Location A' in Figure 2.9). The upstream reach, A–A', has a smooth exponentially declining slope that is graded to this lithologically controlled local base level. The reach downstream from this

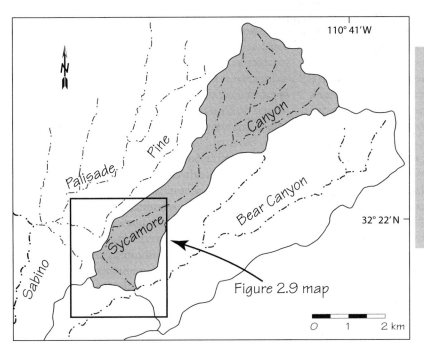

Figure 2.8 Drainage-basin map showing the Sycamore Canyon watershed and the headwaters reaches of the Bear Canyon watershed. The box refers to the Figure 2.9 inset map, which shows the locations of the Figure 2.10 topographic profiles.

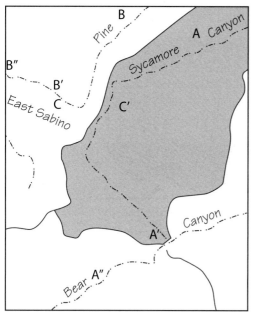

Figure 2.9 Map of locations of topographic profiles in Figure 2.9 that describe landscape elements conducive for stream capture.

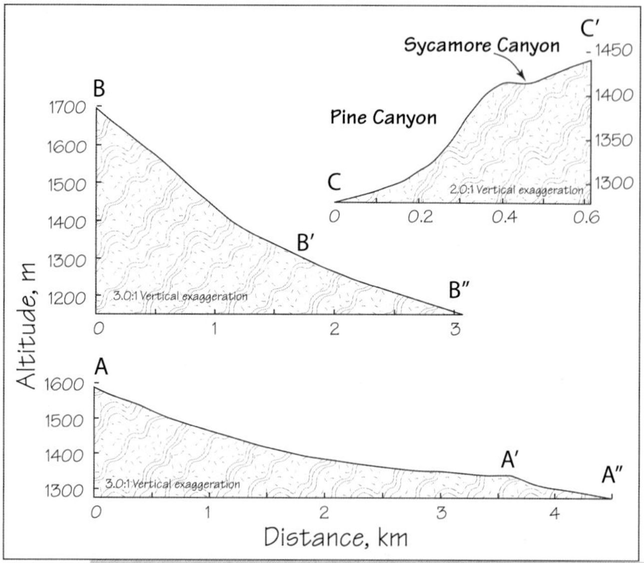

Figure 2.10 Longitudinal profiles showing the low gradient of the stream about to be captured A–A' (Sycamore) and the capturing stream B–B' (Pine). C–C' contrasts the small size of the Sycamore Canyon valley with one side of the steep Pine Canyon valley.

lithologic control , A'–A", is much steeper despite Bear Canyon increasing streamflow discharge by threefold. Location A is rather obscure at the larger scale of Figure 2.5, where it is labeled "Hard Rock".

Pine Canyon (B–B') is steeper for two reasons. It is smaller, and downcutting by the Sabino Canyon master stream is continuing to move upstream through the East Fork to Pine Canyon.

The marked contrast in valley side slopes is shown in the C–C' topographic profile (Fig. 2.10), which crosses the ridgecrest separating Pine and Sycamore Canyons. Rates of hillslope erosion should be faster on the steep, long hillside on the Pine Canyon side, as compared to the short, gentle slope on the Sycamore Canyon side of the divide between the two watersheds.

In conclusion, differences in rock mass strength, and perhaps watershed area, between Sabino and Bear Canyons have led to different responses to tectonic deformation that occurred long ago. The dominate Sabino Canyon fluvial system continues

to capture adjacent subordinate watersheds. Stream capture would be less likely after 5 My of landscape evolution if the Santa Catalina Mountains were being eroded at a spatially uniform rate. Key points include how locally hard rock prevents attainment of equilibrium stream channels, and that stream-capture events may occur during brief episodes of stream-channel aggradation.

I use the Figure 2.1B Mojave Desert drainage net, in alluvium, to further define dominant and subordinate watersheds. One tributary of an assigned dominant watershed was arbitrarily picked to be the hypothetical stream-capture avenue of the headwaters of an assigned subordinate watershed.

Flow from the subordinate watershed is abruptly diverted (avulsion) into the dominant watershed when degradation lowers the divide until it is lower than the trunk stream channel of an adjacent watershed. Stream power is immediately changed in all reaches of either fluvial system that undergo either abrupt increases or decreases of stream discharge.

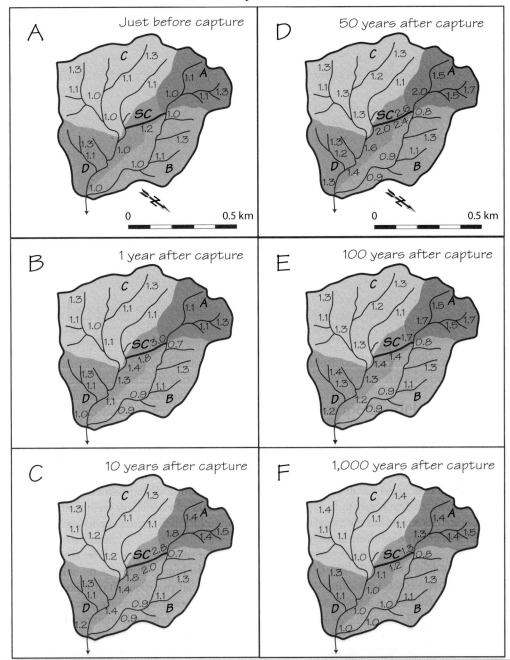

Figure 2.11 Hypothetical changes in ratios of stream power to resisting power in four subbasins A, B, C, and D after stream capture at the headwaters divide of SC. Reaches at equilibrium have a ratio of 1.0. Magnitude of departures from equilibrium characterize tendencies for degradation (ratios >1.0) and aggradation (ratios <1.0). This scenario uses the alluvial drainage basin shown in Figure 2.1B.

The resulting accelerated downcutting or aggrada-tion will vary in time and space until new equilib-rium longitudinal stream profiles are established. I use stream power/resisting power ratios in Figure 2.11 to estimate the magnitude of departures from assumed initial equilibrium.

Stream capture can result from locally un-usual conditions in a fluvial system. Factors that favor stream capture by a dominant watershed include:
1) A tributary stream that is responding to a down-stream base-level fall by rapid downcutting of its stream channel,
2) Divide hillslopes that are underlain by soft, easily erodible earth materials,
3) The rate of headwaters divide migration is in-creased by local tectonic deformation.

Accelerated erosion of a headwaters valley floor promotes ever steeper hillsides adjacent to the headwaters divide. Divide migration into the adja-cent watershed occurs because of relatively steeper headwaters hillslopes of the dominant basin as com-pared to those of the subordinate basin. Divide migration may be quite local; other sections of the dominant basin divide that are underlain by more re-sistant rocks may remain unchanged. Expansion of the headwaters basin area is a minor self-enhancing feedback mechanism. Both the increase of water-shed area and steepening of headwaters hillslopes gradually increase stream discharge. Such increas-es in unit stream power, although small, contribute to further stream-channel entrenchment and divide migration.

Of course the trunk stream channel of the subordinate watershed must be above the altitude of the capturing stream channel in order for streamflow to be diverted. This implies that the trunk stream channel of the subordinate system is not downcut-ting at a rate comparable to that of the capturing stream.

Let us assume that both drainage nets are being affected by a 4 m surface rupture on the range-bounding normal fault. The reach of the subordi-nate stream subject to imminent capture might be slower to respond to this base-level fall because:
1) It is isolated from downstream reaches by a local base-level control such as resistant rock that favors rapids or a waterfall.

2) The earth materials under the reach soon to be captured are more resistant to erosion than the ma-terials under the capturing tributary of the dominant fluvial system.
3) The critical reach has a blanket of bouldery stream-bed armor that inhibits downcutting even where the underlying materials are soft.
4) The reach about to be captured is much more dis-tant from the range-bounding fault than the captur-ing tributary and thus has a longer time lag of re-sponse to a mountain-front base-level fall.

The following case history for a hypothet-ical stream capture focuses on four subbasins and emphasizes the contrasts between the dominant and subordinate basins upstream and downstream from the stream-capture location. Capture is initiated by headward erosion of a small tributary stream (SC) of the dominant watershed. Subbasins A, B, C, and D in Figure 2.11 adjust to initial and consequent pertur-bations induced by stream capture. Migration of the headwaters divide of the SC tributary continues until this topographic saddle intersects the trunk stream channel of the subordinate watershed. New water-sheds A and B are upstream and downstream from this capture point, respectively. Subbasin D includes all parts of the dominant watershed through which the augmented stream flows. Subbasin C is that por-tion of the dominant watershed that is upstream from the junction with flow derived from capturing tributary SC.

Changes range from profound to minor in these four subbasins. Upstream migration of mul-tiple knickpoints in alluvium was patterned after responses to historical surface ruptures in Nevada and California. These personal observations were a rough guide for assigning rates of change used in Figures 2.11B–F.

Pre-capture conditions (Fig. 2.11A) portray the downstream reaches of both trunk stream chan-nels as being at equilibrium (the threshold of critical power ratios are 1.0) in a nearly circular single water-shed. Upstream reaches are slightly on the degra-dational side of the threshold of critical power. As a general rule, the lowest order headwaters streams continue to deepen their stream channels slowly be-cause of minimal annual stream power compared to downstream reaches.

Subbasin A is the headwaters of the subordinate drainage basin. Its flow is diverted into the dominant drainage basin by stream capture induced by tributary SC. Subbasin A suddenly has a new base level as its flow is abruptly switched into a much lower stream channel. This initial base-level fall propagates upstream from the capture point into Subbasin A, but only as fast as knickpoints migrate upstream in the fairly soft materials underlying this semiarid Mojave Desert watershed. The upstream reaches may not feel the effects of the perturbation until more than 10 years after capture and it will take much longer for the perturbation to migrate slowly up the hillslopes and headwaters reaches of Subbasin A. Subbasin A will remain on the degradational side of the threshold of critical power.

Let us apply the general axiom that streams of small basins tend to have steeper gradients than streams of large basins. A 23% increase in the watershed area in the dominant watershed occurs after flow from subbasin A is added to it.

The area of the SC tributary is increased by ~430%. Stream capture promotes rapid stream-channel downcutting in the SC tributary – a situation of a major increase of stream discharge into a steep stream channel. Downcutting also occurs to a lesser extent in the trunk-channel reaches further downstream where departures to the erosional side of the threshold of critical power are not as large.

The consequences of stream capture in subbasin B are equally profound but much different than for subbasin A. Capture initiates an opposite sequence of changes. The headwaters of subbasin B formerly received all of the discharge from subbasin A and had established an equilibrium longitudinal profile and stream-channel pattern. The stream channel in the new reach just downstream from the capture point is now only a headwaters rill. Its basin-position coordinate has been changed from 24 to close to 0. Post-capture stream discharge in this reach is reduced to being miniscule. It no longer has sufficient slope to transport bedload supplied from the adjacent hillslopes, even after tributaries have joined it farther downvalley. All reaches of the trunk stream channel of subbasin B aggrade, but not at the same rate. The tendency for aggradation is strongest in the new headwaters reach where the disparity

between the former and present stream power is greatest (indicated in a relative sense by a ratio of 0.7 in Figure 2.11B). This tendency for aggradation becomes progressively less downstream as tributaries increase trunk stream discharges to amounts closer to pre-capture amounts.

In a general sense the loss of 40% of watershed area is partly compensated for by increases in trunk-channel gradient that would be appropriate for a smaller drainage basin. But such aggradation is slow in subbasin B because long time spans are needed to raise the headwaters stream channel of the new subbasin. It has minimal flux of sediment as well as water. Such headwaters aggradation is an inefficient way to increase stream power. Nevertheless, the stream attempts to establish a steeper gradient by deposition. It may also alter stream-channel pattern, in order to convey some of the bedload with the drastically less stream discharge. The new headwaters reach remains on the aggradational side of the threshold of critical power after 1,000 years, but downstream reaches are presumed to have returned to equilibrium conditions (Fig. 2.11F) because of a postulated change to a more efficient stream-channel pattern.

The consequences of the stream-capture perturbation on hillslopes depend on position within the subbasin B fluvial system. Footslopes adjacent to the aggrading trunk stream channel become less steep and less concave as a result of the base-level rise (Section 3.5.1). So, flux of sediment to the trunk stream channel is decreased by the amount of perturbation-induced deposition on the adjacent footslopes.

In contrast, the watersheds tributary to the trunk stream channel continue to function in the same manner as before the stream-capture event because base-level rises are transmitted only a surprisingly short distance upstream from where deposition has raised a stream channel (Leopold and Bull, 1979). Base-level fall can be transmitted far upstream only where all reaches are degrading (stream power/resisting power > 1.0).

Subbasin C does not respond directly to the stream-capture event. Base-level fall at the mouth of the subbasin is secondary, a consequence of increased discharge in the trunk stream channel of subbasin D. The perturbation migrates upstream in much the same manner as in subbasin A. However, the available unit stream power to erode stream channels in reaches upstream from a basin-position coordinate of 0.2 is exponentially less than at basin-position coordinates of 0.5 or 1.0 It may take much of a thousand years for the perturbation to arrive in the distant headwaters tributaries of subbasin C, which has a larger watershed area than A.

The consequences of stream capture in sub-basin D are most extreme in the catchment of the SC tributary. Gradual migration of the headwaters divide of the SC tributary provided a rather small increase in discharge for the SC catchment, but had minimal impact on the remainder of the dominant watershed. Then the watershed area of the SC catchment increased by two orders of magnitude with the abrupt capture of all subbasin A streamflow. The SC catchment, which was already undergoing accelerated erosion, moved even farther from equilibrium conditions. I illustrate this in Figure 2.11B–D by assigning stream power/resisting power ratios of between 1.8 and 2.4 for inferred intervals of time. However, this degradation rate does not persist for long because this tributary watershed is underlain by soft materials, which was the primary reason for it being the site of stream capture. The ratio is decreased to about 1.4 in only 100 years and after a 1,000 years the rate of downcutting is controlled by, and similar to, the rate of downcutting in reaches downstream from the SC tributary.

The stream power/resisting power ratio in subbasin D downstream from the junction of the SC tributary with subbasin C never acquires the extreme values attained in the SC tributary. The perturbation is not as strong because of the "dilution affect" created by the junction of streamflow from relatively large subbasin C. This downstream reach of watershed C–D is already adapted to large stream discharges downstream from a basin-position coordinate of 66.

The stream power/resisting power ratio decreases systematically in subbasin D downstream from the capture point for two reasons. First, magnitudes of base-level falls induced by stream-channel downcutting decrease exponentially with distance. Second, accelerated downcutting of stream channels

in subbasins A and C increases watershed sediment yield. This increase of bedload transport rates increases resisting power. The increase in sediment flux continues as the influences of valley floor base-level fall slowly migrate up the adjacent hillslopes. It is possible for the ratio to be reversed to the aggradational side of the threshold of critical power in any part of the valley floor of subbasin D.

The effects of stream capture in subbasin D are primarily along the trunk stream channel and consist of stream channel downcutting. In contrast to subbasin B base-level rise, base-level fall in the trunk stream channel migrates up the tributaries to influence the entire watershed. But tributaries in the downstream reaches are subject to only minor amounts of base-level fall so they undergo only negligible change during the 1000 year time span in my postulated scenario. The overall effect increases total sediment yield from the hillslopes in subbasin D.

Stream capture may mimic characteristics associated with tectonically active terrains or climate-change induced changes in stream discharge because capture involves base-level changes. Gradient may not have changed but local increase in stream discharge renews stream-channel downcutting. The hillslopes and active channel of subbasin A develop characteristics that reflect the base-level fall induced by the stream-capture event. Subbasin A landforms will appear more tectonically active than before capture. Valley floors will become narrower, footslopes steeper, and hillslope mass movements will be more common.

In contrast the headwaters of the trunk stream channel in post-capture subbasin B develop characteristics of tectonically inactive terrain. The broadening disequilibrium valley floor becomes a rising base level that decreases hillslope relief, thus promoting changes in hillslope processes opposite to those of subbasin A.

A dramatic example of the consequences of stream capture occurred after 1941 when ephemeral streamflows were diverted through a low drainage divide in Death Valley National Park, California. Occasional flash floods in Furnace Creek Wash threatened tourist facilities, so the Park Service diverted the stream into the headwaters of Gower Gulch. Gower Gulch is a much smaller watershed and its stream channel is steeper because it takes a much more direct route to the regional base level provided by the Death Valley playa. The increase in stream power caused rapid stream-channel downcutting. An incised stream channel migrated upstream from the capture point (Dzurisin, 1975), at a surprisingly fast rate considering the extremely arid, hyperthermic climate. Hillslopes graded to the stream channel have been slow to respond to the base-level fall. Even the very soft, barren badlands hills near Zabriskie Point have retained most of their pre-capture event shapes. Only the footslope portions have become steeper as a result of 8 m of base-level fall that occurred during a 60 year time span.

2.3.2 Drainage-Basin Evolution in a Fold-and-thrust belt

2.3.2.1 Wrench-Fault Tectonics

The preceding discussions summarized several important nontectonic factors associated with stream capture. Next, we add the independent variable of spatially variable uplift in order to examine tectonic constraints on stream capture. Unruh and Sawyer's (1997) splendid example is used here. Their study area is the foothill belt of Mt. Diablo in northern California. Late Cenozoic and historical earth deformation here results from complex interactions between parallel strands of the San Andreas transform boundary. Such strike-fault systems (also called wrench faults) have distinctive structures and landforms, so let us start by summarizing key aspects of wrench-fault tectonics (Fig. 2.12).

Strike-slip faults are steeply inclined and slip mainly along the direction of fault strike. **Wrench** (or transcurrent) **faults** are strike-slip faults that involve basement rocks. Wrench-fault systems form where a regional shear couple tends to rotate basement rocks and overlying sediments. Wrench faults can be depicted as simple shear because the orientations of maximum compressional and extensional tectonic stresses are both parallel to the surface.

Each tectonic shear couple creates diverse structures. A shear couple consists of a primary wrench fault and domains of extension and compression that result in normal faults and in thrust faults

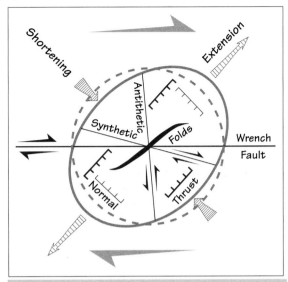

Figure 2.12 Map view of earth deformation associated with a single, straight right-lateral strike-slip fault in homogeneous materials. The shear couple shown by the top and bottom half arrows has distorted a circle (dashed gray line) into an ellipse (solid gray line). Secondary structural domains adjacent to the primary fault include right-lateral (synthetic) and left-lateral (antithetic) strike-slip faults, folds and thrust faults perpendicular to the axis of compression shown by the short stubby arrows, and normal faults perpendicular to the axis of extension shown by the long skinny arrows. Axial traces of folds have a sigmoidal pattern (bent in opposite directions) because of rotation caused by the shear couple. A left-lateral strike-slip fault would be a mirror image of this diagram. From Harding (1974) as modified by Sylvester and Smith (1976). AAPG © 1974, reprinted by permission of the AAPG.

and folds. Secondary strike-slip faults form too. These may have the same (**synthetic**) or opposite (**antithetic**) sense of lateral displacement as the primary strike-slip fault. The style of vertical tectonic deformation of a wrench-fault system depends on the sense of lateral slip on the primary fault and types of departures from a nonlinear trace of a nearly vertical strike-slip fault. *Right-lateral slip* occurs where the sense of relative motion on the opposite side of the fault is towards the right, and *left-lateral slip* where the opposite side appears to have moved left.

Deformation approximates the model of Figure 2.12 where strike-slip faults are straight. But one should keep in mind that progressively more of the tectonic deformation is concentrated along the primary wrench fault as the rocks become a broad weak zone of pervasively sheared and fractured rocks. While total cumulative displacement along dip-slip faults generally is less than 5 km, strike-slip faults commonly have displacements of 20 to 200 km. Such large amounts of cumulative shear reduce the resistance the earth materials to further tectonic shearing and to fluvial erosion. Thus slip may continue on older faults even where they no longer have an optimal orientation to the regional stress field.

Each shear couple has a center of rotation. The entire complex rotates, so secondary structures become less favorably oriented with respect to the driving force, which here is the relative movement between the Pacific and North American plates. Both folds and faults have sigmoidal termination bends (Fig. 2.13) as a result of progressive shearing about centers of deformation (Schreurs, 1994; Eusden et al., 2005b).

Deviations from linear fault traces are common. Bends and sets of en-echelon jogs are typical of strike-slip faults, even in homogeneous materials. Lateral slip along a nonlinear fault trace causes local vertical tectonic deformation. Basins form where crustal blocks are pulled apart and uplift occurs where one side of a strike-slip fault impinges on the other (Crowell, 1974, 1982).

Shear couples created by horizontal movements rotate secondary folds and faults between wrench faults (Fig. 2.12). This leads to further local compressional and extensional deformation at orientations that change with time.

Wrench faults commonly occur in parallel sets. A new direction of strike-slip faulting is created when limits of rotation, 20°–45°, are reached and the primary strike-slip fault is no longer oriented favorably with respect to the plate-tectonics shear couple. Cumulative long-term rotation of a block between parallel wrench-faults can exceed 45°, but limits to rock mass strength (Mohr circle analyses) restrict total rotations for a block to modest values. A new strike-slip fault commonly forms after total rotations of only 20° to 40°. This is less than the maximum theoretical rotation limit of 45° (Nur et al., 1986).

So, one should expect persistent sets of parallel strike-slip faults to have great variety, ages, and scales of earth deformation ranging from 10 m to >10,000 m. Newly formed incipient folds and faults are created in the vicinity of structures that may be old or may have only recently become inactive. Small secondary structures may be temporary as deformation shifts to a nearby structure.

Fascinating distinctive landforms characterize the interactions of lateral tectonic displacements with landscapes whose erosive potential energy ultimately depends on vertical tectonic displacements. This book concentrates mainly on mountains whose fluvial systems have ever more potential energy as a result of vertical uplift. Discussions about the much different suite of landforms produced when uplifted watersheds are torn apart by lateral earth deformation is deferred to another book "Fluvial Tectonic Geomorphology".

2.3.2.2 Mt. Diablo Fold-and-Thrust Belt

The San Francisco Bay region is a diverse arena for tectonic geomorphologists with its historic earthquakes and San Andreas transform boundary landscapes. We summarize one study here and present more in Sections 4.2.1.2.2, 5.2, and 7.3.

Contractional deformation of the Mt. Diablo fold-and-thrust belt is the result of a restraining step from the Greenville fault to the Concord fault, as well as the dextral shear couple between the Greenville and Calaveras faults (Fig. 2.13). Paleoseismologists now realize that it is best to assume that anticlines created in such contractional settings are linked to concealed thrust faults of wrench-fault systems at depth (Bull, 2007, p. 162). Unruh and Sawyer (1997) estimate total horizontal shortening across the Mt. Diablo anticline since Ma to be about 13.5 ± 3.5 km. This equates to an average slip rate for the thrust fault in the core of the Mt. Diablo anticline of 4.1 + 0 – 1.4 m/ky. The Tassajara anticline is secondary to the primary Mt. Diablo anticline and presumably formed as the Mt. Diablo blind thrust rose and propagated to the southwest. Such wrench-fault tectonics has resulted in continual modification of the fluvial landscape, including stream capture.

The drainage net of the Mt. Diablo foothills has undergone many changes. At first, streamflow was directed northwest along tectonic depressions such as the Sycamore Valley syncline. A geomorphic consequence of the progressive growth of the adjacent Tassajara anticline was the creation of small watersheds eroded into the soft rocks on the southwest flank of the new fold. Local tectonic steepening of these stream gradients increased stream power sufficiently to promote accelerated entrenchment in headwaters reaches. This unusual location for rapid downcutting is demonstrated by the fact that three entrenching channels extended their headwaters divides past the axis of the Tassajara anticline (Figs. 2.13, 2.15).

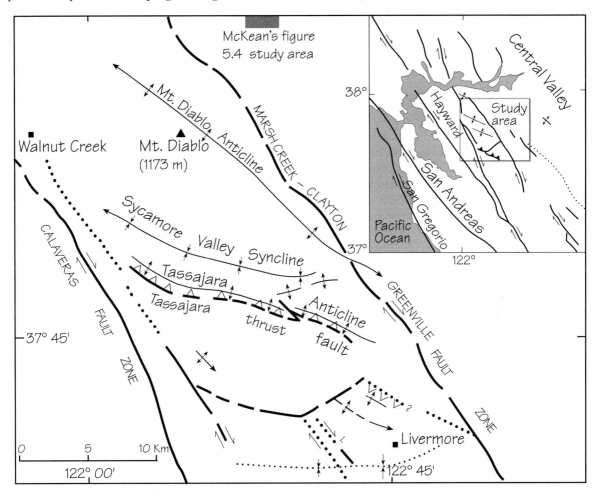

Figure 2.13 Structural characteristics of the Mt. Diablo fold-and-thrust belt. A. Active geologic structures between the Marsh Creek–Clayton–Greenville and Calaveras dextral fault zones. From figure 2-1, provided courtesy of J.R. Unruh, T.L Sawyer, and William Lettis and Associates.

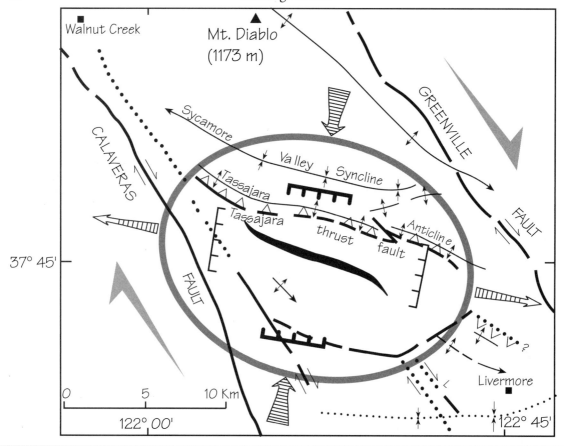

Figure 2.13 Structural characteristics of the Mt. Diablo fold-and-thrust belt.
B. Application of the wrench-fault model of Figure 2.12. The orientation of contractional faults and folds agrees with the model and the propagating Tassajara
anticline has sigmoidal terminations.

Drainage-basin area increased abruptly whenever tributaries of Sycamore Creek were captured by the extending streams of Tassajara, Alamo, and West Branch (Fig. 2.14). This example of multiple stream captures shows that local tectonic steepening of stream gradients increased stream power enough to promote aggressive expansion of headwaters reaches of drainage basins. Headwater's rills of most drainage basins typically are strongly to the degradational side of the threshold of critical power, and exceptionally so in this case. Lithologic control (soft Neogene sedimentary rocks) may be a critical variable here that favored rapid migration of headwaters divides.

An additional factor is obvious in this shifting landscape. Rates of tectonic steepening of valley gradients would be greatest near the anticline axis and should decrease as the headwaters of an expanding watershed moved into the adjacent syncline. One might expect that headward extension into the Sycamore Valley syncline would terminate headward migration because active synclinal downwarping in the headwaters reaches would tend to decrease stream power. Instead headwaters of drainage basins continued to expand. This could occur only if the entire foothill belt were being raised. Regional tectonic base-level fall kept the headwaters reaches of all streams ever more strongly on the degradational

Figure 2.14 Oblique view of the digital photography of the dynamic landscape of the Mt. Diablo foothills. Uplift of the foothill belt, and especially along the Tassajara anticline, has favored headward migration of the drainage basins of Tassajara, Alamo, and the West Branch of Alamo creeks. Figure provided courtesy of J.R. Unruh, T.L Sawyer, and William Lettis and Associates.

side of the threshold of critical power. Maximum incision of the valleys coincided with the axes of anticlines, but uplift of the entire foothill belt was sufficient to allow drainage nets to propagate across the axis of the Sycamore Valley syncline and then into the flank of the Mt. Diablo anticline.

Tom Sawyer's analysis provides additional insights about these fluvial systems. Tectonically induced downcutting resulted in stream terraces as much as 60 m above Tassajara and Alamo Creeks. Longitudinal profiles of these Quaternary streams are locally convex, instead of having concave profiles suggestive of attainment of equilibrium conditions.

Uplift continues to be sufficiently rapid to prevent attainment of equilibrium stream channels despite the significant increases of stream discharge resulting from stream captures. Estimates of late Holocene uplift are >3 m/ky for Mt. Diablo thrust fault and >1–2 m/ky for the Tassajara anticline. He notes that drainage-basin area, stream order, and valley-floor width decrease from Tassajara Creek to Alamo Creek to the west branch of Alamo Creek. This spatial progression of drainage integration suggested to Tom Sawyer that the tip of the Tassajara anticline propagated northwest during the Quaternary. The resulting set of geologic structures

(Fig. 2.13A) matches the basic wrench-fault model (Fig. 2.12) well, including the orientation and sigmoidal terminations of the Tassajara anticline (Fig. 2.13A).

The importance of stream capture in young, tectonically active landscapes is illustrated by the profound changes in the watershed of Sycamore Creek (Figures 2.14, 2.15). This was a large drainage basin in the early Quaternary, extending to the southeast along a synclinal trough. Then the adjacent Tassajara anticline was breached by small subsequent streams on its southwest flank. Three of these streams now have tenfold larger watersheds as a result of capturing portions of the Sycamore Creek watershed and other streams. Note that even Sycamore Creek has a tributary that captured part of an adjacent watershed called ancestral Green Valley Creek on Figure 2.15.

Unruh and Sawyer's study underscores the benefits of geomorphic analyses in assessing Quaternary tectonic deformation, and the potential seismic hazard in this part of the San Francisco Bay region. One should expect magnitude Mw ~6.5 earthquakes in this fold-and-thrust belt, such as have occurred during the past 40 years at several similar California tectonic settings.

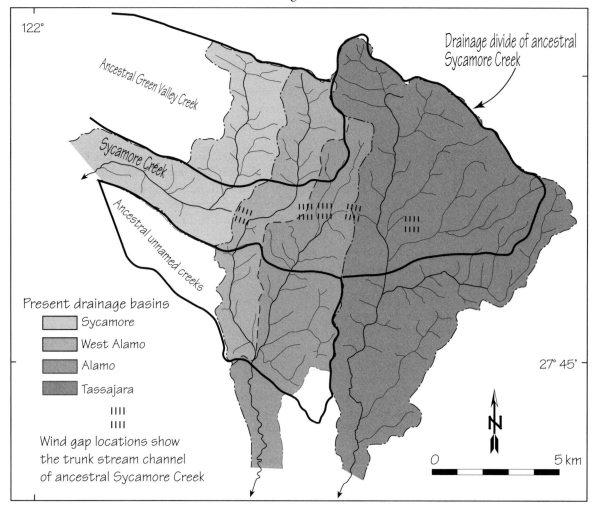

122°

Ancestral Green Valley Creek

Sycamore Creek

Ancestral unnamed creeks

Drainage divide of ancestral
Sycamore Creek

Present drainage basins

Sycamore

West Alamo

Alamo

Tassajara

||||
||||

Wind gap locations show
the trunk stream channel
of ancestral Sycamore Creek

27° 45'

N

0   5 km

Figure 2.15 Three decreases and one increase in the area of the Sycamore Creek watershed
caused by stream capture. The principal tributaries are shown for the present watersheds, and
a heavy line indicates the approximate position of the ancestral Sycamore Creek drainage basin.
Stream captures have greatly reduced the extent and area of the Sycamore Creek drainage
basin. Its ancestral valley floor is indicated by the trend of stream deposits now preserved in
wind gaps between the drainage basins of the three capturing streams of Tassajara, Alamo, and
West Alamo Creeks. From figures 4-6 and 4-7, provided courtesy of J.R. Unruh, T.L. Sawyer, and
William Lettis and Associates.

2.4 Tectonically Translocated Watersheds

The emphasis so far has been transformation
of drainage nets by watershed divide migration.
Equally profound reorganization of drainage basins can occur where strike-slip faulting moves part
of a drainage basin to where the stream can flow
in a new direction. The example of drainage avulsion used here describes the consequences of right-
lateral faulting that slices through the middle of a
presently circular drainage basin of the Kahutara

River watershed of New Zealand (Fig. 2.16). The downstream half of the watershed is rising and continues to enter the Pacific Ocean at the same place. The set of tributaries in the upstream half changes as dextral faulting shifts some tributaries out of the drainage basin while bringing new watersheds in to feed the downstream half of the drainage basin. At the present rate of strike-slip displacement, it would take only ~500 ky for the Kahutara River to receive a completely new set of upstream-half tributaries.

Late Quaternary right-lateral displacements by the Hope fault play a key role in local tectonics and topography. A former major tributary of the Kahutara River watershed, the Kowhai River, has not remained in the same place because each surface rupture on the Hope fault shifts its watershed further northeast. The Hope fault is one of the world's most active fault zones. Van Dissen (1989) used offset stream terraces at Sawyers Creek (Fig. 2.16) to estimate rates of right-lateral displacement on the Hope fault of 28 ± 8 and 33 ± 13 m/ky. I assume a slip rate of 30 m/ky for the following approximate time spans. At only 230 ka floods from the Kowhai

watershed entered the Kahutara River at location A on Figure 2.16. By 110 ka this shifted to location B, 50 ka to location C, and only 3 ka to location D.

The Kowhai River has flowed in a more direct route (Fig. 2.17A) to the Pacific Ocean for only a short time. Stream-channel entrenchment kept the river in a fairly circuitous route to the Pacific Ocean down the present anomalously broad valley of Bellbird Creek (Fig. 2.17B) to join the Kahutara River. Then a recent aggradation event (presumably the result of earthquake-induced landslides in the watershed) backfilled the gorge and the Kowhai River was free to take a much more direct route to the sea. An exceptionally active stream quickly laid down an 8 km wide blanket of alluvial-fan and fan-delta deposits on both sides of Kaikoura Peninsula (Fig. 2.18) in only 3 ky.

The timing of the recent stream channel aggradation event coincides with the time of a dramatic change in landscape characteristics. The reaches of the ancestral Kowhai River downstream from the Hope fault became progressively more gentle (0.025 to 0.021 m/m) as tectonic translocation lengthened

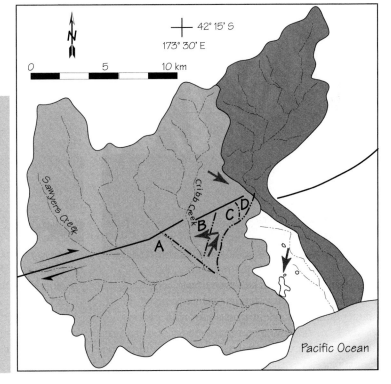

Figure 2.16 Tectonic shift of the Kowhai River (dark gray drainage basin) by right-lateral displacements of the Hope fault, so that it is no longer part of the Kahutara River drainage basin (light gray). Seaward Kaikoura Range, New Zealand. Prior courses of the Kowhai River are shown at 230 ka (A), 110 ka (B), 50 ka (C), and 3 ka (D). Four stubby gray arrows mark the locations and directions of the photographic views of Figure 2.17.

Figure 2.17 Views of the landscape associated with tectonic translocation of the Kowhai River drainage basin. A. View looking at the braided active channel of the Kowhai River, the piedmont of the Seaward Kaikoura Range. Kaikoura Peninsula and the Pacific Ocean in the distance.

Figure 2.17 Landscape associated with tectonic translocation of the Kowhai River drainage basin. B. View looking up ancestral streamcourse D. Lack of a stream channel shows that this valley floor was being filled by an aggradation event at the time of avulsion of the Kowhai River to its new location at the base of the distant mountain.

distances along the ancestral river valley from route A to route D of Figure 2.16. This would bring the stream-channel behavior closer to the threshold of critical power and increase the probability of switching to a mode of valley-floor aggradation.

There is evidence for a more profound and abrupt perturbation. Aggradation of the valley floor was so rapid that it dammed valleys tributary to the Kahutara River. Small tributaries, such as those along Figure 2.17B have small swampy areas upstream from low alluvial dams. Larger tributaries have more impressive dams such as for Lake Rotorua further downstream on the Kahutara River (Fig. 2.17C). Aggradation was so rapid that tributary streams were not able to maintain incised channels. This occurred on both sides of the hills between the ancestral and present routes of the Kowhai River. So the aggradation event, although brief, continued after the Kowhai River switched to its present course. Then modest stream-channel entrenchment prevented flows from going back down route D (Fig. 2.16) to the Kahutara River.

The intensity of soil-profile development on the aggradation event surface is one way of estimating how long the Kowhai River has been in its present course. Phil Tonkin and Peter Almond (2005 written communication) dug and augered soils at sites shown in Figures 2.17B and C. Soil-profile development is only incipient, appearing to be only 2,000 to 4,000 years old. This is my basis for labeling this as the "3 ky aggradation event".

A prominent high stream terrace just downstream from the Hope fault is next to Cribb Creek (Fig. 2.17D), so one might associate this landform as being part of the geomorphic history of Cribb Creek. Not so! This fill terrace was created by the Kowhai River at ~160 ka. It now is being removed by encroaching Cribb Creek whose large watershed is also moving northeast. Yes, tectonically active landscapes require us to be alert for alternative scenarios of landscape evolution.

The post A.D. 1840 behavior of Kowhai River events is so varied as to be unpredictable. The stream has flowed in slightly entrenched channel

Figure 2.17 Landscape associated with tectonic translocation of the Kowhai River drainage basin.
C. View across Lake Rotorua in a tributary of the Kahutara River that was dammed by the 3 ka aggradation event.

Figure 2.17 Landscape associated with tectonic translocation of the Kowhai River drainage basin.
D. View looking west at the high stream terrace next to Cribb Creek. This terrace was formed by the Kowhai River but is being removed by Cribb Creek, which is being shifted northeast by movements along the Hope fault. Low fence at right side of view is 1.4 m high.

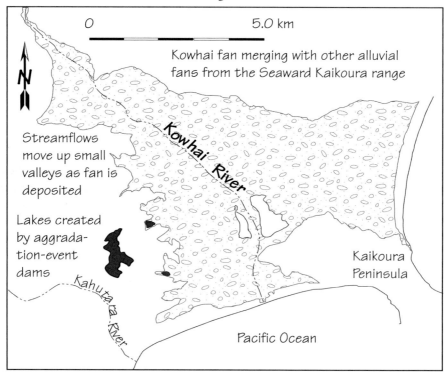

O   5.0 km

Kowhai fan merging with other alluvial
fans from the Seaward Kaikoura range

N

Kowhai River

Streamflows
move up small
valleys as fan is
deposited

Lakes created
by aggrada-
tion-event
dams

Kahutara River

Kaikoura
Peninsula

Pacific Ocean

Figure 2.18 Map showing the extent of post 3 ky alluvial fan and fan delta deposition by
the Kowhai River. Kaikoura floodplain flood hazard map for Kowhai River, Stoney Creek, lower
Kahutara River based on a January, 2,000 map survey by R. I. McPherson Associates and
drafted by R. Elley and S. Hamseed of Environment Canterbury, New Zealand.

part of the time, only to shoot off in a new direction whenever the bedload-transport rate is insufficient to prevent backfilling of the active channel. Such changes in stream behavior occur over time spans ranging from months to several hours. Figure 2.18 shows the preferred course of the Kowhai River, but who knows when it might shift to the east side of Kaikoura Peninsula, flooding farmlands and the floodplain business section of the town of Kaikoura. In just 140 years, floods have surged through the town 15 times, despite expensive engineering works intended to restrain flow directions of the Kowhai River.

Why can't this river be tamed? A detailed Environment Canterbury map shows that a broad swath along the entire north side of the area shown in Figure 2.18 was flooded in 1868.

The present steep course of the Kowhai River may be unstable because insufficient time has passed to establish the more systematic behavior we associate with flow in single-thread stream channels. Geomorphologists make logarithmic plots of 'bankfull' discharge and channel slope to identify domains of meandering and braided streams. Leopold and Wolman (1957) and Schumm (1985) pointed out that some streams (such as the Wairau River upstream from Blenheim, farther north in the South Island) are close to the threshold between stream-channel-pattern domains. This means that humans can install structures to encourage the stream to flow in a slightly sinuous single channel instead of the less stable anastomosing braided channel pattern. The Kowhai River, however, appears to be distant from being a manageable river (Fig. 2.19).

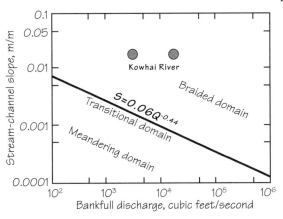

Figure 2.19 Logarithmic stream channel slope (S) vs discharge (Q) plot showing regression line separating domains of braided stream-channel pattern and more stable meandering patterns. Domains regression is from Leopold and Wolman (1957). Kowhai River data points for bankfull and a recent large flow are courtesy of Tony Oliver, Environment Canterbury, Christchurch, New Zealand.

The mean gradient along the present braided stream channel is 0.017 m/m as compared to 0.014 m/m when floods from the Kowhai River catchment flowed down the Kahutara River to the Pacific Ocean. This puts the Kowhai River even farther from the regression line separating more manageable rivers. Large Kowhai River floods have defeated flood-control measures in the past and can be expected to do so in the future.

The problem is compounded by the exceptionally large sediment yield of the drainage basin. Times of insufficient bedload transport rate produce episodes when the streambed is raised quickly by deposition of cobbles and boulders. Flood-control structures are quickly overtopped. The frequency and magnitude of piedmont flooding by the Kowhai since 1868 is partly the result of extreme overgrazing of the Kowhai watershed. Only small patches of the formerly thick soils remain on the ridgecrests. The watershed is no longer grazed but the next Hope fault earthquake should greatly increase landsliding, sediment yield, and flooding by an uncontrollable Kowhai River.

2.5 Summary

This chapter emphasized larger scale aspects of mountainous landscapes such as drainage-basin-shapes and changes in fluvial systems caused by stream capture or tectonically induced avulsion. Highly elongate basins characterize actively rising escarpments. Such juvenile watersheds become more circular through the processes of divide migration and stream capture as noted by early workers such as Horton (1945). Drainage basins change their planimetric shapes and areas when unequal denudation rates in adjacent watersheds cause drainage divides to migrate. Factors that favor diversion of streamflow into a different watershed include:

1) A tributary stream that is responding to a downstream base-level fall by rapid downcutting of its stream channel. This results in steeper hillslopes on one side of a watershed divide, which then shifts laterally.

2) Divide hillslopes that are underlain by soft, easily erodible earth materials accelerate the rate of divide migration. Low rock mass strength is typical of many rising mountains.

3) Aggradation events that briefly raise a valley floor so that flood stream discharges can pour over a low point in a watershed divide to initiate rapid, permanent stream-channel downcutting.

The resulting more circular basins, with stationary drainage divides, may be the most efficient drainage-net configuration for conveyance of sediment from the hillslopes of fluvial systems.

Abruptness of stream capture involves situations of abnormal concentrations of stream power in unusual locations – a stream flowing at the level of a ridgecrest is remarkable. Stream capture may occur in tectonically inactive settings, but is most common in rapidly evolving, rising, mountain ranges. Dominant watersheds slowly (or abruptly) expand their area at the expense of the adjacent subordinate watersheds. Their steeper ridges erode faster, given equal rock mass strength, and migrate into adjacent relatively passive fluvial systems. The trunk stream channel of the subordinate basin gradually becomes closer to the watershed divide instead of flowing down the middle of its watershed. The subordinate stream then is beheaded when its streamflow is diverted into the dominant stream.

Stream power is immediately changed in all reaches of either the dominant or subordinate fluvial system in those reaches that undergo either abrupt increases or decreases of stream discharge. The resulting accelerated valley-floor downcutting or aggradation will vary in time and space until new equilibrium longitudinal stream profiles are established. Rates of change will be a function of the magnitude of departure from the threshold of critical power.

Dimensionless hydraulic coordinates are useful for describing the consequences of stream capture. These ratios minimize the factor of size in comparisons of fluvial landscapes by describing relative positions on hillslopes and along valleys.

Unruh and Sawyer (1997) conclude that the tip of the Tassajara anticline propagated northwest during the Quaternary in their rising study area near Mt. Diablo in central coastal California. Widespread stream captures were associated with local tenfold increases in drainage-basin areas as a result of multiple stream captures affecting most watersheds. Headward erosion of valley headwaters and divide migration were greatly facilitated by the presence of soft Cenozoic marine sediments and rocks.

Tectonic translocation by a strike-slip fault bisecting a watershed gradually replaces the upstream portion of the watershed with a new set of stream channels and associated hillslopes. Prolonged tectonic lateral shifting of the trunk stream channel makes it increasingly difficult to maintain a distorted flow path. This leads to avulsion to a shorter new streamcourse that transmits flow directly downslope. The connection to the former watershed is abandoned and preserved as a broad underfit valley.

Either style of dramatic shift of stream channels – avulsion or stream capture – may occur during a time of alluvial backfilling of valley floors. Such temporary reversal of long-term downcutting may be a climate-change aggradation event, or brief aggradation resulting from an intense seismic-shaking event that causes landslides and increases sediment yield.

We focus next on smaller scales and shorter time spans by examining tectonic controls on hillslopes in tectonically active settings, with an emphasis on the ridgecrest and footslope segments.

Chapter 3

Hillslopes

Infinite combinations of earth materials, climate, uplift rates, and erosional processes influence fluxes of water and sediment moving down hillslopes to the stream channels of fluvial systems. I constrain this discussion mainly to soil-mantled hillslopes where erosion does not exceed weathering of bedrock. Think of these as being transport-limited hillslopes, that react to perturbations more readily than weathering-limited hillslopes. Soil-mantled hills are easier to consider as single-system landforms than where outcrops or inselbergs disrupt creep of colluvium or flow of water.

   Two classes of processes erode hillslopes at different rates, magnitudes, and locations. *Concentration* pertains to the geomorphic processes in areas of converging water flow lines that notch hillslopes, localize sediment flux, and result in mass movements such as debris flows. *Diffusion* pertains to slowly acting processes in areas of divergent flow lines that tend to disperse sediment. Diffusion prevails mainly on ridgecrests and concentration becomes progressively more important near valley floors. Steepness and curvature of hillslopes reflect the blend of concentration and diffusion processes, a mix that reflects local rock uplift rates and climate.

3.1 Hillslope Model Boundaries

I further simplify my model by constraining hillslopes between two limiting boundaries. The upper boundary is the ridgecrest formed by convergence of two opposing hillslopes of adjacent watersheds. Water and sediment fluxes begin at ridgecrests. Changing the shape of a ridgecrest requires large masses of rocks to be weathered and eroded. Soil and colluvium typically are removed slowly by several geomorphic

Smooth soil-mantled hillslopes underlie incipient drainage basins notched into an ~80 ka marine terrace sea cliff on Kaikoura Peninsula, New Zealand. Steep slopes underlain by soft Cenozoic mudstone and marl bedrock favor generation of landslides such as the slump at the lower right side of this view. Grazing sheep and cattle created the miniature hillside terraces.

Tectonically Active Landscapes, 1st edition. By W. B. Bull. Published 2009 by Blackwell Publishing, ISBN 978-1-4051-9012-1

processes from this highest landscape element of fluvial systems. Ridgecrest altitude increases in rising mountains may be similar to the rate of rock uplift. Ridgecrests slowly become lower in tectonically inactive mountains.

The lower boundary is the valley floor formed where fluxes of water and sediment from hillslopes are conveyed along the drainage net of a watershed. Valley floors are local base levels to which footslopes are graded. Nontectonic geomorphic processes of erosion and deposition can raise, lower, or maintain valley floors at the same position with the passage of time. Unlike ridgecrests, the lower hillslope boundary can change rapidly because small areas and powerful streamflow processes are the norm. We might expect response times to tectonic perturbations to be several orders of magnitude faster for valley floors than for ridgecrests.

Hillslope shapes adjust to changes in either boundary. Convex shapes and diverging flow lines are typical of crestslopes, while concave shapes and converging, or parallel, flow lines are typical of footslopes. The transition from convex to concave may be gradual, passing through a straight midslope. Proportions of these slope elements reflect **surface uplift rates** (rock uplift minus denudation rate).

Before the hallmark paper by Molnar and England (1990), we simply noted the magnitude of mountain-range uplift in m/ky – how fast a particle of rock was being raised by mountain-building forces. They pointed out that the rate of surface uplift also depended on a second variable of long-term rate of degradation of the land surface. Emphasizing this geomorphic component of landscape uplift encourages consideration of landform shapes where surface uplift is either less than, more than, or equal to rock uplift rate. Situations of "scrunch and stretch" tectonics (Bull, 2007, Chapter 1) create local departures from the regional rock-uplift rate.

I will conclude that convex ridgecrests prevail in landscapes where bedrock uplift is rapid, and that concave footslopes are conspicuous in tectonically inactive hillslope settings. Midslope lengths can be longer in rising mountains. Midslope steepness may not be significantly greater for active than inactive settings with the same climate and rock type. My discussion focuses mainly on the characteristics of

ridgecrests and footslopes where both steepness and degree of curvature are influenced by magnitude and rate of base-level fall (Penck, 1953). In this way we assess the relative importance of rock uplift and surface uplift in different climatic and tectonic settings.

3.2 Late Quaternary Tectonic Deformation of the Diablo Range

Let us use a study area with soft rocks and fairly rapid uplift rates to start our quest for tectonic signatures in hillslopes, and to further illustrate the usefulness of dimensionless hydraulic coordinates (Section 2.1). The setting is part of a 600 km long tectonic domain that includes the Mt. Diablo fold-and-thrust belt (Sections 1.2.3 and 2.3.2.2).

The main divide of the Diablo Range rises to altitudes of 1100 to 1600 m and is closer to the San Andreas fault (Fig. 1.3) than to the western margin of the San Joaquin Valley. About two million years ago the present foothill belt was the western margin of the San Joaquin Valley. The main Diablo Range was only a "chain of low hills" in the late Pliocene (Davis and Coplen, 1989). The large watersheds that now head in the main range pass through a young, emerging foothill belt that borders the present San Joaquin Valley. Discussions here focus mainly on the hillslope characteristics in the small watersheds of young and rapidly rising mountains (Fig. 3.1). These include the Laguna Seca, Panoche, Tumey, Ciervo, and Kettleman Hills, and Monocline and Anticline Ridges (Fig. 3.2). This fold-and-thrust belt continues to encroach into the San Joaquin Valley. Oil is pumped from juvenile anticlines hidden beneath the valley-floor alluvium.

This new range front greatly changed the width and depth of the San Joaquin Valley (Fig. 1.6). Deposition of basin fill was accelerated by sediment-yield increase from an ever higher Diablo Range, and by a shrinking area for accumulation of basin fill.

This post-4-Ma uplift of the old range and the creation of new range fronts changed the climate. Both factors altered rates of hillslope geomorphic processes and outputs of fluvial systems that promoted accumulation of basin fill in the San Joaquin Valley. Tectonically induced topographic changes are summarized in Figure 3.1 together with climatic changes produced by the resulting rain shadow.

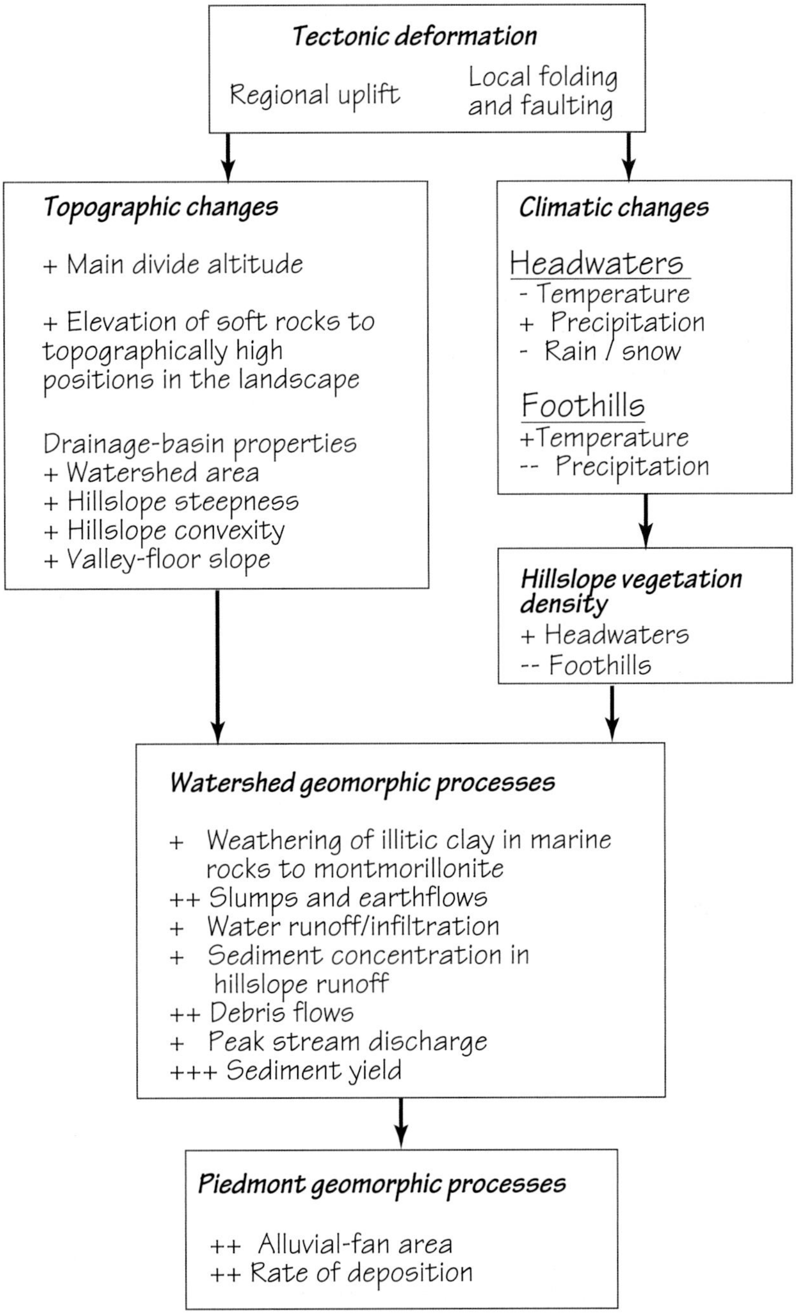

Figure 3.1   Flow diagram showing tectonic controls on landscape evolution of the eastern flank of the Coast Ranges of California, including 1) the consequences of tectonically induced climatic changes by a rising Diablo Range and 2) how creation of new foothill-belt mountains affected basin-fill characteristics.   Changes are shown as increases (+) and decreases (-).

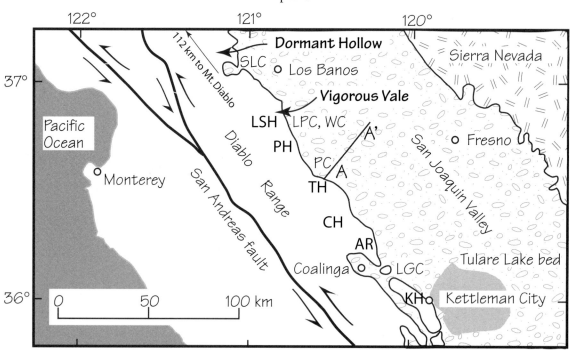

Figure 3.2 Tectonic setting of the Diablo Range, central California. Geologic cross section for A–A' is shown in Figure 1.6. Two unnamed watersheds are Vigorous Vale, which is tectonically active compared to Dormant Hollow. LSH, Laguna Seca Hills, PH, Panoche Hills, TH, Tumey Hills, CH, Ciervo Hills, AR, Anticline Ridge and KH, Kettleman Hills. Names for streams are – North to South – SLC, San Luis Creek; WC, Wildcat Creek; LPC, Little Panoche Creek; PC, Panoche Creek; and LGC, Los Gatos Creek.

Climatic changes occurred as the rising Diablo Range intercepted more precipitation from winter storms moving down the California coast from the Gulf of Alaska. Headwaters reaches on the lee side of the range received progressively more precipitation. Streamflow characteristics changed from ephemeral to intermittent. A higher main divide promoted warm and dry airflow that descended the east flank of the mountain range during passage of storms.

A rain shadow developed in the foothill belt. Present mean annual rainfall locally is only 180 mm. Hillslope plant cover decreased as the foothill belt climate became more arid. This favored flashy, ephemeral streamflow and debris flows. Such changes in independent variables changed watershed processes as the foothill belt rose out of the San Joaquin Valley. The result was increases in area and rate of accumulation of Diablo Range basin fill depicted in Figure 1.6 along the line of section A–A' in Figure 3.2. Concurrent tectonic subsidence created space for rapid deposition of valley alluvium.

We focus on the youthful foothill belt using the Figure 3.2 study sites here and in Chapters 4 and 5. Foothill belt hillslopes tend to be convex and steep from the divides to the stream channels as a result of rapid tectonically induced stream-channel downcutting into soft rocks. Longitudinal profiles of streams become steeper where surface uplift of headwaters is more than at the mouth of a watershed. Comprehensive discussions about landscape signatures in tectonically active and inactive watersheds are based on data from two unnamed streams that I call Vigorous Vale and Dormant Hollow.

We need estimates of rates of rock uplift in order to evaluate landscape responses to tectonic controls. The first impression is that sufficient quantitative data are available. A nasty complication is that uplift rates vary substantially over short distances in a tectonic environment characterized by propagating folds and blind thrust faults. That is not a significant problem here because we merely seek information regarding average rate of surface uplift for large topographic units such as the Panoche Hills. Some young tectonic elements of the rising foothill belt are capped by remnants of pre-uplift basin fill and/or by erosion surfaces of former pediments. Rock uplift equals surface uplift where a landform, such as an uplifted pediment, has undergone little erosion. Our goal is just to compare and contrast drainage basins that are rising rapidly, relative to their denudation rates, with basins that are virtually tectonically inactive and where erosion is lowering the landscape.

The hills of the foothill belt rose through the edge of the San Joaquin Valley basin fill. This deformed and uplifted alluvium provides us with both stratigraphic and geomorphic time lines for assessing rates of tectonic deformation. These include:

1) Fairly good paleomagnetic control on the age of the Tulare Formation. Lettis (1982) used radiometric and paleomagnetic data to conclude that the Tulare Formation was deposited from 1,900 to 250 ka. The significance of the Tulare Formation is that it marks the creation of new elements of the Coast Ranges that were sources of new alluvial fans. This tectonic evolution continues today, so the time-transgressive nature of the Tulare Formation can be applied to many sites on the western and southern sides of the San Joaquin Valley.

2) An extensive lake clay is so widespread that it has been given formal member status in the stratigraphy of the region. The Corcoran Clay Member of the Tulare Formation has a 10 cm bed of volcanic ash near its top. Sanidine crystals from pumice fragments in the correlative Friant ash provide an excellent potassium–argon age estimate of 620 ± 22 ka (Janda, 1965). Corcoran lacustrine silts have normal magnetic polarity so the estimated time span of deposition is between 700 and 615 ka (Lettis, 1988).

3) A soils chronosequence for stream terraces and erosion surfaces is post-Corcoran in age and dates back to the mid-Pleistocene (Lettis, 1982, 1985, 1988; Anderson and Piety, 2001).

Late Quaternary damming of the Central Valley of California at Carquinez Straits prevented the combined flow of many large rivers from reaching the Pacific Ocean through San Francisco Bay. The resulting 700 km long lake persisted for ~85 ky and was the site of deposition of thick lacustrine silty clays. Subsequent tectonic deformation is nicely shown by a structure contour map on the top of the Corcoran Clay. The Corcoran Clay is strongly warped next to the Panoche, Tumey, and Laguna Seca Hills (Figs. 1.6, 3.2), and Tulare Formation paludal or lacustrine beds are present in the flanks and on the crests of these rising folds.

A map of Corcoran-Clay thickness constrains times of formation of foothill-belt structures (Fig. 3.3). Lake beds in a tectonically passive environment should thicken away from the shoreline, especially if lake level varies. This appears to be the situation northwest of the city of Fresno. The Corcoran Clay laps onto the piedmont alluvial fans downslope from a tectonically inactive Sierra Nevada mountain front and it becomes progressively thicker to the west.

Corcoran Clay thickness is highly variable next to the Diablo Range. Alluvial fans were already being shed from the rising folds of Anticline Ridge and the Big Blue Hills so these structures pre-date the Corcoran Clay. Thicknesses are only in the <20 to 60 feet range (6 to 18 m), and the area of thickest lake beds coincides with the present axial trough of the San Joaquin Valley. Corcoran Clay thickness of more than 120 feet next to the Tumey and Panoche Hills records substantial tectonic deformation during deposition of the Corcoran Clay. Folding depressed the Corcoran to below sea level while the foothill belt was being raised. Initiation of foothill belt uplift is still younger for the Laguna Seca Hills. Undeformed Corcoran Clay occurs in mountain-front pediment embayments northwest of the town of Los Banos. Tectonic deformation has yet to occur here.

Maximum Corcoran Clay thickness occurs just east of the Kettleman Hills at the site of the modern Tulare Lake bed. This basin is still subsiding and its tectonic deformation is related to a thrust-fault-propagation shift from the Diablo Range mountain

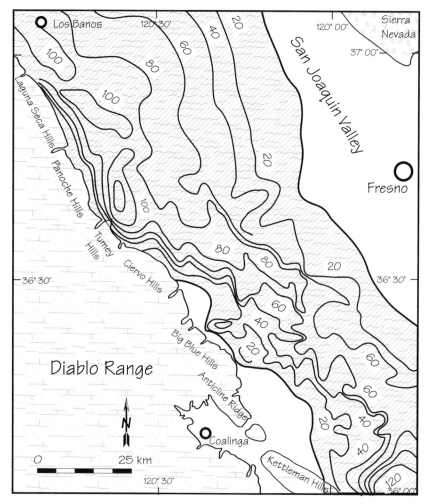

Figure 3.3 Thickness of the Corcoran Clay Member of the Tulare Formation in western Fresno County, California. Contour interval is 20 feet (6.1 m). From figure 11 of Miller et al. (1971).

front to a newer structure in the fold-and-thrust belt – the Kettleman Hills (Bull, 2007, p. 160–163).

Spatial variations in uplift on range-bounding folds are reflected in thicknesses of the subsequent basin fill. I assume that sediment yields were similar from drainage basins with similar climatic, lithologic, and geomorphic characteristics. So, thicknesses of alluvial-fan deposits above the Corcoran provide clues as to when the various tectonic elements of the foothill belt rose and began to provide sources of alluvium to bury the Corcoran.

The 250 m of post-Corcoran Diablo Range alluvial-fan deposits opposite the Ciervo hills suggests that parts of those hills were already rising at 600 ka. The Corcoran is little deformed here but has sunk below sea level. Alternatively, the uplift that induced erosion was concentrated on structures that do not presently border the San Joaquin Valley, such as Monocline Ridge. The strongly deformed Corcoran opposite the Panoche Hills (Fig. 1.6) suggests that the Panoche Hills were raised well after 600 ka. But one needs to allow at least 300–500 ky to accumulate the 300 m of post-Corcoran fan deposits. The Laguna Seca Hills have only about 50 m of fan deposits above the Corcoran, suggesting that this structure may have become active only since 200 ky. Still further to the northwest, Tectonic rejuvenation has yet to begin at San Luis Creek and other pediment embayments where the Corcoran is almost at the land surface.

The progressive decrease in thickness of post-Corcoran Lake Clay alluvial-fan deposits towards the northwest suggests tectonic propagation towards the northwest. Creation of Anticline Ridge front was followed by the Big Blue Hills, Ciervo Hills, Panoche Hills, and then the Laguna Seca Hills. This rather simple model for times of creation of new tectonic elements has several dilemmas.

It does not take much uplift for tectonically induced downcutting to terminate deposition on a piedmont surface. Minor warping may put the fanhead reach of a stream permanently on the degradational side of the threshold of critical power. An example is the present-day Little Panoche Creek fan between the Panoche and the Laguna Seca Hills. The present *threshold-intersection point* (where the entrenched stream channel ends and fan deposition begins) is 4 km downslope from the mountain–piedmont junction. The entrenched reach has been developing a soil profile for more than 600 ky as shown by Corcoran Clay onlap onto this piedmont. This might be a case where local warping began and then ceased as uplift shifted to a new blind thrust fault at a nearby location. The Panoche and the Laguna Seca Hills continue to rise rapidly.

Uplift rates decrease as folding becomes tight. New hilly landscapes should be created quickly if 36% of potential uplift has occurred after only 4% horizontal shortening (Bull, 2007, fig. 1.15; after Rockwell et al., 1988). Propagation of an active thrust fault to the surface would further change the style, distribution, and rate of bedrock uplift.

Post-Corcoran rates of basin-fill accumulation have increased. Aggradation rates doubled in the axis of the valley near the Tulare Lake bed from about 0.3 m/ky for the 600 ky time span to about 0.6 m/ky for the 26 ky since deposition of the "A" lake clay of Croft (Atwater et al., 1986; Croft, 1968,1972). The pronounced stratigraphic overlap

(Fig. 1.6) indicates that the rates of fan deposition at 500 ka were substantially less than the 600 ky mean rate, perhaps only 0.1 m/ky.

Tectonically induced increases of mean hillslope steepness in new, rising Diablo Range watersheds may account for both increases in area and aggradation rate of the expanding Diablan piedmont. Slower erosion rates in the early stages of a new mountain range (low hills) leads me to double my estimate of the time of initial uplift of the Laguna Seca Hills to roughly 400 ka. Having only general information about the changes in the rates of deposition means that estimates of the times of initial mountain range uplift are only approximate.

Conveniently, we have a cross-check provided by flights of dated geomorphic surfaces. Hall (1965) named an extensive pre-uplift surface between Little Panoche Creek and San Luis Creek the "Las Aguilas land surface". Subsequent studies by Lettis (1982, 1985, 1988) and by Anderson and Piety (2001) used radiometric ages of a soils chronosequence to estimate the age of this raised pediment and the several younger stream terraces inset below it.

An example from the southern part of the Laguna Seca Hills (Fig. 3.4) shows the Las Aguilas surface of Hall (Q1) and the inset younger Los Banos stream terrace of Lettis. Tectonically induced downcutting by Wildcat Creek provides a minimum estimate of the amount of rock uplift since Q1 time. It is about 230 m, which is less than my 550 m estimate for the more deeply notched Vigorous Vale, which is 13 km NNW in the Laguna Seca Hills. The Anderson and Piety age estimate for Q1 is 300 to 500 ka (400 ± 100 ka). An age estimate of about 400 ka is useful for estimating when pedimentation was terminated by initial uplift of the Laguna Seca Hills.

The Q2 stream terrace is inset well below the level of the Q1 surface and post-dates the initiation of the Laguna Seca Hills uplift. Anderson and Piety's age estimate of about 250 ± 50 ka for this inset surface does not conflict with the premise that these new drainage basins began to form at about 400 ka. So, using the rate of stream-channel downcutting, it appears that mean rock uplift at Wildcat Creek is about 0.6 m/ky. This is at the upper end of the 0.3–0.6 m/ky range of Anderson and Piety (2001, table 3.3). Their estimate is based on the

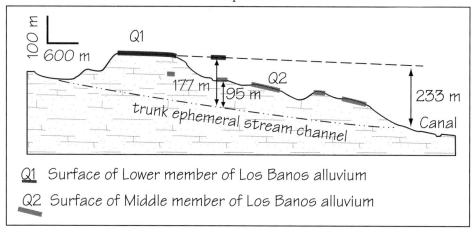

Q1 Surface of Lower member of Los Banos alluvium

Q2 Surface of Middle member of Los Banos alluvium

Figure 3.4 Topographic cross section of the Laguna Seca Hills in the Wildcat Creek watershed. Q1 is old pediment surface created prior to formation of the Laguna Seca Hills. Q2 is a stream terrace that is younger than the time of creation of the Laguna Seca Hills and Wildcat Canyon. From Figure 3-10 of Anderson and Piety, 2001.

estimated rate of slip of the Laguna Seca Hills thrust fault, assuming a 20° to 30° west dip of the fault plane. My 0.6 m/ky estimate should be larger than theirs because it includes tectonic scrunching components of rock uplift (Chapter 1, Bull, 2007), and slip on the thrust fault.

The more active part of the Laguna Seca Hills at Vigorous Vale seems to have been rising at about 1.4 m/ky since 400 ka. Even that estimate may be a minimum because deposition that raises the fluvial system downstream from the zone of maximum differential uplift would tend to decrease tectonically induced downcutting within the mountains. Uplift rates may exceed 2 m/ky, but for the purposes of discussions in this book I assign an uplift rate for Vigorous Vale of 1.5 ± 0.5 m/ky.

Dormant Hollow is tectonically inactive. However, one should allow for the possibility of initial incipient mountain-front folding at some localities as tectonic activity propagates ever farther northwest. In any case long-term uplift rates are much less than 0.1 m/ky. Dormant Hollow is used here to evaluate the importance of stable base levels on evolution of hillslope morphologies.

We can use three base-level processes to estimate the total tectonic deformation across this margin of the Diablo Range.

$$\Sigma td \approx \Sigma cd + \Sigma pa \qquad (3.1)$$

where td is total tectonic deformation, cd is trunk stream-channel downcutting within the mountains, and pa is the vertical-space needed to accumulate basin fill in response to tectonically induced erosion of the mountains.

Equation 3.1 can be used to estimate a mean rate of uplift for the young, rising mountain front. The Panoche Hills (Fig. 3.2), a surprisingly young 24-km-long dome, is another recently emerged thrust-fault cored fold. Total uplift exceeds 600 m (Lettis, 1982, 1985; Davis and Coplen, 1989). Level-line surveys show that uplift continues (Bull, 1975b). Folded Corcoran Clay laps onto the adjacent Tumey Hills (Fig. 1.6). Reduced paludal and lacustrine deposits are exposed in the upwarped edge of the Laguna Seca Hills monocline. Total tectonic deformation = 600 m + 300 m = 900 m. The Corcoran lake would not have occupied this part of the San Joaquin Valley if the foothill belt was already rising at 600–700 ka. The average uplift rate for this part of the fold-and-thrust belt approximates 2 m/ky if we assign an age for initial uplift of 450 ka.

Soft marine sandstone and mudstone underlie these soil-mantled hillslopes and promote landslides during infrequent wet winters. Landslide types include slumps, earthflows, and debris flows (Section 4.2). Gravity-driven diffusion includes creep of soils when wet, shrink and swell of montmorillonite-rich soils that is enhanced by the highly seasonal climate, shifting of soil particles by burrowing rodents, rainsplash of unprotected particles, and dry granular flow on surfaces without vegetation.

The result is a systematic style of erosion that creates distinctive hillslopes. Ridgecrests are smooth and rounded despite the arid, highly seasonal, climate (Appendix A). Midslopes have rills that become stream channels on the footslopes. Flux of sediment is a function of size of ephemeral streamflow events, as well as mass-movement processes. Annual unit stream power varies greatly and most work is done during the occasional exceptionally wet winter. Footslope morphologies are determined by a combination of diffusion and concentration geomorphic processes.

3.3 Sediment Flux and Denudation Rates

G.K. Gilbert's hallmark paper "The Convexity of Hilltops" was published in 1909. His concept is that hillslope gradient should progressively increase downslope from the ridgecrest divide. Steeper slopes are needed to provide sufficient gravitational force to shift an ever increasing sediment flux downslope from a ridgecrest. Hillslope sediment flux is delivered to the valley floor where channelized streamflow sweeps it downstream. All else being equal, the rate of increase of slopes below a ridgecrest is constant.

Two-dimensional studies of hillslopes model the relation between sediment flux and topography (Davis, 1892; Gilbert, 1909; Bakker and Le Heux, 1952; Culling, 1960, 1963, 1965; Hirano, 1968; Scheidegger, 1961; Kirkby, 1971; Ahnert, 1967, 1970, 1973). Three-dimensional models take advantage of recent advances in obtaining and analyzing of detailed topographic information and dating with terrestrial cosmogenic nuclides (Troeh, 1965; Ahnert, 1976, Dietrich et al., 1995; Howard, 1994, 1997; Heimsath et al., 1997).

The following set of equations from Roering et al. (1999) define important variables and processes

for tectonic geomorphologists to consider. Gilbert's simple linear diffusion model has sediment flux, q_s, as being proportional to hillslope gradient, where K_{lin} is linear diffusivity and z is altitude.

$$q_s = K_{lin} \nabla z \qquad (3.2)$$

Roering et al. relate sediment flux and denudation rates though a continuity equation

$$-\rho_s \frac{\delta z}{\delta t} = \rho_s \nabla . q_s + \rho_r C_o \qquad (3.3)$$

where ρ_s and ρ_r are bulk densities of soil and rock, z is the rate of change of surface altitude and C_o is bedrock uplift rate. $\delta z / \delta t = 0$ where dynamic equilibrium has been attained (denudation rate = bedrock uplift rate).

$$-\rho_r C_o = \rho_s \nabla . q_s \qquad (3.4)$$

They combine equations (3.2) and (3.4),

$$-\frac{\rho_r C_o}{\rho_s K_{lin}} = \nabla^2 z \qquad (3.5)$$

In this model, the ratio of denudation rate to linear diffusivity is equal to the rate of increase of hillslope curvature. Equilibrium hillslopes that erode by linear diffusion should have a constant curvature.

But such was not the case in the Oregon Coast Ranges study area of Roering et al. (1999). Sediment flux increases are almost linear on ridgecrests with low gradients, but flux increases rapidly as hillslope gradient approaches a critical angle.

A key point for soil-mantled hillslopes is that hillslope gradient becomes increasingly uniform with increasing distance from the divide. Both slope and curvature become increasingly insensitive to increase of erosion rates. Linear diffusion cannot explain such hillslopes that become progressively more planar. They conclude that a nonlinear theoretical expression best explains how sediment flux varies with hillslope gradient.

$$q_s = \frac{K \, \nabla z}{1 - (|\nabla z| / S_c)^2} \qquad (3.6)$$

where K is diffusivity and S_c is a critical hillslope gradient. This nonlinear transport law is similar to those of Andrews and Bucknam (1987) and Howard (1997). S_c is different from threshold slope angles such as the angle of repose (Strahler, 1950; Young, 1972; Burbank et al., 1996). S_c is the slope at which sediment flux becomes infinite, and thus is an abstraction. Diffusivity in their model varies linearly with the power per unit area exerted by geomorphic processes, and inversely with soil strength. Thickness of soil mantle does not appear to affect hillslope sediment flux rates according to Heimsath et al. (1997, 2001).

Roering et al. (1999, Fig. 2) compared linear and nonlinear models with their field data. Figure 3.5 summarizes results that are as applicable to the dry, grassy eastern margins of the California Coast Ranges as to the wet forests of the Oregon Coast Ranges. Both are soil-mantled slopes with minimal complications of protruding outcrops. Note that both the topographic profile (Fig. 3.5A) and gradient (Fig. 3.5B) depart from the linear model only 20 m from the ridgecrest divide. Additional variations of nonlinear equations are sure to be developed, as are hillslope models that analyze landscapes from a non-steady-state perspective.

This introduction presents the flavor of intriguing work being done by many geomorphologists on hillslopes. Modelers portray complicated pictures for well-studied areas. They provide much food for thought, but are problematic even after making numerous simplifying assumptions.

Two major goals of this book are to discern tectonic signatures in hillslopes (Section 3.7), and to assess the validity of the common assumption in models that steady state is typical of hills that have been eroding for long time spans under a given set of conditions (Section 5.4). First, we need to compare the characteristics of ridgecrests and footslopes in tectonically active and inactive study areas that have the same climate and lithology.

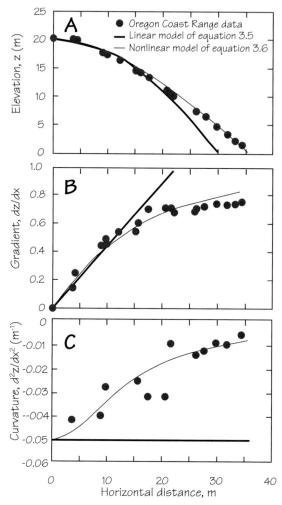

Figure 3.5 Changes in altitude (A), gradient (B) and curvature (C) along a ridgecrest profile using a linear model (equation 3.2), a nonlinear model (equation 3.6), and data from humid, forested slopes in the Oregon Coast Ranges. Both modeled hillslopes assume steady-state erosion conditions, and use a diffusivity of 0.003 m²/yr, an erosion rate, C_o, of 0.075 mm/yr, and ρ_r/ρ_s (the ratio of bulk densities of rock and sediment) is 2.0. S_c is 1.2 for the nonlinear hillslope. Figure 2 of Roering et al. (1999).

3.4 Ridgecrests

3.4.1 California Coast Ranges

The effects of stream-channel base-level change are transmitted from footslopes to midslopes and eventually to ridgecrests along lines that are perpendicular to the contours.

Profile convexity is the subject here and Sections 3.7 and 5.4 examine planform convexity. A simple power function (Hack and Goodlett, 1960) describes two-dimensional configurations along the flow lines of hillslope runoff. It can be used to compare topographies of hillslopes that are subject to base-level rise and base-level fall. Slope length, L, is the horizontal distances for a sequence of hillslope-position coordinates and slope fall, H, is the vertical distances for a sequence of hillslope-fall coordinates. Downslope changes in cumulative slope length and cumulative fall are described by

$$H = bL^f \qquad (3.7)$$

The ways in which slope steepness and curvature vary with changes in the constants of equation 3.7 are shown in Figure 3.6. The coefficient, b, describes slope steepness for a given exponent. Coefficients for different slopes may be compared, but only where exponents are similar. Base-level change, erodibility of surficial materials, climate, and geomorphic processes influence the values of coefficient b and exponent, f, which describes the type and amounts of curvature of hillslope topographic elements.

Ridgecrests were studied in two areas of rapid base-level fall. Base-level change during the Late Quaternary at the Panoche Hills site in the Diablo Range of central California is tectonic. It is nontectonic during a one-decade time span at Downpour Gulch badlands site in southeastern Arizona.

Hillslopes in the Panoche Hills tend to be strongly convex (Fig. 3.7), some from the divides to the stream channels, as a result of persistent tectonically induced stream-channel downcutting. A decade of measurements at 200 spike-and-washer benchmarks on three ridgecrests indicate that mudstone is downwasting five times faster than sandstone.

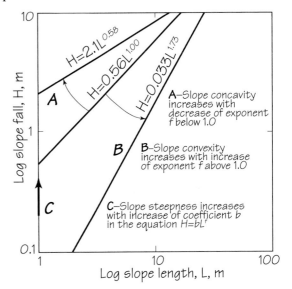

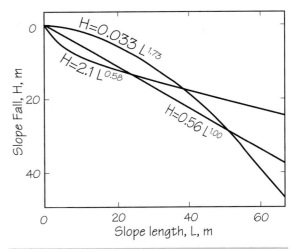

Figure 3.6 Interrelations of slope length, L, and fall, H, for hypothetical hillslopes; $H = bL^f$. Figure 5 of Bull (1975a). Upper. Logarithmic slope length–fall plots. Lower. Arithmetic topographic profiles

The slope length–fall interrelations of Figure 3.7 are representative of soil-mantled ridgecrests in soft marine Panoche sandstone in a semiarid climate and where rapid uplift is occurring. More than one equation generally is needed to describe slope morphology between the divides and the stream channels even where the entire hillslope is convex. Figure 3.7 data are from surveys made normal to contour lines between hillslope position coordinates of 0.0 and 0.25–0.40.

The cumulative nature of both variables results in absurdly high correlation coefficients of 0.999. Tectonic control that results in unique exponents for power-function relations (Wilson, 1968) may be difficult to demonstrate because spatial variations in rock strength and microclimate also affect ridgecrest morphologies. However, the constants of equation 3.7 are much different for the rapidly rising Panoche Hills and the tectonically inactive Panoche sandstone ridgecrests of the Howard Ranch quadrangle 63 kilometers to the northwest. The mean coefficients and exponents for the two families of curves are:

$$H = 0.07L^{1.78} \qquad (3.8)$$

for ridgecrests in the rising Panoche Hills, and

$$H = 1.32L^{1.41} \qquad (3.9)$$

for ridgecrests in the tectonically inactive Howard Ranch quadrangle.

The exponents suggest that ridgecrest convexities in the tectonically inactive Howard Ranch area are less – a difference of 0.37.

Statistical significance is low (1σ is 0.4) because other variables affect ridgecrest convexity. Panoche sandstone beds vary in width along strike in both areas, as does abundance of hard concretions in the soft sandstone. One should expect spatial variation in convexity along ridges that vary in width and rock mass strength. Dispersion of the ten exponents for the Howard Ranch ridgecrest dataset reflects lithologic and base-level controls.

These two study areas have contrasting tectonic and nontectonic styles of base-level change. Stream channels are cutting down into the rising Panoche Hills, whereas erosional widening of valley floors in the stable Howard Ranch area creates a

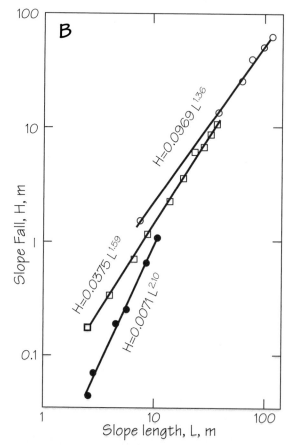

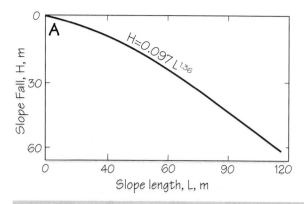

Figure 3.7 Convexity of sandstone ridgecrests in the Panoche Hills, central California.
Figure 6 of Bull (1975a).
A. Topographic profile of ridgecrest described by $H = 0.095L^{1.36}$.
B. Logarithmic fall-slope length plots for three ridgecrests.

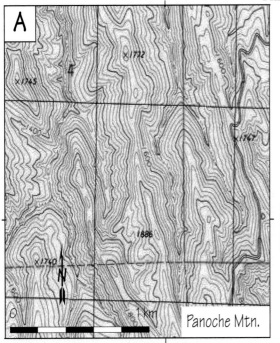

Panoche Mtn.

120° 46' W

36° 44' N

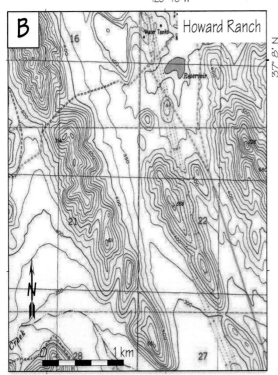

Howard Ranch

37° 8' N

121° 5" W

base-level fall induced by lateral erosion for the adjacent hillslopes. Stream-channel downcutting creates steep hillslopes and concave ridgecrests, and subsequent lateral erosion of the valley floor maintains these steep hillslopes.

Topographic maps illustrate contrasts in the terrains of rising and stable landscapes (Fig. 3.8). The Cretaceous Panoche Formation dips 35° to 45° in both study areas. Headward erosion and stream-capture processes direct the initial stages of drainage-net development along soft beds. Concretionary sandstone has a much greater rock mass strength and these beds tend to form ridgecrests even in the early stages of erosion (Fig. 3.8A). Steep hillsides may consist of either hard or soft rock, but only materials with a low rock mass strength underlie the broad valley floors of tectonically inactive settings (Fig. 3.8B).

3.4.2 Badlands

Complications resulting from spatial variations in rock mass strength characteristics can be largely avoided in studies of badlands or in laboratory erosion models. Miniature erosional landscapes in badlands generally lack vegetation and have fairly uniform fine-grained lithologies with low resistance to erosion.

Entrenchment of the San Pedro River of southeastern Arizona resulted in rapid dissection of silty clay along one tributary to form the highly intricate miniature topography of a study site named "Downpour Gulch" (Fig. 3.9). Downcutting by stream channels is a base-level fall and local valley-floor aggradation is a base-level rise. Measurements

Figure 3.8 Concretionary sandstone beds of the Panoche formation rise above the intercalated softer strata to form long parallel ridgecrests in the Diablo Range of central California.
A. Tectonically active (Class 1) landscape north of Panoche Mountain in the Panoche Hills.
B. Tectonically inactive (Class 5) landscape southwest of Howard Ranch near the Merced–Stanislaus county line.

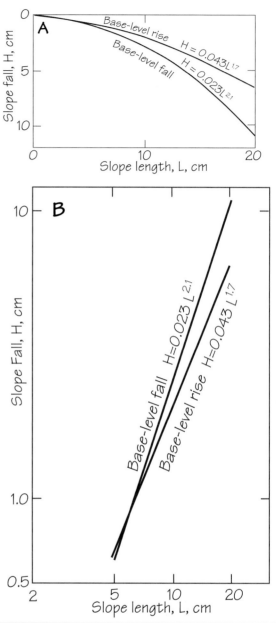

at 400 spike-and-washer benchmarks reveal a ten-fold variation of hillslope erosion rates. Steady-state denudation is not approximated here. I used the Downpour Gulch data to examine the influence of base-level change on the exponent, *f*, of equation 3.7. The hillslope-position coordinates for ridgecrest data for these miniature hillslopes ranged from 0.00 to 0.4–0.5.

The variables of lithology, slope exposure, and slope distance (non-horizontal length) were held constant at ten transects at the Downpour Gulch site for the purpose of comparing convexities of ridgecrests in areas of base-level rise and fall (Fig. 3.10A). Figure 3.10B shows the mean regression lines for both of the five ridgecrest groups. The mean coefficients and exponents of the two families of curves are:

$$H = 0.023L^{2.1} \qquad\qquad (3.10)$$

Figure 3.9 Intricate fluvial landscape eroded into the clayey silt of the Downpour Gulch badlands site, southeastern Arizona. Spike-and-washer line installed for measuring erosion between the two squares. Individual spikes and washers are circled. Protrusion of spike heads indicates the amounts of post-installation erosion. Rock hammer for scale.

Figure 3.10 Mean fall–slope length plots for ten ridgecrests in the Downpour Gulch badlands, southeastern Arizona that are affected by base-level rise ($H = 0.043L^{1.7}$) or base-level fall ($H = 0.023L^{2.1}$). Figure 7 of Bull (1975a).
A. Topographic profiles of ridgecrests.
B. Logarithmic fall–slope length plots.

for ridgecrests in areas of base-level fall, and

$$H = 0.043L^{1.7} \qquad (3.11)$$

for ridgecrests in areas of base-level rise.

Although the silty clay is highly conducive for the development of smooth, rounded ridgecrests, the ridgecrests affected by base-level fall are markedly more convex than those where base-level rise is occurring in the adjacent valley floor. The difference in the means of the two sets of exponents is significant at the 0.995 confidence level. The much greater convexity of ridgecrests responding to stream-channel downcutting, in these badlands and in the Panoche Hills, is a distinctive signature of ongoing base-level fall. The power-function exponent, f, is about 0.4 greater in both study areas.

Ridgecrest analyses of the Panoche Hills and Downpour Gulch data imply orderly adjustments of fluvial systems to base-level changes, even though steady state has not been achieved. One way to examine such systematic interrelations is to use the allometric change conceptual model (Bull, 1975a; 1991, section 1.9).

Total hillslope relief is a function of base-level fall, which is largest near the mouths of the drainage basins of mountain ranges with active range-bounding faults. The rest of the system must adjust first in order for the ridgecrests of the headwaters divides to reflect a portion of the range-front tectonic activity. Each ridgecrest morphology is related to its open system – the sequence of interconnected slopes of its drainage basin. Topographic studies of ridgecrests indicate how static (landscape interrelations at a point in time) or dynamic (landscape interrelations of changes with time) allometric analyses provide insight into how tectonic perturbations influence topographic evolution of hills and streams. In the next section we use the concept of allometric change to evaluate the effects of base-level change on footslopes in the canyonlands of southeastern Utah.

3.5 Canyonlands

We now shift to a much different tectonic setting than the California Coast Ranges. First we need to describe the rock types and the styles and magnitudes of recent tectonic base-level falls of part of the Colorado Plateau in the southwestern United States.

The canyonlands of Arizona and Utah in the southwest United States are scenic wonders that provide opportunities to study hillslope shapes in a much different lithologic setting than the soil-mantled Diablo Range of California. Local base-level change is generated by normal faults and by folding of monoclines and anticlines. Rates of tectonic deformation are much slower than for the Diablo Range study region, and knickpoints created by Late Quaternary faulting have migrated more than 200 km up large rivers.

Before we examine the characteristics of footslopes at one site in southern Utah we need to assess how much of the relevant tectonic perturbation is locally generated and how much consists of base-level falls generated by displacements in distant downstream reaches. Active normal faulting in the transition zone between the Colorado Plateau and the Basin and Range Province is the primary source of knickpoints that migrate upstream. Compressional deformation, such as folding of the Monument Upwarp near the San Juan River, is a more local tectonic perturbation.

Gently dipping beds of sandstone, shale, and limestone vary considerably in strength. This results in hillslopes with many outcrops. The resulting "cliff-and-ledge" topography is typical of the canyonlands.

The Colorado River drains much of the western United States. Two of its larger tributaries (Figs. 3.11, 3.12) are the Little Colorado River and the San Juan River – both have drainage-basin areas about 65,000 km^2.

The Colorado River has a drainage area upstream from Lees Ferry that is 264,000 km^2. River boatmen measure distances in miles downstream from Lees Ferry, which gives us a convenient standard way to note distances along the trunk stream channel. Lees Ferry also divides the upper and lower basins of the Colorado River for water-allocation purposes. The great power of the Colorado River is demonstrated by the ease with which it removed successive 100 to >300 m high lava dams at river mile 179 (288 km). Dams were created by occasional outpouring of basaltic lava from cinder cones on the

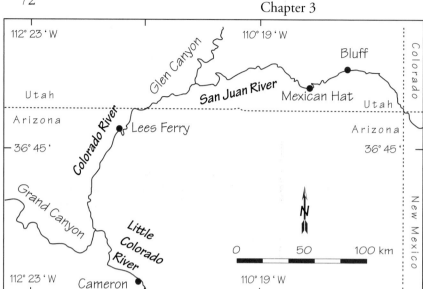

Figure 3.11 Map showing places in the Grand Canyon–Four Corners region, southwestern USA.

adjacent canyon rim (Hamblin, 1994; Dalrymple and Hamblin, 1998). The active Toroweap and Hurricane normal fault zones cross the Canyon at river miles 179 and 191 (288 to 307 km).

This is the transition zone between Colorado Plateau and the Basin and Range tectonic provinces. Normal faulting began at about 3.5 to 2 Ma (Billingsley and Workman, 2000). Displacement magnitude for each normal fault was approximately 70–170 m/My since 3–5 Ma for a combined normal fault displacement of roughly 580 m (Fenton et al., 2001). About 380 m, or 65% of this displacement seems to have occurred during the past 2 My (Jackson, 1990; Stenner et al., 1999; Fenton et al., 2001). Cosmogenic ^{3}He dating by Fenton et al. (2001) reveals uniform displacement rates during the

past 200 ky of 0.111 ± 0.009 m/ky and 0.081 ± 0.006 m/ky, respectively, for the Toroweap and Hurricane faults.

The reach just downstream from this pair of faults may also have been affected by minor faulting. Lucchitta et al. (2000) estimate that river downcutting rates increase from 0.07–0.16 m/ky in the downstream reach to about 0.40 m/ky in the reach upstream from the Toroweap fault.

These Quaternary tectonic base-level falls caused the river to cut ever deeper into the Grand Canyon reach, here considered as being from the junction of the Little Colorado River and 185 km to the west (river miles 65 to 190). Active normal faulting at the western end of the Canyon resulted in a mean rate of tectonically induced downcutting of

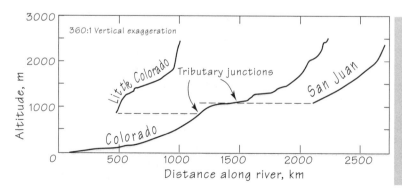

Figure 3.12 Large-scale longitudinal profiles of the Colorado River and its larger tributaries, the Little Colorado and San Juan Rivers in the Colorado Plateau region of the southwestern USA. Profiles based on U.S. Department of Interior, 1946.

~0.4. m/ky, but 100 km upstream the rate is only 0.2 m/ky. Remnants of straths capped with river gravel record this recent deepening of the Grand Canyon. The 185 km reach downstream from the junction with the Little Colorado River has been steepened an impressive 400 m during the past 2 My.

One might expect the mighty Colorado River to quickly re-establish an equilibrium longitudinal profile after local faulting creates a trivial waterfall knickpoint that is only 3 to 5 m high. Such a large river with an unlimited supply of bouldery cutting tools should quickly re-establish a smooth exponentially decreasing gradient characteristic of type 1 dynamic equilibrium. Surprisingly, this has not occurred. The longitudinal profile of the Grand Canyon reach is convex, not exponentially concave, and half of the fall in this reach occurs abruptly in a series of rapids (Leopold, 1969).

Two processes retard the process of tectonically induced downcutting into bedrock by this large river. One is climate-change induced aggradation events. Rivers cannot degrade their stream channels and cut broad straths indicative of the base level of erosion when the streambed is being raised by depositional processes during long time spans of full-glacial climate. Such valley-floor aggradation occurs when glacial erosion furnishes more bedload than the river can shift further downstream.

Second, debris-flow fans at the mouths of tributaries have effectively decoupled the Colorado River during times of interglacial climate such as the Holocene (Bull, 2007, section 2.5.2 and figs. 2.19 C, D). Tributary stream alluvial fans have boulders large enough to prevent continuity of behavior by the Colorado River. The river lacks the power to sweep clumps of huge boulders downstream. The result is a series of hydraulic jumps – rapids and short reaches of pools. The result is a Grand Canyon reach that has an irregular longitudinal profile. The San Juan River appears to have a more conventional longitudinal profile at the regional scale used in Figure 3.12.

The Colorado River can create the tectonic landform of strath terraces only during brief time spans (Pederson et al., 2006, fig. 4). Times of strath cutting are dictated by size of stream discharge upstream from Lees Ferry, and influxes of boulders from local tributaries.

The Little Colorado River is anomalously steep upstream from the Colorado River. A 400 m departure from the projection of the adjacent upstream reach of this stream probably records long-term tectonically induced downcutting generated by movement on the normal faults at the western end of the Grand Canyon. But the consequences of this distant base-level fall have either arrived fairly recently or the river has been slow to respond to the large cumulative tectonic perturbation.

Oblique aerial views of the Little Colorado River (Fig. 3.13) illustrate striking changes in the visual appearance of this landscape. The steep notch at the head of the terminal reach is upstream from the town of Cameron (Fig. 3.12), but valley-floor entrenchment at Cameron is modest (Fig. 3.13A). A narrow incised canyon is present at 16 km downstream from Cameron (Fig. 3.13B). It becomes much deeper further downstream (Fig. 3.13C) and then is both broad and deep closer to the junction with the Colorado River (Fig. 3.13D).

This background introduces how far upstream the perturbations generated by slip on the Hurricane and Toroweap faults are felt. We now discuss hillslopes along the present and past course of the San Juan River at a site called "The Loop". It is 15 km downstream from the town of Bluff, Utah, and 270 km upstream from the Little Colorado River (Fig. 3.11).

3.5.1 The Loop of the San Juan River

The San Juan River is a classic meandering stream that is entrenched deeply into bedrock. Best known is the "Gooseneck" reach (Fig. 3.14). A new, short and steep, reach will be created when the neck of this meander eventually is breached. Hillsides adjacent to the abandoned river channel then will be graded to a new base-level(s). The gradient of the abandoned river channel will be much too gentle for the new, much smaller, stream to flush away detritus derived from adjacent hillslopes. The resulting aggradation will raise and steepen the valley floor. Such meander abandonment gives us a setting of base-level rise with which to assess footslope morphology.

The San Juan River abandoned the long course around a meander when it eroded through

Figure 3.13 Aerial views of the Little Colorado River showing rapid change in canyon width and depth downstream from the town of Cameron, Arizona. Photos by William Tucker.
A. Slightly entrenched stream channel at Cameron, Arizona.
B. Incised valley 15.6 km northwest of Cameron, Arizona.
C. Incised valley 27.6 km northwest of Cameron, Arizona.
D. Incised valley 36.3 km northwest of Cameron, Arizona.

the narrow neck at a site called "The Loop" (Figs. 3.15, 3.16). Then it rapidly cut down through the moderately resistant mudstone, limestone, and sandstone of the Comb Ridge monocline as flow plunged down the new short, steep route.

Here we estimate the time of meander abandonment, and analyze footslope morphologies that reflect base-level rise and base-level fall. This provides insight about rates of tectonic base-level change, and responses of the river and adjacent hillslopes. Late Quaternary base-level fall includes local folding and perhaps the effects of distant normal faulting in the western Grand Canyon.

Abandoned meanders of the Colorado Plateau seem to be preferentially located on the crests of folds. If so, they might be used to identify tectonically active geologic structures. Besides The Loop, other examples include the Rincon of the Colorado River on the Waterpocket anticline in southern Utah; Labyrinth Canyon where the Green River crosses an anticline in northern Utah, and Antelope Wash on the east Kaibab monocline in northern Arizona.

Actively rising anticlines across alluvial valleys promote increased stream-channel sinuosity downstream from their axes (Ouchi, 1985; Schumm et al., 2000). Response times to base-level fall should be much longer for channels in bedrock. Fold axes have tension fractures and joints that may promote abnormally large meander amplitudes of stream-channel meandering (Hack and Young, 1959).

Figure 3.14 "Goosenecks" reach of meander bends on the San Juan River located 7 km downstream from the town of Mexican Hat. Photo by William Tucker.

The substantial height of The Loop above the present level of the San Juan River led me to anticipate a substantial age for its time of abandonment. Sand and gravel left by the San Juan River should be carbonate cemented by now. A valley of great age would have a flight of fill terraces reflecting the pervasive influences of Quaternary climatic changes. Instead, the few terraces are surprisingly young. Meander abandonment seems to have occurred in the Late Quaternary, even though the river now flows 61 m below its former level (Fig. 3.17).

I had expected the deposits left by the San Juan River to be strongly cemented by pedogenic calcium carbonate derived from the local limestone and abundant atmospheric dust. Instead, I found loose sand that is unweathered and river cobbles that lack calcium carbonate coatings.

Pockets of loose, unoxidized magnetite-bearing sand between cobbles and boulders were impregnated with plastic (see fig. 4 of Leopold and Bull, 1979) and tested for remnant magnetism by Robert DuBois at the University of Oklahoma. Normal polarity and high-latitude poles indicate an age of less than 790 ka, the time of the Brunhes-Matuyama polarity reversal (Johnson, 1982).

Only one prominent fill terrace occurs in The Loop. None of the post-abandonment alluvial surfaces in The Loop have strongly developed

Figure 3.15 The San Juan River used to flow counter-clockwise around the central island in a valley that is now drained by two ephemeral streams with a common drainage divide in the backfilled part of the valley floor. The present incised canyon of the San Juan River is at the left side of this aerial view. Photograph by Peter L. Kresan ©.

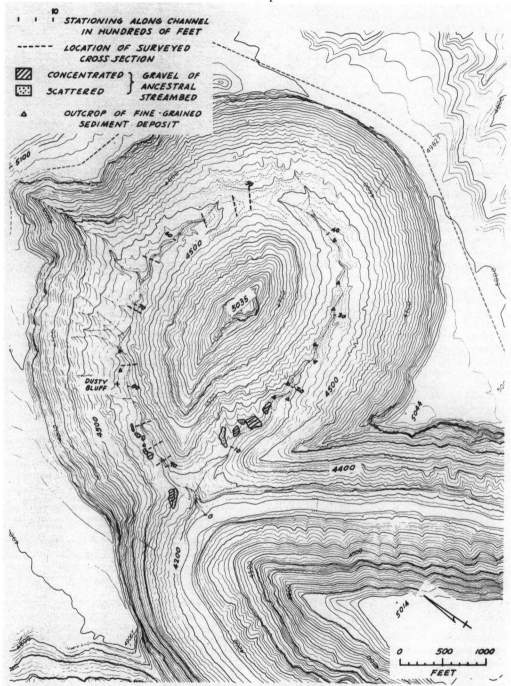

Figure 3.16 Footslope study sites in the aggrading valley of an abandoned meander adjacent to a downcutting reach of the San Juan River in southeastern Utah. This topographic map of The Loop of the San Juan River is figure 2 of Leopold and Bull (1979).

Figure 3.17 View upstream to the east from of the Loop of the San Juan River. The two ephemeral streams in the foreground drain the east and west sides of the Loop. They are incised below the former level of the San Juan River and end now abruptly at this dry water-fall, the top of which is 37 m above the present river and 24 m below the level of the river at the time of meander abandonment. About 10% of total canyon downcutting occurred since meander abandonment.

soil profiles. The highest alluvial surface has a Late Pleistocene soil resembling those of the of lower Colorado River region (typic haplargids). Soils of this age are described in detail in Chapter 2 of *Geomorphic Responses to Climatic Change* (Bull, 1991). This Loop soil profile has pedogenic clay and carbonate charac-teristics suggestive of a younger age than the 120 ka soils of the lower Colorado River region. It appears to record an aggradation event at ~ 60 ka, which is widespread in the semiarid to arid American south-west (table 1.2, Bull, 2007). Leopold and Bull (1979) concluded that meander abandonment occurred in the Late Pleistocene (Table 1.1).

Meander abandonment occurred sufficient-ly before 60 ka for accumulation of as much as 45 m of deposits in the broad valley floor. I assign an age of meander abandonment of at least 100 ka. This implies an average downcutting rate by San Juan River of about 0.6 m/ky.

Incision of the San Juan River could be partly the result of Late Quaternary normal faulting at the downstream end of the Grand Canyon reach of the Colorado River about 570 km downstream from The Loop. The cumulative magnitude of this perturbation emanating from the Basin and Range Province since 2 Ma is about 800 m.

Let us presume that increments of this tec-tonic perturbation decrease in an exponential manner as knickzones created by normal faulting events mi-grate upstream. Analyses of fluvial landforms in the eastern Grand Canyon suggest a downcutting rate of only 0.2 m/ky (Luchitta et al., 2000; Pederson et al., 2002, 2006). So, the effects of the tectonic perturba-tion seem to be halved in just the first 180 km.

Estimating rates of canyonlands downcut-ting has been a challenging and profitable playground for teams of geochemists and tectonic geomorphol-ogists. The times of formation of various land-forms have been estimated by dating of basalt lava flows, gravels capping strath terrace surfaces, and re-gional soil profile chronosequences. Paleomagnetic characteristics constrain possible ages. Geochemical procedures include potassium–argon dating, terrestrial cosmogenic nuclides, and uranium-series dating. The broad plateau provides a reference surface for estimating the amounts of stream-channel downcutting that has created the slot-like canyons (Figs. 3.13–3.17).

Several as yet unresolved complications arise when one tries to assess the rates and implications of regional canyon downcutting at a particular site. These include:
1) The amount of stream-channel downcutting that has migrated upstream from the active normal faults in the distant transition zone.
2) Late Quaternary tectonic deformation of monoclines upstream from the Grand Canyon reach. Active structures, such as the Monument Upwarp near the San Juan River, need to be included in estimates of tectonic base-level fall. This includes uplift of the Comb Ridge monocline at The Loop.

Appraisal would be much simpler if these big rivers quickly restored smooth, concave equilibrium longitudinal profiles as anomalously steep sections migrated quickly upstream. We now know that much of the mighty Colorado River downstream from its junction with the Little Colorado River is a series of rapids on a convex longitudinal profile. What about the reaches further upstream? Are they in equilibrium or disequilibrium? Or do we have a spatially variable mix of equilibrium and disequilibrium reaches?

Surely Quaternary canyon cutting of the terminal reach of the Little Colorado River (Fig. 3.13A–D) must have also proceeded up the Colorado River. The 250 m deep slot of Glen Canyon was created since 500 ka (Hanks et al., 2003; Garvin et al., 2005). They estimate rates of Colorado River downcutting to be 0.4 m/ky between 500 and 250 ka, accelerating to 0.7 m/ky since 250 ka. But how far up the San Juan River did such stream-channel downcutting migrate from Glen Canyon? It is 190 km downstream from The Loop. Wolkowinsky and Granger (2004) estimated that the average rate for 140 m of incision near Bluff to be only 0.1 m/ky, but Hanks and Finkel (2006) believe that the same data indicate a rate of 0.2 m/ky since 600 ka.

I assume that the regional tectonically induced downcutting (the sum of broad upwarping and upstream knickpoint migration) since 100 ka was between 6 and 12 m at the Loop. The remaining 49 to 55 m is best ascribed to two types of local base-level fall.

The style and magnitude of local nontectonic base-level fall is straightforward. Suppose the reach of the San Juan River at The Loop is tectonically inactive – no local uplift, and no downstream sources of base-level fall. Breaching of the meander neck would shift the threshold of critical power from equilibrium conditions (1.0) to degradation (1.1–1.3) because of the steeper flow gradient along the much shorter route. The magnitude of this nontectonic base-level fall is equal to the drop in altitude where the ancestral river flowed around The Loop. It is only 3 m.

The remaining 46 to 52 m of base-level change is best ascribed to uplift of the Comb Ridge monocline. Uplift at about 0.4 to 0.5 m/ky contributed to abandonment of the former streamcourse and caused most of the rapid incision below the abandoned meander.

Did the San Juan River achieve equilibrium conditions when it was flowing around The Loop reach? Both the prior and present stream channels have flat bedrock floors. However, the present river flows in a gorge, and the prior valley (Figs. 3.16, 3.17) was wide enough to allow creation and preservation of low strath terraces. So the prior San Juan River achieved type 1 dynamic equilibrium. The present river has characteristics of a type 2 dynamic equilibrium landscape. This change from type 1 to type 2 suggests a Late Quaternary acceleration of base-level fall caused by 1) local and distant base-level fall perturbations and 2) streamflow being able to take a shorter route.

3.5.1.1 Footslopes

Footslope characteristics may be even more indicative of relative tectonic activity than ridgecrests – convex footslopes in fast-rising mountains and concave footslopes where base-level fall has ceased or is slow.

Rapid stream-channel downcutting across the neck of the meander loop quickly made it difficult for large streamflow events to enter The Loop. The Loop became isolated from the river. This

Figure 3.18 Characteristics of footslopes subject to base-level rise in the Loop of the San Juan River. Outside of bend of the abandoned meander that was used for evaluation of the effects of base-level rise is at the left middleground. Note the relatively greater amounts of colluvium compared to the slopes above the actively downcutting river shown in Figure 3.17.

Foreground is near the north end of the bedrock central island where aggradation has created a watershed divide between the two ephemeral streams. About 45 m of base-level rise has occurred here since meander abandonment. Rockfalls, debris flows, and water flows from the bedrock hillslopes contributed to aggradation of the valley-floor.

provided a site for us to study footslopes being affected by depositional base-level rise and to compare them with footslopes next to present-river where base-level is continuing to gradually fall.

Streamflow dynamics changed greatly in The Loop. Consider the major changes in the components of the threshold of critical power. Stream power decreased greatly when flows of the second largest tributary of the Colorado River were replaced by highly ephemeral local runoff from the adjacent hillsides. Resisting power changed too; bedload transport rate decreased greatly, but sediment size, and hydraulic roughness increased, and the stream-channel pattern changed from meandering in bedrock to braided in gravelly alluvium. The net result was a huge decrease in stream power and major increases in resisting power variables. The abandoned-meander reach shifted far to the aggradational side of the threshold of critical power.

Two new watersheds were created in The Loop (Fig. 3.16). Maximum valley-floor aggradation of about 45 m coincided with the divide that separates the two ephemeral streams (Leopold and Bull, 1979, fig. 9). As one might expect, the location of this divide is at the maximum distance from the meander-neck cutoff. One small ephemeral stream flows in the same direction as the prior river, but the other flows up the former canyon (Fig. 3.16). Aggradation in the headwaters combined with degradation in the

downstream reaches of The Loop nearer the present San Juan River steepened the gradient of both ephemeral streams until stream power was sufficient to transport the new sediment flux and size of bedload washed from the adjacent cliffy hillsides. The sum of ephemeral-stream downcutting of 24 m at the mouth (Fig. 3.17) and abandonment-induced aggradation of 45 m at the new watersheds divide (Fig. 3.18A) is an impressive 69 m. The San Juan River dropped only 3 m when it flowed around The Loop, a gradient of 0.0012. The downstream reaches of the two ephemeral streams have a gradient of 0.10.

Two sets of footslopes were described using equation 3.7, one influenced by 40 m of base-level rise in the aggrading part of the abandoned meander and the other influenced by 61 m of base-level fall adjacent to the present San Juan River (Figs. 3.16, 3.18). Both sites are on the outside of a bend of the prior or present river, face towards the southwest, and are underlain by similar sedimentary strata. Distances of starting points for the topographic profiles were constrained to 212–227 m from the edge of the valley floor in order to compare coefficients for equation 3.7. The hillslope-position coordinates for these footslope data were 0.6–0.7 to 1.0. Plots of slope length and fall were made of footslopes for eight topographic profiles; four were affected by base-level rise and four were affected by base-level fall. The representative graphs shown in Figure

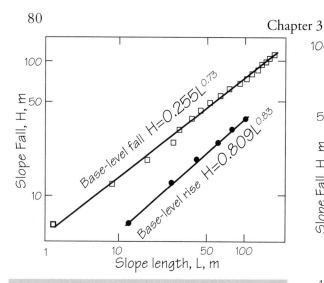

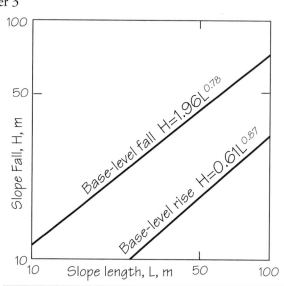

Figure 3.19 Concavity of footslopes adjacent to the San Juan River, Utah. Figure 8 of Leopold and Bull (1979).
A. Logarithmic fall–slope length plots for footslopes in areas of base-level rise and base-level fall.

Figure 3.19 Concavity of footslopes adjacent to the San Juan River, Utah. Figure 8 of Leopold and Bull (1979).
C. Mean logarithmic fall–slope length plots for five footslopes in areas of base-level rise and base-level fall .

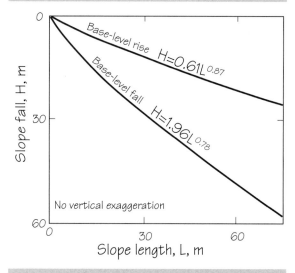

Figure 3.19 Concavity of footslopes adjacent to the San Juan River, Utah. Figure 8 of Leopold and Bull (1979).
B. Mean topographic profiles of five footslopes in areas of base-level rise and base-level fall.

3.19A have slightly more scatter than for the Panoche Hills ridgecrests (Fig. 3.7B) because of the cliff-and-ledge topography that is typical of stratified rocks in the canyonlands. Footslopes affected by base-level rise consist of bedrock mantled with colluvium and talus. Footslopes affected by base-level fall consist mainly of bedrock with a few patches of colluvium and talus. Mean equations are

$$H = 1.96L^{0.78} \qquad (3.12)$$

for the footslopes affected by base-level fall, and

$$H = 0.61L^{0.87} \qquad (3.13)$$

for the footslopes affected by base-level rise.

The topographic profiles (Fig. 3.19B) of the mean equations (Fig. 3.19C) for the footslopes are much different. Means of the two classes of exponents are significant at the 0.975 level and suggest contrasting styles of allometric change for footslopes affected by base-level rise and base-level fall.

Footslopes at the base-level fall site are – perhaps unexpectedly – more concave than those at the base-level rise site (0.78 departs much more from a value of 1.00 than does 0.87). Footslopes above an aggrading valley floor maximize sediment-transport efficiency by undergoing minimal decreases in slope – the result is a low concavity (equation 3.13). A coefficient of 1.96 compared to 0.61 suggests (as predicted by Penck, 1953) that much steeper slopes prevail in the base-level fall site. Increase of total relief creates a situation of surplus power to transport sediment across footslopes, which become highly concave (equation 3.12).

The significant differences in footslopes and ridgecrests caused by either base-level fall or rise demonstrate the migration of the effects of base-level perturbations through a series of interrelated slopes as first proposed by Gilbert (1877). Rupture along a range-bounding fault induces headcutting that migrates rapidly up the trunk stream channels because stream power is concentrated into a small area. Migration of the effects of a base-level fall up hillslopes is much slower because of lesser competence of processes such as raindrop splash, overland flow, and rillwash combined with a variety of episodic mass movements.

In an overall sense, mountain fronts respond to rapid tectonic base-level fall by forming convex or straight hillslopes that rise abruptly above the degrading valley-floor base level as relief is progressively increased. Erosional retreat of slopes in an inactive terrain is accompanied by conversion to gentle concave hillslopes that are graded to valley floors, which either are at equilibrium or are aggrading. Steeper concave footslopes are maintained where rates of base-level fall are slow.

3.6 Cross-Valley Shapes

3.6.1 Lithologic Controls
Relief and slope steepness depend on rock mass strength (Selby, 1982a,b; Burbank and Anderson, 2001; Brook et al., 2004) as well as on rates of base-level fall and how geomorphic processes deepen valleys. Hard and soft rocks are raised together where uplift rates are fast and peaks and ridgecrests may be composed of soft or fractured rock types. Even erosional base-level fall may allow soft materials to form steep slopes with a high drainage density – the badlands of Bryce Canyon National Park, Utah, USA are an example.

Rock mass strength and widths of valleys are closely related to slope steepness in tectonically inactive mountains. Soft rocks form gentle slopes and occur in topographically low positions and resistant rocks form summits and ridgecrests. Distributions of slope steepness and drainage-net patterns are obviously different when compared with tectonically active landscapes (Fig. 3.8). Such contrasts in tectonic signatures are quantified in the Section 3.7.

Two images illustrate this notion. The Basin and Range orogeny was largely concluded by 8 Ma in southern Arizona, USA, allowing ample time for erosion to create Baboquivari Peak, which has a landscape with a distinctive signature indicative of tectonic inactivity (Fig. 3.20A). Quartz monzonite underlies the lower slopes, but silicification of part of the intrusion near dikes created locally resistant rocks. Spatial differences in denudation rates are controlled by rock type here. Several million years of differential erosion has resulted in peaks composed only of highly resistant rocks in a class 5 landscape.

An even longer period of tectonic inactivity began along the tectonically passive coastline of Brazil after South America drifted away from Africa (Fig. 3.20B). The coastline is highly embayed and numerous inselbergs rise above a pedimented coastal plain. All of the high peaks are underlain by massive resistant lithologies that have exceptionally high rock mass strength.

A sensitive index describing recent deepening of valleys is the valley floor width/valley height ratio, or for brevity the V_f ratio (Bull, 2007, section 4.2.1.2). If V_{fw} is the width of a valley floor and A_{ld}, A_{rd}, and A_{sc} are the altitudes of the left and right divides (looking downstream) and the altitude of the stream channel, then

$$V_f = \frac{V_{fw}}{\dfrac{(A_{ld} - A_{sc}) + (A_{rd} - A_{sc})}{2}} \qquad (3.14)$$

Figure 3.20 Topographic characteristics of tectonically inactive mountains.
A. Baboquivari Peak in southern Arizona illustrates how resistant dike rocks and plugs coincide with the highest parts of tectonically inactive mountains.   Photograph by Peter L. Kresan ©.

In a parallel discussion, Bull (2007, section 4.2.1.2) discusses the relative merits of several valley-shape equations and their application for defining classes of relative tectonic activity of mountain fronts for short and long time spans.

Valley-floor widths increase with watershed size, erodibility of rock type, and with decrease of uplift rates. Valley heights decrease with the passage of time after cessation of uplift, but not nearly as fast as valley floors widen.   A key point is that

Figure 3.20 Topographic characteristics of tectonically inactive mountains.
B. View of rugged mountains around a 40 km long estuary, the Baia de Paranaguá, 200 km southeast of Sao Paulo, Brazil. Pico Paraná at the left edge of the image rises to an altitude of 1922 m. This perspective view was created from a side-looking radar digital elevation model and was provided courtesy of the German Remote Sensing Data Center and NASA. Compare this class 5 landscape with the class 1 landscape of Figure 2.2A.

flowing-water processes capable of rapid landscape change are concentrated mainly along valley floors. The valley floor width/valley height ratio is especially good for recording to Late Quaternary tectonism because narrowing and incision of a valley floor is accomplished quickly by most streams.

Lithologic control of V_f ratios is illustrated by the rugged, but tectonically inactive, mountain front north of Sheep Mountain (Fig. 3.21A). Southwestern Arizona has a strongly seasonal, arid, hyperthermic climate (Appendix A). The range-bounding fault has not slipped for more than 700 ky (Bull, 2007, p. 146). Two contrasting rock types are responsible for the dark and light tones, and for the differing textures of Figure 3.21B.

The lighter toned rock type is biotite-quartz diorite and monzonite. It is not strongly foliated except near the contact with the dark rocks, which are hornblende-plagioclase amphibolite, foliated mafic gneiss, and biotite schist. Joint orientations in the diorite are N 80° W and N 5° W and joint spacing at about 100 m upstream from the range-bounding fault is about 15 per m. The diorite–monzonite weathers mainly to gruss and large blocks – typical weathering products of coarsely crystalline plutonic rocks. Two-thirds of the hillslopes are bare rock and

the remaining third has less than 0.3 m of soil, which decreases rock weathering rates.

A common joint orientation in the mafic rock complex is about N 12–18° W and the abundance of joints at about 100 m upstream from the range-bounding fault is about 30 per m. About a quarter of the hillslopes consists of highly fractured outcrops with low relief and the remaining three-quarters consists of poorly sorted colluvium with a surface lag of rock fragments. These differences in rock-weathering products of the plutonic and mafic rock types are sufficient to affect erosion rates and valley shapes.

The effects of lithology and structure controls include valleys in terrain underlain by the mafic rocks that have U-shaped cross-valley shapes (Figs. 3.21C, D). Pediment embayments are typical of the large valleys. Values of V_f ratios measured at basin-position coordinates of 0.9 average about 1.8 for basins underlain by rocks of the mafic complex, and about 0.6 for basins underlain by quartz diorite. The markedly different landforms of this mountain front are the result of different rates of weathering,

Figure 3.21 Influences of contrasting lithologies on valley morphologies, Sheep Mountain, southwestern Arizona.
B. Aerial photo showing contact between light toned biotite-quartz diorite and dark toned mafic metamorphic rocks. Note the wider valley floors associated with the mafic rocks.

Figure 3.21 Influences of contrasting lithologies on valley morphologies, Sheep Mountain, southwestern Arizona.
A. A rugged tectonic activity class 5C mountain front rises above a pediment with a thin veneer of alluvial-fan deposits.

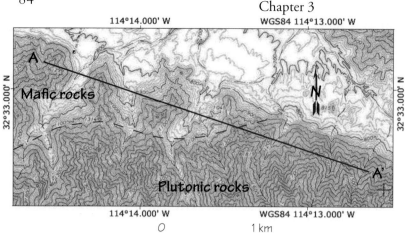

114°14.000' W   WGS84 114°13.000' W

32°33.000' N

A

Mafic rocks

N

A'

Plutonic rocks

114°14.000' W   WGS84 114°13.000' W

32°33.000' N

0   1 km

Figure 3.21 Influences of contrasting lithologies on valley morphologies, Sheep Mountain, southwestern Arizona.
C. Topographic map and two classes of rock types. A–A' is the line of topographic profile for Figure 3.21D.

and different erosional processes, of these two rock types. Weathering rates are extremely slow for the plutonic rocks in this hot desert, in part because exposed outcrops are continually dry. The sheared and jointed mafic rocks have downwasted faster. Their markedly different cross-valley shapes demonstrate the importance of lithologic control on landscape evolution in this dry, tectonically inactive setting.

3.6.2 Tectonic Controls

The previous sections show that rates and magnitudes of base-level change influence curvature and steepness of ridgecrests and footslopes. Valley shapes may also tell us how fast the mountains are rising. Here we examine topographic profiles across valleys in diverse tectonic settings at a constant basin-position coordinate of 0.9, and at different basin-position coordinates along two valleys.

Wide valley floors are significant because they suggest the attainment of equilibrium conditions for sufficiently long time spans to allow lateral cutting by the stream into adjacent hillslopes. V-shaped valleys gradually are transformed to U-shaped valleys

with concave footslopes by conditions of prolonged stable base level.

Cross-valley shape is different for streams that are aggrading, degrading, or are at equilibrium. Downcutting dominates over lateral cutting in degrading valleys. Valley floors commonly are only as wide as bankfull stream discharges. Such narrowing of a stream channel increases unit stream power and may initiate a self-enhancing feedback mechanism that promotes continued downcutting. The resulting valley topographic cross sections have pronounced V shapes.

Lateral cutting dominates over vertical cutting in valleys with protracted stream-channel equilibrium conditions. Such streams have remained at their base level of erosion for >50 ky, being interrupted only by climate-change induced aggradation–degradation events. This lack of incision is because further stream-channel downcutting would reduce gradients to the point where not all of the bedload can be transported. Short-term valley-floor aggradation occurs as stream power and bedload-transport rate fluctuate. From a landscape-change perspective

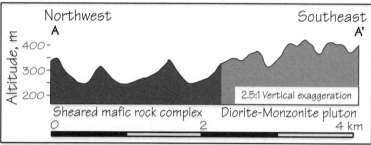

Northwest

A

Altitude, m

400–
300–
200–

Southeast

A'

2.5:1 Vertical exaggeration

Sheared mafic rock complex   Diorite-Monzonite pluton

0   2   4 km

Figure 3.21 Influences of contrasting lithologies on valley morphologies, Sheep Mountain, southwestern Arizona.
D. Topographic profile along line A–A' showing the effects of spatial changes in rock type on valley-floor width.

these are merely fluctuations about a long-term equilibrium condition. Such valleys commonly have moderately U-shaped cross sections. Broad, concave footslopes are typical of the adjacent hillsides. Valleys may be extremely U-shaped where climate-change induced aggradation has raised the valley floor and buried the lower hillsides to create an abrupt topographic transition.

Valley shape and size is a function of both rate and duration of uplift. Uplift rate is rapid at the Laguna Seca Hills (Fig. 3.22A), but the magnitude of uplift is modest. The rate of uplift at Happy Canyon (Fig. 3.22B) is only half that at Laguna Seca Hills

but has been sustained for so long that valley relief is large. Both have a tectonic activity class rating of 1. Valley cross sections for Fall and Eaton Canyons (Figs. 3.22C, D) appear similar to Figures 3.22 A, B but these fluvial systems have a tectonic activity class of 2 because their fanheads are permanently entrenched.

The class 3 Hanuapah Canyon valley cross section is much different. Prolonged widening of the valley floor has created the present strath, and the bench above the valley floor may be a strath terrace. Valley floors are even wider for the class 4 Wilson Canyon (Fig. 3.22F) and class 5 Fortuna

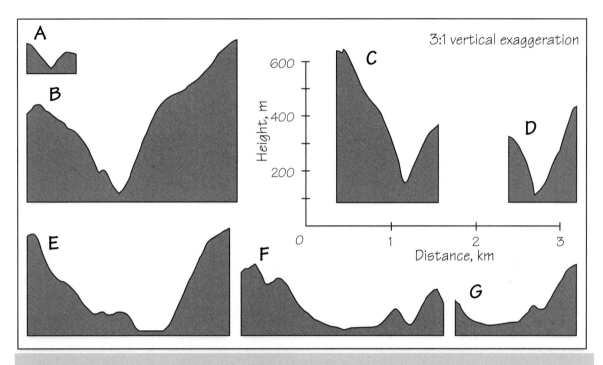

Figure 3.22 Cross-valley shapes of arid and semiarid drainage basins of different tectonic activity classes at basin-position coordinates of 0.9. Each has the same height and distance scale and the same vertical exaggeration of 3:1.
A. Class 1 unnamed basin of "Vigorous Vale", Laguna Seca Hills, Diablo Range, California.
B. Class 1 Happy Canyon, Panamint Range, southeastern California.
C. Class 2 Fall Canyon, Grapevine Mountains, Death Valley, southeastern California.
D. Class 2 Eaton Canyon, San Gabriel Mountains, southern California.
E. Class 3 Hanuapah Canyon, Panamint Range southeastern California.
F. Class 4 Wilson Canyon, Argus Range, southeastern California.
G. Class 5 Fortuna Wash, Gila Mountains, southwestern Arizona.

Wash (Fig. 3.22G). Although steep slopes are present locally in tectonically inactive terrains, maximum valley side slopes are not as consistently steep as the slopes of tectonically active landscapes.

A great variety of valley shapes are shown in Figure 3.22, but this approach is a rather crude measure of the degree of tectonic activity of these landscapes. Plots A–D can be classed as active and E–F as inactive. This lack of precise identification can be attributed to long response times to large, prolonged tectonic perturbations. Class 1 and class 2 hillslopes still are convex. The valley shapes of classes 3, 4, and 5 could just as well be lumped together as being inactive. Uplift is either not present or so slow that streams easily maintain their wide channels and equilibrium longitudinal profiles.

Tectonic activity classes were assigned on more sensitive landscape signatures. These include the degree of sinuosity of the mountain–piedmont junction, stage of degradation of triangular facets, and whether the adjacent piedmont is aggrading or degrading. Only two classes of tectonic activity are apparent – active or inactive – if one uses only cross-valley shapes. Channel downcutting, if present in tectonically inactive settings, such as Australia and Arizona, is slow and restricted to crustal rebound in response to gradual erosional unloading (Bishop and Brown, 1992; Goldrick and Bishop, 1995).

Even the power of episodic glacial degradation (Montgomery, 2002) may not be sufficient to permit tectonically induced downcutting to equal rapid uplift rates (Fig. 3.23). Stirling Creek in the southwestern part of the South Island of New Zealand is now an ice-free watershed. Former great depths of recent glacial ice are indicated by the smooth U-shape of the cross-valley shape, which is truncated by Milford Sound. The combined flows of glaciers scoured resistant high-grade gneiss to 300 m below mean sea level to create Milford Sound (Bruun et al., 1955). Stirling Creek is now a hanging valley; its mouth is 150 m above sea level in the fiord. This tributary glacier may have had slower valley-floor downcutting rates during the Pleistocene. This setting has yet to return to the characteristics of a fluvial landscape. Glaciation is a major complication for geomorphologists who study behavior of fluvial systems (Section 8.2.1).

Valley floor width/valley height ratios vary within a fluvial system. Reaches close to the mountain–piedmont junction are more likely to be closer to base level of erosion conditions, but headwaters reaches are nearly always strongly degradational.

This is demonstrated by comparing two drainage basins. Both are arid, are underlain by granitic rocks, and have areas of about ~7 km². V_f ratios in the headwaters reaches are similar for the two basins, but increase much more rapidly in the downstream direction in the tectonically inactive basin (Fig. 3.24). The entire class 1 basin is on the degradational side of the threshold of critical power. The middle and downstream reaches of the class 5 basin have been at the base level of erosion for ~1 My; sufficient time to create broad valley floors.

Some valleys have footslopes that are steeper than the adjoining midslopes, which results in an inner gorge type of cross-valley shape. This is presumed to be the result of accelerated uplift that enhanced valley-floor downcutting (Kelsey, 1980, 1988; Kelsey et al., 1995). This process can be considered as a tectonically induced knickpoint that is migrating up the hillslopes from the valley floor.

Tectonically induced steepening of footslopes in the San Gabriel Mountains of Southern California (Fig. 3.25) is responsible for hillslopes that are convex from the ridgecrest to the valley

Figure 3.23 View of the U-shaped cross section of the valley of Sterling Creek where it was truncated by a Pleistocene glacier in Milford Sound, New Zealand.

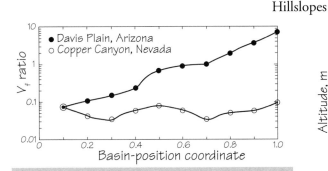

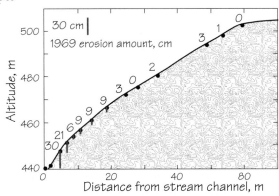

Figure 3.24 Downstream changes of V_f ratios in tectonically active and tectonically inactive streams of arid-region granitic landscapes. Copper Canyon is a tectonic activity class 1 valley in the Wassuk Range of west-central Nevada. The unnamed stream flows onto the Davis Plain of the Gila Mountains in southwestern Arizona. It has a tectonic activity class rating of 5.

Figure 3.25 Steepening of footslopes in tectonically active mountains. Convex hillslope underlain by weathered gneiss and schist along a headwaters stream of a class 2 mountain front, San Gabriel Mountains, southern California. Figure 2 of Scott (1971).

floor. Kevin Scott measured amounts of slope degradation after two major winter of 1969 rainstorms; maximum hillslope denudation occurred at the footslope. Such nonuniform slope erosion is tectonically induced. It decreases slope stability and increases the likelihood of mass movements.

These are but several of many geomorphic consequences of rock uplift. The landforms of tectonically active and inactive landscapes are indeed profoundly different. These distinctive characteristics are quantified in the next section.

3.7 Tectonic Signatures in Hillslopes

The preceding sections of this essay about hillslopes show that class 1 landscapes have more convex and steeper ridgecrests, more concave and steeper footslopes, and narrower and steeper valleys. This section steps up from landform to landscape scale by providing watershed-size examples that quantify the profound influence of tectonic base-level change on hillslopes of small fluvial systems. We return to the soil-mantled hillslopes on the eastern margin of the Diablo Range whose tectonic setting was described in Sections 1.2.3 and 3.2.

First let us compare the overall topography of two unnamed Diablo Range drainage basins

to which I have assigned informal names (Fig. 3.2, Table 3.1). Tectonically active Vigorous Vale is in the Laguna Seca Hills, and inactive Dormant Hollow is about 33 km to the northwest on the Merced–Stanislaus County line (Howard Ranch topographic map quadrangle). Both watersheds have ephemeral streams and the same arid to semiarid, mesic, strongly seasonal climate (Appendix A). They are at about the same altitude and both flow into the San Joaquin Valley (Table 3.1). Marine sedimentary strata underlie these watersheds; bedding parallels the mountain fronts and strikes to the northwest. Soft sandstone, mudstone, and shale are eroded quickly where uplift has created steep hillsides. Valleys tend to be underlain by these soft rocks and trend SE–NW (Fig. 3.8). The more resistant concretionary sandstone beds of the Panoche Formation form the ridgecrests and hilltops. Soft mudstone may occur on ridgecrests and in valley floors during the initial stages of uplift.

Hills and knobs in Dormant Hollow (Fig. 3.26A) are underlain by harder sandstone. Erosion over a time span of >1 My has lowered areas underlain by soft rocks faster than the adjacent concretionary sandstone. A key point here is that hard rocks underlie the higher portions of this tectonically inactive landscape because of nonsteady-state landscape evolution as discussed in Section 5.4.

	Area km^2	Uplift rate m/ky	Relief and altitude range	Mean slope	Climate	Rock type
Vigorous Vale	1.30	~2	550 m 320 to 870 m	0.54 28°	Semiarid, mesic, strongly seasonal	Soft, NW strike, concretionary sandstone
Dormant Hollow	3.04	<0.1	680 m 450 m to 1130 m	0.49 26°	Semiarid, mesic, strongly seasonal	Soft, NW strike, concretionary sandstone

Table 3.1 Geomorphic characteristics of the watersheds of tectonically active Vigorous Vale and tectonically inactive Dormant Hollow.

Topography does not match rock mass strength so nicely in the rapidly rising Laguna Seca Hills. Rock uplift is so fast that ridgecrest surface uplift nearly matches the rock-uplift rate. Rock uplift is much faster than surface uplift in the valley floors.

Shapes of watersheds reveal their stage of tectonic evolution (Section 2.2.1). The young, rapidly rising sections of the foothill belt typically have elongate watersheds (Fig. 3.27) as compared to more circular watersheds (Fig. 3.26) in the tectonically stable parts of the Diablo Range.

A valley floor that remains at the base level of erosion allows lateral erosion to create concave footslopes. Such widening of the valley floor maintains steep midslopes on the adjacent valley sides, which in turn slows the rate of decrease of ridgecrest convexity.

Areal distributions of convex and concave hillslopes were identified and compared for Dormant Hollow and Vigorous Vale. Nonuniform rock strength and variable spacing of small headwaters stream channels introduce noise into the hillslope

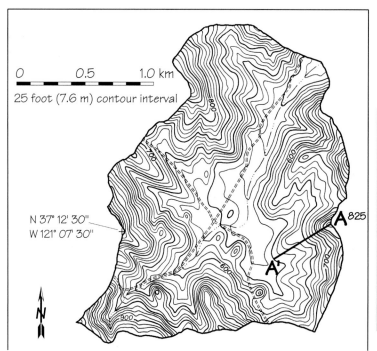

0   0.5   1.0 km

25 foot (7.6 m) contour interval

N 37° 12' 30"
W 121° 07' 30"

A'

A 825

N

Figure 3.26 Topographic characteristics of a tectonically inactive (class 5) Diablo Range watershed.
 A. Topographic contours in a 3.04 km^2 drainage basin (rising at <0.1 m/ky), which I have named "Dormant Hollow". Double dashed lines show locations of single lane dirt roads. The topographic profile for A–A' is shown in Figure 3.28.

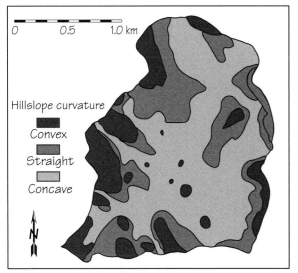

Figure 3.26 Topographic characteristics of a tectonically inactive (class 5) Diablo Range watershed.
B. Map showing the distribution of convex, straight, and concave hillslopes in Dormant Hollow.

ridgecrests. However, comparison of the topographic maps shown in Figures 3.26A and 3.27A shows that broad hilltops are typical of Dormant Hollow and relatively narrow ridgecrests are typical of most of Vigorous Vale. Maximum hillslope steepness is similar in the tectonically active and inactive settings.

The two topographic profiles of Figure 3.28 can be used to illustrate marked contrast in values of the exponent f of equation 3.7. The Dormant Hollow topographic profile has a fairly narrow ridgecrest that is quite convex, having a power-function exponent f value of 1.95. The Vigorous Vale ridgecrest is extremely convex, having an exponent of 3.15. Ridgecrests in Vigorous Vale are so strongly convex that they grade into straight midslopes much sooner than in Dormant Hollow.

A huge difference in footslope characteristics is visually apparent, with the inactive basin having long, gentle footslopes that are highly concave. The exponent f is only 0.39. The extensive concave character of the footslopes in Dormant Hollow suggests that they are equilibrium surfaces of detrital transport. The much different hillslopes of Vigorous Vale plunge down to the trunk stream channel. Concave footslopes are lacking as shown by an exponent f value of 0.96 (1.00 would be straight hillslope). The virtual lack of footslopes in Vigorous Vale also should contribute to a steeper drainage basin mean slope (0.535 or 28°) than for Dormant Hollow (0.489 or 26°).

Vigorous Vale would have an even steeper mean slope were it not for the substantial time lag for tectonic perturbations to migrate upstream. Mountain-front uplift has begun to steepen mid-basin hillslopes but has yet to substantially affect headwaters hillslopes.

Small areas of concave slopes in the headwaters of the Laguna Seca Hills watersheds may be inherited from former times of tectonic inactivity. The accordant summits of the hills mark the level of a former dissected pediment estimated to be ~400 ± 100 ka. Uplift appears so recent that it has yet to reverse the characteristics of hillslopes in the headwaters from those that prevailed during an initial phase of much slower uplift rates. More valley-floor downcutting will be needed to complete the

topographic data. Even when using closely spaced contour intervals of 6.1 and 7.6 m, various workers may class areas of convex and concave slopes differently on these 1:24,000 scale topographic maps.

I made visual assessments of contour spacings starting at hilltops and ridgecrests. Progressively closer contour lines are indicative of convex slopes, and progressively greater distances between contour lines were classed as concave slopes. Transition zones between these convex crestslope and concave footslope end members were classed as straight.

The overall topographic characteristics of the two study watersheds are greatly different, so important generalizations can be made. I focus on the location of change from convex to straight hillslope morphology, which I call the *convex–straight transition*. Straight slope elements range from amazingly uniform to irregular.

Examples of how I classed slope morphologies are shown in Figure 3.28. Crestslope width also affects ridgecrest convexity, with broad hilltops increasing in slope at a slower rate than narrow

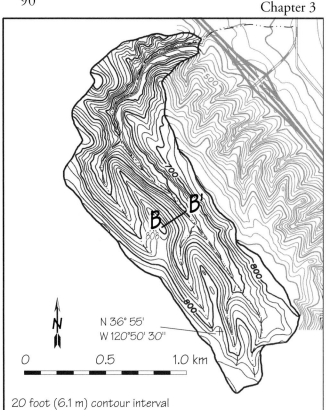

N 36° 55'
W 120°50' 30"

0   0.5   1.0 km

20 foot (6.1 m) contour interval

Figure 3.27 Topographic characteristics of a tectonically active (class 1) Diablo Range watershed.

A Topographic contours of a 1.30 km² drainage basin (rising ~2 m/ky), which I have named "Vigorous Vale". The freeway interchange is adjacent to the abrupt mountain–piedmont junction. Hillslopes become progressively steeper downstream to the source of the tectonic perturbation, which is a monocline with a concealed thrust fault. The topographic profile for B–B' is shown in Figure 3.28.

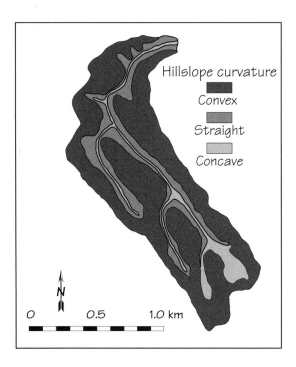

Hillslope curvature

■ Convex

■ Straight

■ Concave

0   0.5   1.0 km

transition to a tectonically active landscape in these headwaters areas.

Large mid-basin areas in Dormant Hollow have achieved maximum steepness, which augments the mean slope value for the watershed. In both these representative fluvial systems, tectonic controls and response times to rock uplift greatly affect types, magnitudes, and rates of geomorphic processes and the resulting hillslope morphologies.

Models that simulate hillslope evolution commonly assume that increases of ridgecrest convexity approach zero as a straight midslope is

Figure 3.27 Topographic characteristics of a tectonically active (class 1) Diablo Range watershed.

B. Map showing the distribution of convex, straight, and concave hillslopes in Vigorous Vale.

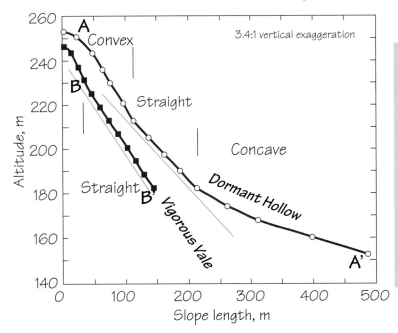

Figure 3.28 Topographic profiles of representative hillslopes, from ridgecrests to trunk stream channels in the tectonically active and inactive drainage basins, at locations A-A' (Fig. 3.26A) and B-B' (Fig. 3.27A). The thin straight lines below the profiles are parallel to the straight hillslope segment and show the subtle nature of the convex–straight transition.

approached whose steepness is regarded as a stability threshold (Kirkby, 1985; Anderson, 1994; Howard, 1997; Roering et al., 1999). It is postulated that small increases in hillslope steepness above this critical value will increase mass movements to the point that steeper soil-mantled slopes cannot be maintained. Such models assume attainment of steady-state hillslope morphology through increase in the frequency of both deep-seated and shallow landslides.

I doubt if steepness of planar hillslopes in the Laguna Seca Hills is limited by critical hillslope angles. Steepness of planar slopes in Vigorous Vale varies greatly (19° to 46°), even though bedrock lithology is the same.

An alternative model is that magnitude and rate of tectonically induced downcutting of a valley floor largely controls steepness of adjacent hillslopes. For example, the 23° planar footslope of section B-B' (Figs. 3.27A, 3.28) would become steeper, perhaps to >40°, if relief became greater as a result of erosional lowering of the valley floor. Note that my model assumes lack of attainment of steady state.

Mass-movement processes do become more common with increase of footslope steepness. But it is the continued tectonically induced lowering of the lower boundary of hillslopes – the valley floor

of the trunk stream channel – that causes spatially variable denudation rates in tectonically active watersheds. Exceptionally slow typical response times of hillslopes to tectonic perturbations leads me to believe that even tectonically inactive landscapes are forever changing.

Concave slopes clearly dominate the tectonically inactive landscape of Dormant Hollow (Fig. 3.26B). Only the headwaters tributary streams have retained large areas of convex hillslopes. Concave slopes comprise about 51% of the basin, convex slopes only 20%, and straight slopes occupy about 29% of the watershed area. The trunk stream channel has been at the base level of erosion for so long that the wide valley floor is beginning to have characteristics suggestive of a pedimented terrain. Broad embayments extend into even minor tributaries, and erosional knobs of concretionary sandstone rise above the broad valley floor much like inselbergs in the pedimented terrains of the Mojave Desert of California (Bull, 2007, figs. 4.14, 4.15).

In marked contrast, Vigorous Vale has virtually no concave slopes. Steep hillsides plunge down to narrow valley floors. The narrow zone of convergent flow next to the stream channels borders on being too narrow to map at the scale of Figure

3.27B. Convex slopes occupy about 68% of the watershed, and straight slopes about 22%. Only 10% of the basin has concave slopes, and that percentage will decrease as tectonically induced downcutting extends more completely into the headwaters reaches.

Comparison of Figures 3.27A, B with topographic maps having a 2 m contour interval in the humid Coast Ranges of Oregon (Roering et al., 1999, plate 1) shows several similarities. The Oregon mountains rock uplift may be an order of magnitude slower. Concave hillslopes in Figure 3.29 occur only adjacent to valley floor debris-flow channelways (Stock and Dietrich, 2003). Uplift rates near the Oregon coast are less than 0.25 m/ky (Kelsey and Bockheim, 1994; Kelsey et al., 1994; Personius, 1995). Short-term geodetic uplift rates are an order of magnitude faster (Mitchell et al., 1994). Uncertainty about the Late Quaternary rates of rock uplift in the Figure 3.29 study area includes lack of information about how regional uplift rate varies from the coastal zone where the published rates were estimated, and the degree of activity of geologic structures adjacent to the study area. A radiocarbon dated 40 ka strath terrace of the nearby Siuslaw River provided Almond et al. (2007) a tectonically induced downcutting rate of 0.18 m/ky, which supports the premise of a slow regional uplift rate.

The similarity of the tectonic signatures in the Oregon Coast Ranges hillslopes when compared with the Laguna Seca Hills suggests several possible models of landscape evolution.
1) I tentatively favor the idea that low, as well as rapid, uplift rates create and maintain similar slope morphologies in the headwaters reaches of tectonically active fluvial systems. As for my Dormant Hollow site, the upstream reaches of a fluvial system adjust very slowly to rock uplift. I say "tentatively" because I am surprised that the Oregon site, with its much wetter climate and greatly slower uplift rates, has not achieved equilibrium valley floors for 3rd-order stream channels. Perhaps the valley floors of "Vigorous Vale" are even farther from equilibrium than I had supposed.
2) Not as likely, but perhaps the headwaters reaches at the Oregon site have yet to degrade to their respective base levels of erosion because hillslope sediment yields are high enough, and sufficiently

Figure 3.29 Airborne laser-altimetry relief map of small soil-mantled watersheds in the Oregon Coast Ranges. Regular spacing of first-order stream channels reflects uniform rock mass strength of the underlying greywacke sandstone. Note how narrow ridgecrests change quickly to roughly planar slopes that plunge down to narrow valley floors that have virtually no stream terraces. This study area was clear-cut logged about 7 years before the laser-altimetry flight, except for area south of the major ridgeline that has the rectangular patch of 25-year old trees. From Figure 4 of Roering et al. (1999).

bouldery, to inhibit rapid downcutting by the headwaters streams. If so, ratios of stream power to resisting power would be between 1.1 and 1.3 instead of being about 1.8 (the higher the value, the stronger the tendency to incise valley floors). Note that the 4th-order stream (Sullivan Creek) that parallels the top of the Figure 3.29 view has a wider valley floor. Sullivan Creek has sufficient unit stream power to achieve the base level of erosion – stream power/resisting power = 1.0 – under the prevailing uplift and lithologic constraints.
3) Debris flows, although frequent, may not be efficient in degrading resistant bedrock beneath headwaters stream channels at the Oregon site.

In any case the Oregon valley floors are being lowered faster than the ridgecrests to create a topography indicative of a rising mountain range.

Locations of convex–straight transitions are not constant in either Vigorous Vale or Dormant Hollow (Figs. 3.30, 3.31). Hillslope-position coordinates for the convex–straight transitions in tectonically inactive Dormant Hollow decrease progressively downstream from 0.7 in upstream reaches to 0.25 near the basin mouth. Convex ridgecrests quickly become straight midslopes and then concave footslopes.

This steadily decreasing trend reflects the dynamics of drainage-basin evolution in the absence of uplift. Downstream reaches have been at base level of erosion >1 My, but upstream reaches have yet to erode to base level of erosion. The effects of a prolonged stable base level in the downstream reaches has provided longer time spans for base-level change to migrate up the sides of valleys. A combination of greater stream power and longer time at the base level of erosion near the basin mouth accounts for the dominance of concave hillslopes at basin-position coordinates of 0.7 to 0.9.

Vigorous Vale has the opposite trend. Hillslope-position coordinates for convex–straight transitions increase slightly from 0.5 to 0.65 in the downstream direction. The change to straight hillslopes is farther from the ridgecrest as one gets closer to the basin mouth – and the location of the active thrust fault hidden in this monocline. The basin-mouth hillslope-position coordinate is 0.65 as compared to only 0.25 in Dormant Hollow. The two headwaters points at basin-position coordinates of 0.2 and 0.3 (Fig. 3.30) may be anomalous because they have hillslope-position coordinates that are less than for similar basin-position coordinates in Dormant Hollow. They reflect characteristics inherited from the pedimented landscape. Without these two points, the trend for Vigorous Vale would be constant, or only slightly rising, in the downstream direction.

Changes in the vertical locations (hillslope-fall ratio) for convex–straight transitions (Fig. 3.31) have more scatter about the control points but the trends are similar to those for convex–straight transitions in the horizontal dimension. Hillslope-fall ratios

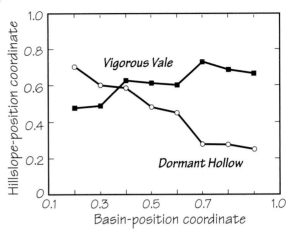

Figure 3.30 Graphs comparing trends of hillslope-position coordinates for convex–straight transitions at a sequence of basin-position coordinates in tectonically active and inactive drainage basins.

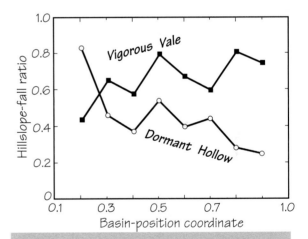

Figure 3.31 Graphs comparing trends of hillslope-fall ratios for convex–straight transitions at a sequence of basin-position coordinates in tectonically active and inactive drainage basins.

(Section 2.1) rise downstream in Vigorous Vale and decline downstream in Dormant Hollow. Maximum degradation below summits and ridgecrests has occurred in the downstream reaches. Figures 3.30 and 3.31 clearly show that convex–straight transitions

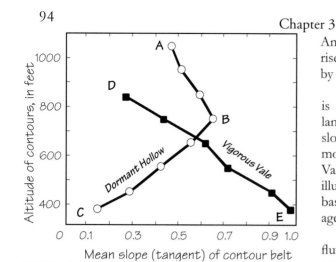

Figure 3.32 Trends in the distribution of mean slope at altitude increments of 100 feet (30.5 m) in Vigorous Vale and Dormant Hollow.

are both progressively farther from, and farther below, the divides in downstream reaches of Vigorous Vale.

Comparison of mean slope plots is an even more distinctive tectonic signature. Figure 3.32 underscores the contrast between tectonically active and inactive drainage basins. I determined slopes of consecutive altitudinal slices in both watersheds using a procedure that is discussed later (equations 4.1 and 4.2). Tectonically active Vigorous Vale slopes are not uniformly steep despite a uniform lithology; instead slopes are progressively steeper nearer the basin mouth. The headwaters hillslopes have a tangent value of only 0.27 (15°), but mean slope progressively increases to about 0.99 (45°) near the basin mouth.

The headwaters hillslopes of tectonically inactive Dormant Hollow have a tangent value of about 0.47 (25°) and mean slope increases to about 0.66 (33°) in the upper part of the basin. The upstream portion of Dormant Hollow has a declining trend as does Vigorous Vale. Note that the slope of AB is steeper than that of DE in Figure 3.32; slope increases more rapidly from the headwaters ridgecrests of tectonically inactive terrains. Of course this partly reflects the relatively greater significance of lithologic control in stable landscapes.

An extreme end result can be spectacular peaks that rise above gently sloping plains such as is illustrated by Figure 3.20B.

The abrupt reversal of trend in mean slope is a defining characteristic of tectonically inactive landscapes. The trend abruptly reverses with mean slope decreasing to only 0.15 (8°) near the basin mouth. These markedly different plots for Vigorous Vale and Dormant Hollow provide one of the best illustrations of the unequivocal impact of tectonic base-level fall, or to define a tectonically stable drainage basin.

Our comparison of the inactive and active fluvial systems would not be complete unless we examined the longitudinal profiles of the trunk stream channels (Fig. 3.33). After all, it is the stream channel that transmits a tectonic perturbation at the basin mouth to progressively more distant upstream footslopes that are graded to the valley floors. Much of the stream profile in Dormant Hollow has a smooth, concave shape that reflects long-term attainment of type 1 dynamic equilibrium. The downstream half of Vigorous Vale has a steep and fairly constant gradient that borders on being convex instead of concave. Continuing rapid uplift may not have allowed ephemeral streamflows to attain even a type 2 dynamic equilibrium, despite the soft watershed lithologies. Once again the headwaters reaches are quite similar, and both have the occasional steeper reach that appears to be associated with locally harder

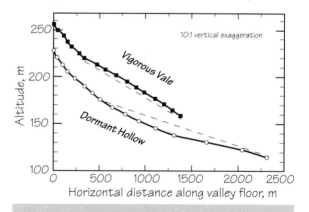

Figure 3.33 Longitudinal profiles of trunk stream channels in Vigorous Vale and Dormant Hollow.

sandstone. Neither headwaters reach is suggestive of attainment of a steady-state longitudinal profile. Annual unit stream power is insufficient to achieve the base level of erosion.

3.8 Summary

Chapter 2 examined ridgecrests in the context of changing drainage-basin areas. This chapter considered ridgecrests as the upper boundary of hillslopes – that most ubiquitous landform of mountains. The lower boundary is the edge of a valley floor created by streamflow. Valley floors are local base levels to which hillslopes are graded.

Understanding how hillslopes evolve in different climatic and lithologic settings is essential for answering important geomorphic questions. Does the height and shape of a hillslope between the upper and lower boundaries maintain a constant relief and shape – a steady-state behavior? Are the crestslopes, midslopes, and footslopes of tectonically active and inactive terrains different?

Convex shapes and diverging flow lines (diffusion) are typical of crestslopes, while concave shapes and converging flow (concentration), or parallel, flow lines are typical of footslopes. Convex ridgecrests extend farther downslope in landscapes where bedrock uplift is rapid, and concave footslopes are more conspicuous in tectonically less active mountains. Patterns of contour lines for soil-mantled hillslopes clearly illustrate such contrasts in fashion.

Both slope and flux of sediment increase downhill from ridgecrest divides. Hillslope gradient becomes increasingly uniform farther downslope from the divide instead of becoming steeper to shift the increasing flux of erosional sediment. Changes in both slope and curvature seem rather insensitive to assumed increases in erosion rates. Linear diffusion processes cannot explain hillslopes that become progressively more planar. Roering et al. (1999) conclude that a nonlinear expression best explains how sediment flux varies with hillslope gradient.

The power-function equation of Hack and Goodlett (1960) describes two-dimensional configurations along the flow lines of hillslope runoff. It nicely compares convexities and concavities of

hillslopes subject to base-level rise and base-level fall. Erodibility of surficial materials, climate, and type of geomorphic processes also influence hillslope steepness (the coefficient) and curvature (the exponent).

Crestslopes affected are markedly more convex where base-level fall is occurring (stream-channel downcutting) than where base level is stable or rising (at equilibrium or the valley-floor is aggrading). The power-function exponent is about 0.4 greater for tectonically active crestslopes. Footslopes subject to adjacent base-level fall are both steeper and more concave than where base level is stable or rising.

Cross-valley shapes differ for streams that are aggrading, degrading, or are at equilibrium. Valley floors may be only as wide as bankfull stream discharges where downcutting dominates over lateral cutting. Valley floor width/valley height ratios record responses to uplift, and position within a drainage basin. Base-level-of-erosion conditions are more likely where fluvial systems generate maximum unit stream power – in reaches just upstream from range fronts. Migration of base-level perturbations up streams and hillslopes was first proposed by Gilbert (1877). Headwaters reaches remain strongly degradational because miniscule unit stream power cannot achieve the base level of erosion even in tectonically inactive settings.

Two contrasting unnamed watersheds in the eastern Diablo Range of central California illustrate contrasting tectonic signatures in hillslopes. I call the tectonically inactive basin "Dormant Hollow" and the rapidly rising basin "Vigorous Vale". These two fluvial systems have unambiguous contrasts of landscapes that record much different rates and responses to past and present rates of tectonic base-level fall.

Distribution of soft and hard rock types in a drainage basin is influenced by Quaternary uplift rates. Soft mudstone occurs on ridgecrests as well as in valley floors in Vigorous Vale. Locally harder sandstone occurs as ridgecrests in the tectonically inactive landscape of Dormant Hollow and mudstone underlies the valley floors. Here, tectonic quiescence that has promoted differential erosion for more than 1 My has lowered areas underlain by mudstone faster than the adjacent concretionary sandstone.

The position on a topographic profile where hillslope morphology changes from convex to straight – the **convex–straight transition** – records how erosion is influenced by rock-uplift rate. It is not constant in either Vigorous Vale or Dormant Hollow. Hillslope-position coordinates for convex–straight transitions in tectonically inactive Dormant Hollow decrease progressively downstream from 0.7 in headwaters reaches to 0.25 near the basin mouth. Downstream from the headwaters reach, convex ridgecrests quickly become straight midslopes, which change to concave footslopes. These hillslope characteristics reflect the dynamics of drainage-basin evolution without uplift. Greater stream power and longer time at the base level of erosion near the basin mouth accounts for mainly concave hillslopes at basin-position coordinates of 0.7 to 0.9. These extensive concave footslopes may be equilibrium surfaces of detrital transport.

Vigorous Vale has the opposite trend. Hillslope-position coordinates for convex–straight transitions increase slightly from 0.5 to 0.65 in the downstream direction. The change to straight hillslopes is farther from the ridgecrest as one gets closer to the basin mouth – and the active thrust fault beneath this monocline. The hillslope-position coordinate of the convex–straight transition at the basin mouth is 0.65 as compared to only 0.25 in Dormant Hollow. Concave footslopes are lacking

in Vigorous Vale where hillslopes plunge down to the trunk stream channel. Steepness of Vigorous Vale planar footslopes, with constant lithology, varies from 19° to 46°. Slopes are steeper near the tectonic perturbation, which is a range-front thrust-faulted monocline.

Comparison of mean slopes of contour belts is striking. Vigorous Vale slopes are progressively steeper nearer the basin mouth. The headwaters hillslopes have a tangent value of only 0.27 (15°), but are 0.99 (45°) near the basin mouth.

Headwaters slopes of tectonically inactive Dormant Hollow have a tangent value of about 0.47 (25°) and mean slope increases to about 0.66 (33°) in the upper part of the basin. This slope increase reflects recent arrival of stream-channel downcutting from downstream reaches. Then the trend abruptly reverses with mean slope decreasing to only 0.15 (8°) near the basin mouth, where the process of valley-floor downcutting ceased long ago. This abrupt reversal of down-valley trend in mean slope is a defining characteristic of tectonically inactive landscapes.

Landscape signatures in tectonically active and inactive drainage basins were a theme of Chapters 2 and 3. We have yet to examine influences of rock uplift on geomorphic processes such as mass movements and sediment yield. Those subjects are introduced in Chapter 4, where you can also expect a discussion of the hydrodynamics of wet, soggy hillsides.

Chapter 4

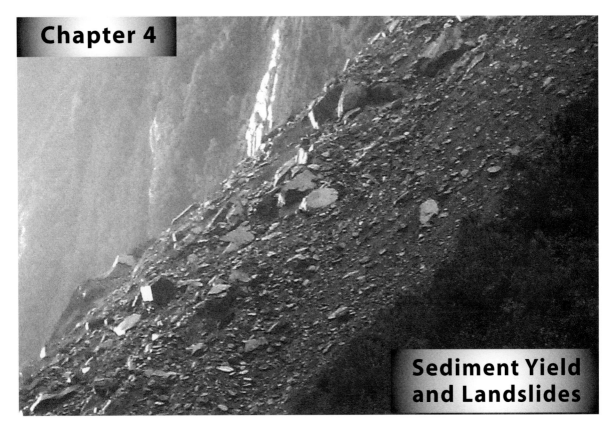

Sediment Yield and Landslides

Uplift of mountains affects rocks, climate, and the geomorphic processes that shape landscapes. Tectonically active landscapes typically have many landslides. Mass movements are especially common where soft, crumbly rocks have been raised recently out of oceans to form steep hills and mountain ranges. Increasingly higher terrain squeezes more rain out of passing clouds. Streams rapidly deepen valleys underlain by soft rocks, making footslopes adjacent to the valley floor steeper. Uplift sets the stage for occasional episodes of widespread slumping of hillslopes and generation of debris flows, especially during rainstorms and earthquakes.

This discussion of tectonic controls on hillslope denudation addresses two essential topics, sediment yield and mass movements. The main focus is on how water promotes landsliding in rising mountains such as those in coastal California and New Zealand. I examine how prolonged rainfall changes driving forces in hillslopes, thus triggering mass movements. Infiltrating rain changes groundwater levels and promotes landslides. We need to consider changes to driving and resisting forces that many refer to simply as "pore pressures".

Rain-soaked schistose debris on a steep debris cone in the Fox Glacier valley. High sediment yields of the Southern Alps, New Zealand are a product of fractured, weathered schistose rocks, an extremely humid climate (5 to 15 m precipitation/year), and powerful rivers and glaciers that cut down into valley floors at rates that match the 5 to 8 m/ky uplift rates.

Tectonically Active Landscapes, 1st edition. By W. B. Bull. Published 2009 by Blackwell Publishing, ISBN 978-1-4051-9012-1

4.1 Sediment Yield

4.1.1 Influences of Rock Uplift

Uplift increases both **sediment yield** (amount of material, per km², eroded from a drainage basin per year) and **denudation rate** (average erosional lowering of the overall drainage basin surface in m/ky). Why do tectonically active landscapes have large sediment yields? Tectonic deformation shatters strong rocks, and uplift creates high, steep mountains that intercept more rain and snow. Let us appraise tectonic control of sediment yield in regard to each of these three factors.

Mountain-building forces squeeze, fold, and shear shallow rocks in the brittle deformation zone of the crust, and metamorphose the deeper rocks. The result of fracturing is delivery of disintegrated fragments of rock to the land surface where they are swept away by hillslope and stream processes (Gilbert, 1877; Dutton, 1882). Molnar et al. (2007) nicely argue the case that such disaggregation of rock is as important an end-product of tectonic deformation as the more publicized raising of the land surface. Massive bare rock sheds only water. Brittle rocks sheared by strike-slip and thrust faults may be reduced to rock chips, gouge and blocks permeated with fracture planes in the process of further splitting.

We examine brittle shearing of rocks on several landscape-size scales in order to underscore its importance as an independent variable controlling sediment yield. Let us start with the wrench-fault shear zone diagrammed in Figure 2.12 where domains of shortening and extension are shown for a point in time. Yes, gouge is pronounced along the primary wrench fault, but the emphasis here is on the variety of other concurrent tectonic deformation processes. The entire packet of faults and folds is rotated with the passage of time. Each structure may be rotated until it is no longer favorably oriented in regard to the stress field that created it. New sets of structures are generated with alignments that better match the orientation of the present primary shear couple. This realignment happens again and again, because primary wrench-fault displacements commonly are 10 to 100 km.

Plate boundaries typically have parallel sets of strike-slip faults (Fig. 2.13 for example). The intervening blocks cannot be rotated more than 45° in response to long-term shearing by the regional shear couple (Nur et al., 1986). Secondary strike-slip faults commonly cut across the blocks between the primary faults after only 25° to 40° of rotation. See figure 4.19 in Bull, 2007 and the related discussion. The blocks between the cross faults are rotated and further sheared as shown in Figure 4.1. A lack of smooth, planar fault planes means that rock shoved along an undulating fault plane will be subject to extension and compression. Brittle rocks, at shallow depths in the crust, are further disaggregated with each earthquake, be it large or very small (Molnar et al., 2007).

Pervasive disaggregation processes are just as common along active thrust faults, including the Hope fault (Fig. 4.1 and Section 2.4). Movement along curving gently inclined thrust-fault planes further batters the rocks. The end result of such scrunch tectonics is "bulldozed" masses of fractured materials of anomalously high fault scarps (Bull, 2007, figures 1.8–1.13).

The South Island of New Zealand is a superb example of such processes. The fractured greywacke and argillite of the Torlesse Group is the dominant rock type underlying much of the Southern Alps. It grades into brittle schist where metamorphosed closer to the Alpine fault. The ongoing Kaikoura orogeny is responsible for much of the sheared and fragmented rock mass now arriving at the surface. Earlier, equally profound tectonic deformations during the Pahau and Rakaia orogenies at 225–200 Ma and 135–100 Ma had already put the Torlesse through the fault-slip-and-crunch mill. It is pretty easy for us, or erosional processes, to dig into such fractured rocks.

Low rock mass strength was illustrated by the aseismic collapse of the uppermost 60 m of Mt. Cook in 1991. Uplift rates are ~8 m/ky (Bull and Cooper, 1986). This rock avalanche (Chinn et al., 1992; Owens, 1992) revealed the inside of the highest peak in the Southern Alps. It resembled a pile of roughly layered bricks instead of solid rock.

Such fracturing can affect the relative importance of fluvial and landslide processes in the

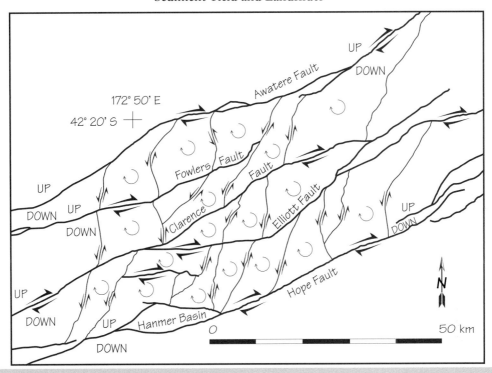

Figure 4.1 Pattern of primary faults (large dextral arrows) and secondary cross faults (small sin-estral arrows) of the Marlborough fault zone. Arrowed circles show sense of fault-block rotation about vertical axes (Eusden et al., 2005a, b). These are the larger strike-slip faults. Fracturing is promoted by strike-slip faults that are too small to show here, and by highly active range-bound-ing thrust-fault complexes. Map supplied courtesy of J.D. Eusden.

Seaward Kaikoura Range (Fig. 4.2). Bedrock uplift is 4 to 6 m/ky. Intense fracturing promotes pervasive frost wedging in such alpine settings.

Intense shattering may result in fluvial dissection (Fig. 4.2B) instead of deep-seated land-slides. Pervasive disaggregation decreases rock mass strength and the relevance of bedding and structural attitudes as controls on geomorphic processes. The result is fluvial erosion instead of landsliding. Debris flows and rock falls are universal, but slumps and deep-seated landslides are rare in these mountains.

Large sediment yields characterize the Seaward Kaikoura Range. Mean annual precipitation ranges from 1.5 to 4 m. Daily rainfall amounts can be as much as 0.5 to 2.0 m. The Kaikoura weather station received 2,001 mm of rain in 24 hours from Cyclone Alison on 12 March, 1975.

These rugged mountains are underlain by highly fractured greywacke sandstone with some ar-gillite and tuffaceous sandstone. Late Quaternary in-creases of relief (4–6 m/ky) have caused hillslopes to crumble instead of forming high, massive cliffs such as one might expect in massive plutonic rocks. Hillslopes in the Kowhai drainage basin (Section 2.4) yield sediment in excess of 5,000 tonnes/km^2/yr (O'Loughlin and Pearce, 1982; Mackay, 1984). This is equivalent to a mean watershed denudation rate of 1.8 m/ky. This magnitude of sediment yield is ~250 times greater than that of undisturbed forested watersheds, with minimal tectonic uplift, in the USA Pacific northwest (Costa, 1994).

Although impressive, this tectonically in-duced erosion of the Seaward Kaikoura Range is only 17% of the sediment yield measured in wetter

Figure 4.2 Typical fracturing of Torlesse rocks. A. Exposure at the summit of Hill 1168 m on east divide of the Hapuku River, Seaward Kaikoura Range, New Zealand. The larger chunks of this rock may have many microscopic fracture planes, but some of the larger blocks will become boulders in streambeds.

Figure 4.2 Typical fracturing of Torlesse rocks. B. A 600 m high, 41° slope hillside above the Snowflake Tributary of the Kowhai River, Seaward Kaikoura Range, New Zealand. The high drainage density is indicative of rapid erosion. The yielded sediment is flushed away by the braided stream in the valley floor.

parts of the Southern Alps (Griffiths and McSaveney, 1983) that receive five times the mean annual precipitation of the Seaward Kaikoura Range. New Zealand landscapes respond rapidly to tectonic base-level falls because of such rapid erosion rates of tectonically shattered rocks.

Other parts of the Southern Alps have a rock mass strength that favors different landslide processes. Deep-seated landslides are a prominent feature of South Westland landscapes (Korup, 2005a,b). Rock avalanches are widespread and many coincided with times of earthquakes (Adams, 1981; Bull, 1996, 2007; Hovius and Stark, 2006). Such mass movements increase sediment yield greatly (Korup and Crozier, 2002; Korup et al., 2004).

Consider how these mountains intercept rain and snow from passing storms. High mountains have a propensity for causing weather fronts to drop their moisture by forcing air to rise, thereby becoming cooler and saturated. But not all mountain ranges are the same in this regard.

Let us compare the Southern Alps of New Zealand with the Alps of central Europe (Rother and Schulmeister, 2006). The vertical scales of Figure 4.3 A and B are the same. The European Alps are much

higher and wider than the Southern Alps but annual orographic precipitation is only a third of that in the South Island of New Zealand.

The orientation, height, and latitudinal position of the Southern Alps are ideal for interception and abrupt orographic forcing of precipitation from a never-ending sequence of low-pressure weather systems. Moist air masses flow eastward from the Tasman Sea, which is an important area for generation of new storms in the southern oceans. Much of the Southern Alps is truly deserving of an extremely humid classification (table 2.1 in Bull, 2007). Maximum annual precipitation is ~16 m and the unit annual runoff measured in one watershed is 11.7 m

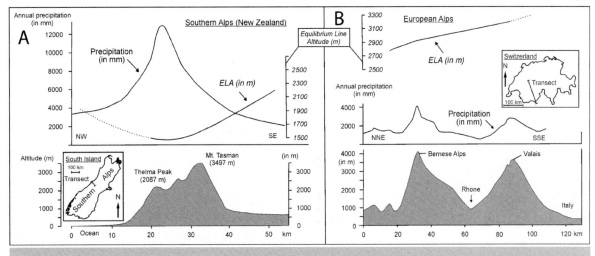

Figure 4.3 Comparison of precipitation and ELA levels across the Southern Alps (A) and European Alps (B) plotted at equal scales (except horizontal distances of the two inset maps). This Figure was furnished courtesy of Rother and Schulmeister (2006) and is their figure 1. It is based on information from Griffiths and McSaveney (1983); Chinn and Whitehouse (1980); Sturman and Wanner (2001); and Mueller, et al., (1976).

(Griffiths and McSaveney, 1983). Even 10 m/year of precipitation is huge from a geomorphic perspective – 1 km per century. Convergence of the Australian and Pacific tectonic plates rapidly raises these crumbly mountains into a fluvial–glacial buzz saw.

The many glaciers of the Southern Alps flow fast, with some sweeping down into the rain forest. They follow the present global pattern of glacier retreat, but even a brief spell of increased orographic precipitation causes a temporary readvance.

Orographic control of precipitation is reflected in the ELA or *glacial equilibrium line altitudes*. An ELA is the altitude on a glacier where annual accumulation equals annual ablation and the net mass balance is zero. ELA varies with latitude, mean annual temperature, and precipitation. The rising trend in the Figure 4.3 ELA for the European Alps reflects warmer temperatures towards the south. The marked depression of ELA for the Southern Alps (Fig. 4.3A) coincides with the phenomenal increase of orographic precipitation. Then the ELA rises steeply on the rain-shadow side of the Southern Alps.

This brief summary makes it clear why sediment yields from New Zealand fluvial systems are amazingly high. Steep mountain landscapes are rising fast, their brittle rocks being disaggregated into small fragments, and the climate is conducive for frequent large floods.

4.1.2 Lithologic Controls

Uplift and consequent erosional steepening of hillslopes greatly increases watershed sediment yields. Steeper slopes erode faster (Schumm, 1963), and should create larger, thicker piedmont alluvial fans (Whipple and Trayler, 1996).

We need to constrain many variables in order to evaluate the importance of increasing watershed mean slopes on sediment yield. I do this for the east-central Coast Ranges of California. It would be desirable to:
1) Use a time span that is sufficiently long to eliminate pervasive human impacts on the environment,
2) Reduce the effects of large glacial–interglacial climatic fluctuations during the Quaternary,
3) Have an efficient trap that retains sediment outputs from basins underlain by specific rock types.

These requirements are largely met in the western San Joaquin Valley by comparing areas of gently sloping alluvial fans derived from drainage basins with a large range of mean slopes. Fan area is used as a proxy for fan volume. Electric logs of the numerous deep water wells and cores from scientific sites show that adjacent fans derived from the sandstone and mudstone watersheds of the Diablo Range have not encroached upon each other during the past 100–200 ky. Areas of adjacent fans have been proportionately constant and record the sediment yielded from their source watersheds. The piedmont became larger by expanding towards the Sierra Nevada side of the San Joaquin Valley (Fig. 1.6). Fan area is largest when derived from large, steep source areas underlain by soft rocks. Sediment yield is also a function of density of plant cover relative to annual precipitation (Langbein and Schumm, 1958).

The fan-source drainage basins have similar climates but have different mean slopes, lithologies, and areas. Two lithologic classes are used in Figure 4.4; basins underlain mainly by 1) mudstone and shale, and 2) soft sandstone.

Ratios of alluvial-fan area to drainage-basin area were used to normalize the variable of basin area. A *sediment-yield index* – the amount of fan area per unit drainage-basin area – approximates the long-term sediment yield from each source watershed. Regressions of sediment-yield index and mean basin slope describe a distinctive increase of sediment yield as the rising Diablo Range source areas became steeper.

The sediment-yield index for all Figure 4.4 points increases from about 0.5 for a mean slope of about 0.2 to about 2.0 for a mean drainage-basin slope of 0.4. Doubling watershed mean slope increases the sediment-yield index by fourfold. The regression line of points for basins underlain by sandstone is above that for the group of mudstone and shale basins. For a given mean basin slope (for example 0.3) about 1.8 times as much sediment is yielded from mudstone basins as from basins underlain by sandstone – the sediment-yield index is increased from 1.0 to 1.85.

Radar digital elevation models are now available for much of planet Earth. Such marvelous data sets will allow estimation of many watershed parameters with a few clicks of a computer keyboard, but resolution is limited (Stark and Stark, 2001). However, the precision that can be obtained by using topographic maps of the quality of those used in the Figure 4.4 analyses is better than most digital elevation models.

Mean slope was measured using the method of Strahler (1952):
1) Select contour interval (C_i) for measurement of consecutive altitudinal slices of a drainage basin.
2) Measure the lengths of each bounding contour line, and compute the mean length for each contour belt (L).
3) Planimeter the areas of each contour belt (A).
4) Divide the area by mean length to obtain the mean width of each contour belt (W).

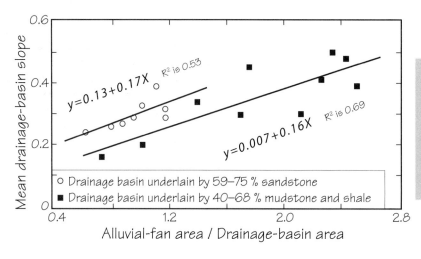

Figure 4.4 Relations between mean drainage-basin slope and sediment-yield index (fan area/source area ratio) for the west-central San Joaquin Valley, California.

$$W = \frac{A}{L} \qquad\qquad 4.1$$

5) Divide contour interval by the mean width to get the tangent of mean slope for each contour belt.

$$\tan \alpha = \frac{Ci}{w} \qquad\qquad 4.2$$

6) Calculate the sum of mean slope–area products for the set of contour belts in the drainage basin to get watershed mean slope

$$\tan \alpha = \frac{\Sigma[(\tan \alpha)(A_p)]_{1-n}}{100} \qquad\qquad 4.3$$

Where A_p is the percent of total basin area within a contour belt. Such data can be used to estimate mean watershed slope and to graph distributions of slope steepness within a watershed. Variable contour belt areas were normalized by multiplying *tan α* by percent of basin area represented by each contour belt, and then dividing the sum for the watershed by 100.

Spatial variations of mean slopes of the contour belts make distinctive mean slope plots (Strahler, 1957) for comparison of drainage basins. See Figure 3.32 for an example comparing markedly different slope distributions for tectonically active and inactive watersheds.

4.2 Mass Movements

Residents of California are devastatingly aware of the coincidence between duration of winter rainfall and the incidence of landslides and debris flows that damage their homes, roads, and utilities. A sampling of the interesting literature on this subject includes Campbell (1975), Nilsen et al. (1976a,b), Cannon, (1988), Cannon and Ellen (1985, 1988), Keefer et al. (1987), Wentworth (1986), Iverson and Major (1987), Ellen and Wieczorek (1988), Reneau and Dietrich (1987), Dietrich et al. (1995), Wilson (1997), Wilson and Jayko (1997), Wieczorek et al. (1998), De Vita and Reichenbach (1998), Crozier (1999), Iverson (2000),

Montgomery and Dietrich, (2002), D'Odorico and Porporato (2004), and D'Odorico et al. (2005). Early European studies that mirrored work being done in California include articles by Starkel (1979), and Govi and Soranza (1980). For a global set of references go to http://rainfall thresholds.irpi.cnr.it/.

Californians believe that rainstorms cause these destructive mass movements, and that the hazard is increased through poor planning and construction practices. It is also widely recognized that steep slopes, clayey rocks, and geologic structures that parallel hillslopes increase mass-movement hazards.

This section starts with clarification of the several ways in which water decreases, or increases, the strength of porous hillslope materials. Separating out such hydrologic controls is necessary before discussing how mass movements are affected by prolonged rain and by tectonic activity – the ultimate cause of most unstable hillslopes.

4.2.1   Rain, Ground-Water Levels, and Landslide Thresholds

4.2.1.1 Ground-Water-Induced Stresses in Hillslopes

A deadly landslide experiment near Kawasaki, Japan in 1975 demonstrated how little we know about the role of water in promoting landslides. Water was being sprayed on a landslide-prone hillslope when the slope collapsed. About 500 m³ of hillslope soil and colluvium liquefied, and rushed downslope to bury scientists, reporters, and the shed used as the headquarters of the failed experiment (Ochiai et al., 2004). In what ways did the application of water trigger this landslide?

We need more than antiquated general statements about landslides being triggered by "pore pressures" in hillslope colluvium. Such loose terminology does not tell us if the intended focus is on hydrostatic pressures, hydraulic head and potential flow gradients, or seepage forces. Examples include "slip episodes were induced by gradually rising pore water pressure" (Iverson, 2000), "role of pore pressure transients" (Stegmann et al., 2007), "slope stability in Hong Kong is primarily controlled by transient pore pressure in response to short intense rainfall process" (Lan et al., 2003), and "analysis of pore pressure fluctuations in a thin colluvium landslide complex" (Haneberg, 1991). These fine papers might have used more appropriate terminology, such as was done in the 1987 hallmark paper by Iverson and Major about groundwater seepage vectors. I like, and will use, their emphasis on the importance of seepage forces. Let us also consider changes in forces and strength of materials associated with rainfall-induced changes in hillslope ground-water tables. See the Glossary for definitions of the terms used in this book.

The simple ratio describing the factor of safety, G_s, in landslides encourages a focus on many variables that influence two opposing forces. Equation 4.4 describes a geomorphic threshold for hillslope materials. Hillslopes are either stable or they are moving downslope as a mass towards the valley floor. As described by Ritter et al. (1995),

$$G_s = \frac{Resisting\ force}{Driving\ force} \qquad (4.4)$$

Stable slopes have a ratio greater than 1.0. Tectonic steepening of hillslopes increases driving forces, thereby moving parts of a valley side closer to a G_s threshold value of 1.0. Seismic shaking or loading of crestslopes increases driving force. Excavation by humans of footslope materials or logging that kills tree roots reduces resisting forces.

Raising or lowering the ground-water table changes both resisting and driving forces, so will be a key part of our discussions here. Such changes cause the factor of safety to locally become less than 1.0. The resulting hillslope failures include slumps with a mix of translational and rotational downslope movement, fast moving debris flows, slow moving earthflows, and debris slides and rock avalanches in more brittle rock types.

Resisting force components include shear strength, S, of material above a landslide plane, and the weight of these materials, called the normal force, N, directed perpendicular to the slope (Fig. 4.5).

Resistance also includes materials further downhill that act as a buttress, which tends to hold upslope soil and rock in place. Undercutting of

footslopes by streams, coastal erosion, or road construction commonly triggers a landslide. Conversely, construction of buttress fills helps stabilize footslopes. Footslopes upstream from a new dam become saturated. The resulting buoyant support of earth materials reduces their effectiveness as a buttress – an common example of the importance of ground-water levels on hillslope stability.

Driving force components include that part of the weight that is directed parallel to the slope, τ. Downslope flow of groundwater may increase driving forces sufficiently to cross to the unstable side of the G_s threshold.

Rainfall amounts and intensity affect both runoff and infiltration rates. Peak stream discharge, Q_p, is proportional to rainfall intensity as defined by

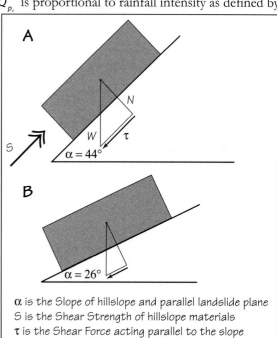

α is the Slope of hillslope and parallel landslide plane
S is the Shear Strength of hillslope materials
τ is the Shear Force acting parallel to the slope
W is the Gravitational Weight of hillslope materials
N is the Normal Force, perpendicular to the slope

Figure 4.5 Comparison of shear force, τ, for gentle and moderately steep hillslopes where the weight of materials, W, and normal force, N, above a potential slide plane is the same. A. A 44° planar slope. B. A 26° planar slope.

the Rational Method equation (Chow, 1964)

$$Q_p = ciA \qquad (4.5)$$

where c is a dimensionless runoff coefficient, i is the rate of rainfall, and A is the watershed surface area. The runoff coefficient is a function of rainfall intensity and amount, watershed geomorphology, and the amount of water already present in the surficial materials. Tectonic setting is important because it affects orographic increase of precipitation and erosion rates.

Tectonic setting also influences infiltration and runoff processes by creating watersheds with steeper hillslopes. Their rubbly surficial materials that have a larger infiltration capacity because soil-profile development generally has much less clay than in tectonically inactive settings.

Slope steepness is a major control on hillslope susceptibility to landsliding. Consider the downstream reaches of Vigorous Vale and Dormant Hollow (Section 3.7), which have similar rock types and climate. The tectonically active setting has a mean slope of 44° and the inactive setting 26°. The slope-parallel component of gravity increases with slope steepness; in the case of Figure 4.5 by a factor of 1.92. Erosion that creates the steep landscapes of rising mountains greatly increases potential shear forces, τ, thus locally reducing the factor of safety (equation 4.4) to values <1.0.

Prolonged heavy rain changes the degree of saturation of surficial materials, and thus leads to landsliding of soil-mantled hillslopes. The ways that rain can decrease resisting forces and increase driving forces are so important that much of this section is devoted to this key topic.

Let us visit a beach composed of no more than sand, water, and air to illustrate the role of water in mass movements. Water can increase, as well as decrease, the strength of granular materials. Dry sand is cohesionless, and cannot stand at slopes steeper than the angle of repose.

The behavior of clay is much different because of its *adsorption* properties. Water is held between the thin layers of microscopic platy clay minerals by negative electrical charges. Clay becomes weaker when wet, because the added water is loosely held. When dry the remaining adsorbed water is tightly held, making clay strong. Small-size clay particles, such as montmorillonite having a larger surface area have stronger adsorption properties.

Wet, but unsaturated, sand has considerable strength; one can drive on it with ease and you may even choose to build a sand castle with it. Why? Capillary air–water interfaces between the sand grains create tensional forces that hold sand together, even in a vertical cut. Water can increase the strength of cohesionless materials. Capillarity height increases with decrease in particle size and can be more than 2 m in silts. So, by increasing moisture-tension forces, water can make surficial materials in hillslopes stronger, not weaker.

This frictional resistance at the boundary between water and air can be regarded as apparent cohesion. Sand castles quickly melt away when the incoming tide saturates the sand and eliminates moisture-tension forces. Surficial hillslope materials also may move across the equation 4.4 threshold when rain raises the ground-water table to near the land surface, thereby reducing the capillary-strength component of resisting force.

While at the beach we might as well play games with a bucket, funnel, and a sheet of plywood. First, pour non-cohesive dry sand through the funnel onto a horizontal plywood sheet. The resulting cone surface has an angle of repose of about 35°, and the tangent of 35° is close to the coefficient of sliding friction of the sand – its value of internal friction. The *coefficient of sliding friction* is a balance between the driving force of gravity and the resisting force of internal friction. It is equal to the ratio between the downslope component of the gravitational weight and the component of gravitational weight acting perpendicular to the slope during movement.

Use the funnel to make another cone of sand, but this time at the bottom of a bucket of water. The angle of repose will be the same as for the dry cone. Remember this the next time you see the superficial statement that water acts as a lubricant in promoting landslides. This experiment also shows that we should not make perfunctory generalizations about the role of pore-pressure increase in triggering landslides. The distribution of pore pressure throughout our saturated cone of sand is directly proportional to the hydrostatic head at that point below the surface of the water. This hydrostatic pore pressure should be regarded as a **neutral stress** because it does nothing to change grain-to-grain stresses or slope stability (Hillel, 1998, p. 353, 360).

Now, let us make another cone of dry sand and then use the funnel to slowly trickle water into the top of the cone. The slope face fails in a mass movement as water emerges from the base of the cone. At last! On our third attempt we have found a way in which water might promote failure of hillslopes — by creating flowage of water in the direction of the sloping ground-water table. This type of pore-pressure increase augments **effective stresses** in the driving force component of equation 4.4. The combination of hydraulic head and flow of water through the sand creates a force along the flow path. Such **seepage forces** are directly proportional to head differentials in granular materials, which can be large. These relations can be summarized with a few simple equations.

The Coulomb equation, where c is stress in units of force per unit area, σ is gravitational stress component perpendicular to the surface, and $tan\phi$ is the coefficient of sliding friction (the internal friction),

$$S = c + \sigma tan\phi \qquad (4.5)$$

which can be used to illustrate resolution of forces for a dry, cohesionless cone of sand and for a cone of sand with seepage forces. The constant c in the Coulomb equation is 0 for non-cohesive materials such as sand, gravel, and talus. Thus the resisting force, S_r, is the shear strength of the material,

$$S_r = \sigma tan\phi \qquad (4.6)$$

σ is a product of the gravitational weight of a mass (W) times the cosine of the slope angle, β. The driving force, S_d, or the stress component parallel to the sloping surface, is described by

$$\sigma = Wcos\beta \qquad (4.7)$$

where S_d is the product of the weight times the sin of β.

$$S_d = Wsin\beta \qquad (4.8)$$

The Coulomb equation does not apply to situations where seepage stresses are present. So let's include seepage stresses ($'$). Equation 4.5 becomes

$$S' = c' + \sigma' tan\phi' \qquad (4.9)$$

which more closely approximates the distribution of stresses in wet landslide-prone hillslopes. The key point here is that outward seepage of water allows the slope of the cone of sand to fail at much less than the angle of internal friction of 35° (the case if seepage stresses were not present).

Seepage stresses are just one way in water promotes crossing of the hillslope stability threshold, in this case by increasing the driving forces of equation 4.4. Sources of ground-water flow include recharge by rain and snowmelt, irrigation by humans, and leaky pipelines.

Before discussing cohesive, plastic landslides we need to consider the effects of stress applied to more complex, cohesive materials. Let us start with water and then mix in some clay.

$$\tau = du/dy \qquad (4.10)$$

τ is the response to a stress applied to water, μ is a coefficient of viscosity, and du/dy is the rate of shear. As described in Rouse (1950, p. 88) μ and β are the molecular and eddy viscosities respectively. Eddy viscosity increases with rate of shear but molecular viscosity is constant. Neither the concept of molecular or eddy viscosity seems ideal in the case of adsorbed water on clay minerals because we are changing the bonding of clay–water systems.

A stress–strain graph (Fig. 4.6) for water shows the distinctive properties of a Newtonian fluid. Note that even a small applied stress results in measurable strain. There is no yield strength as is shown by the straight line passing through the origin of the graph. Shear increases linearly with increase in shear stress.

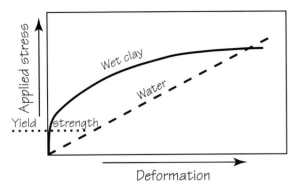

Figure 4.6 Stress–strain plots for a clay–water mixture and of water, which is a Newtonian fluid.

In contrast, a plastic clay has to reach and exceed its yield strength before strain can be initiated.

$$\tau_\mu = \mu(\tau)du/dy \qquad (4.11)$$

The $\mu(\tau)$ notation indicates that viscosity of plastic masses is a function of the applied stress. Roger Hooke (1967) found that viscosity decreased with increasing applied stress after the applied-stress threshold had been exceeded in mudflows. Such viscosity decrease accompanies rapid landslides of cohesive materials – a self-enhancing feedback.

We need to further elaborate on the role of seepage stresses, capillarity (the water of specific retention), and of buoyant support of submerged materials, to better assess the role of water in landslides. Let us start with water flowing through granular materials in the direction of head differentials.

Many geomorphic processes are influenced by seepage stresses. Flow of water into an alluvial streambank during high stages of streamflow creates a seepage stress that temporarily increases bank strength. The outward flow after the flood crest has passed increases the chances of bank failure because reversal of the seepage force decreases streambank strength. Slush avalanches in alpine and arctic snowfields are triggered during the spring when a perched water-table rise creates head differentials within the snow that increases driving forces (Bull et al., 1995). Inflow and outflow of water on sandy beaches occurs with diurnal tidal ranges of ~2 m.

The flow paths of ground water after rain infiltrates into a hill are controlled by the hydraulic conductivity of geologic units. Degree of bedding determines whether flow is constrained (Fig. 4.7A) or dispersed where flux of ground water per unit area is much smaller (Fig. 4.7B). Flow results in seepage forces that are added to the gravitational driving forces, τ, and thus may increase total driving forces sufficiently to reduce the factor of safety (equation 4.4) to a value of less than 1.0.

Seepage forces can be important even in unlikely settings, such as the sheer glaciated granitic cliffs of Yosemite National Park, California (Wieczorek and Snyder, 1999; Bull, 2004). Rain or melting snow washes debris into exfoliation cracks that parallel a cliff face. Freezing temperatures promote partial plugging that reduces ground-water flows out of the cliff. A temporary water-table rise creates seepage forces that may be sufficient to cause part of a cliff face to cross a stability threshold. The resulting landslide ranges in size from a small rockfall to a huge rock avalanche of granitic blocks.

Concave footslopes of hills commonly have colluvial wedges with argillic soil-profile horizons, which create head differentials and potential seepage forces in the direction of ground-water flow. Seepage forces may be large in hillslopes, but even a small increase may augment the driving forces of equation 4.4 sufficiently to cross to the unstable side of the stability threshold. Piezometers (wells) can be used to measure ground-water levels to define head differentials in hillslope aquifers. Even a head differential of only 10 m exerts 1 kg/cm² force downslope, in the direction of ground-water flow.

Rising or falling water tables also change the resisting-force component in the numerator of equation 4.4. Elimination of capillary tension during a water-table rise decreases the resisting force.

Grain-to-grain stresses also change when materials become saturated or unsaturated. A water-table rise provides buoyant support to each grain that is submerged, thereby decreasing its cumulative weight on all underlying connected grains.

To return to our simple model – a beach bucket of sand with water. The hydrostatic weight of the water is transmitted through the interstices between sand grains. It is the bottom of the bucket

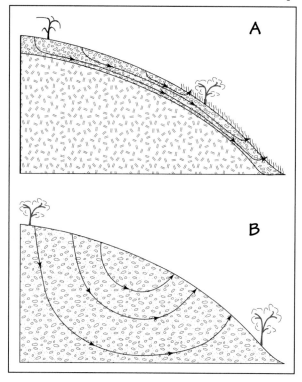

Figure 4.7 Lithologic control of seepage flow paths in two hills during a rainy season. Ground-water flow emerges as hillside springs and contributes to the stream flowing in the adjacent valley floor.
A. Thin grassy colluvium overlying impervious massive granite restricts flow to a small cross-sectional area at shallow depths.
B. Homogeneous sandy gravel promotes deep and diffuse flow of ground water.

Many landslide-prone, soil-mantled hillslopes have both water-table conditions in surficial granular materials and partial confinement of ground water below layers of finer-grained materials that tend to impede and direct ground-water flow.

Ground-water hydrologists study stress-producing factors in aquifer systems to better understand elastic changes and permanent decrease of thickness of alluvial basin fill brought about by the pumping of water by irrigation wells (Bull and Poland, 1975; Poland et al., 1975; Riley, 1998). Increases in applied stress that resulted from pumping ground water caused compaction of silty and clayey alluvial-fan, floodplain, and lacustrine deposits of the San Joaquin Valley. This causes subsidence of the land surface. Pumping of ground water accelerates the basic geologic process of compaction of saturated sediments (Bull, 1973), especially in young, weak basin fill. Compaction occurs as grain-to-grain stresses are changed by several factors in the unconfined and confined depth zones of ground-water systems.

The concepts of *applied stress*, *effective stress*, and *neutral stress* (see p. 106) are used in analyses of the effects of ground-water withdrawal (Terzaghi, 1925; Tolman and Poland, 1940; Terzaghi and Peck, 1948, 1967; Bull and Poland, 1975; Borchers, 1998; Holzer, 1998; Riley, 1998; Galloway et al., 1999; Sneed and Galloway, 2000). Water-level changes in either the confined or unconfined parts of aquifer systems can alter pre-existing stress distributions within each sedimentary bed.

Applied stress is the weight per unit area of sediments and water above the water table, plus the submerged weight of deposits, algebraically plus seepage stresses. Changes in applied stresses become effective stresses only as rapidly as water can be expelled from a given bed. New equilibrium pore-pressure distributions can occur within seconds after a head change in thin beds of permeable sand, but may require centuries in thick clay beds (Poland, 1961; Riley and McClelland, 1971; Bull and Poland, 1975; Poland et al., 1975, Lofgren, 1979). Pore-pressure changes occur quickly in thin, moderately permeable hillslope layers, such as buried argillic soil profiles. As in the case of our hypothetical beach-bucket experiment, hydrostatic stresses are considered neutral because, although hydrostatic stress

that supports the weight of the water. Grain-to-grain increase in the cumulative weight of sand grains increases less rapidly below the water table because buoyant support by water reduces the effective weight of each grain. Effective weight of quartz sand becomes 1.67 (2.67–1.00 = 1.67) if submerged by water-table rise.

Geomorphologists can use basic concepts of mechanics of aquifer systems to gain a better appreciation of the several roles of water in landslides.

tends to compress each grain, it does not change grain-to-grain relationships.

Seepage forces are proportional to the flux of groundwater, per unit area, through porous materials. They are exerted in the direction of flow and are equal to the decrease of hydraulic head. Hydrodynamic viscous drag associated with head loss of flowing water is the connecting link between the fluid and solid phases of a dynamic ground-water system. Groundwater flow directions create hydrodynamic forces that act vertically to compress aquifer systems in basin-fill, and to change downslope driving forces on landslide prone hillslopes.

Changes in internal grain-to-grain stresses – the effective stresses – in aquifer systems are summarized in Figure 4.8. Unsaturated deposits extend to a depth of 100 m, which is the initial position of both the water table and the potentiometric surface of the confined aquifer system. The top of an aquitard at a

depth of 250 m is the reference plane used in this set of computations of applied stresses.

The water table and the potentiometric surface occur at the same level in Figure 4.8A, so a head differential does not exist across the aquiclude and no seepage stress is present. The distribution of neutral pressures is hydrostatic, the same as if no confining bed were present.

A water-table decline of 50 m in Figure 4.8B increases the vector of applied stress due to the weight of the dry deposits above the water table so that it is greater than the sum of the other two vectors. A water-table rise would have the opposite effect. The weight of the submerged deposits is the dominant source of changes in stress applied to the top of the aquitard (Fig. 4.8C).

Changes in the position of the ground-water table in landslide-prone hillslopes influences applied stress in three ways:

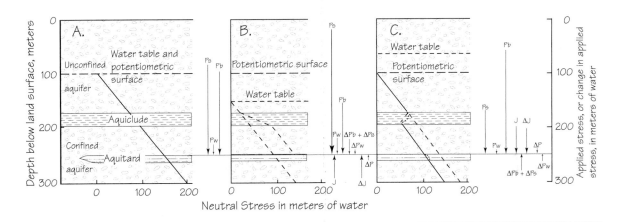

Figure 4.8 How water-table rise or fall changes applied stresses in the underlying confined zone. Position of the potentiometric surface in this hypothetical aquifer system is constant for all three examples.
A. Water table is at the same level as the potentiometric surface of the confined zone.
B. Water table is lowered.
C. Water table is raised.
All stresses are expressed in meters of water, assuming a porosity of 0.40, specific gravity of grains of 2.70, and specific retention of 0.20 by volume. *Ps*, applied stress due to dry weight of unsaturated deposits; *Pb*, applied stress due to buoyant weight of submerged deposits; *Pw*, applied stress due to the weight of water clinging to unsaturated deposits; *J*, seepage stress; Δ*P*, change in total applied stress from condition A. From figure 2 of Bull and Poland (1975).

1) By changing head differentials and therefore the potential for seepage stresses across layered sandy to clayey materials;

2) By removing or supplying buoyant support of grains within the depth interval affected by water-table decline or rise;

3) By eliminating or developing water that clings to grains (specific retention) as deposits become saturated or unsaturated. This water-table change affects both specific-retention weight and capillary forces.

It is convenient to express stress changes in terms of meters of water; 1 cm = 1 g/cm². This notation is used because field measurements of water-level change in hillslope piezometers or water wells compare changes of water level with time. 1 meter of water at 4°C is equal to 0.1 kg/cm².

Water-table rise tends to increase seepage stresses in an underlying confined zone. Each meter of water-table rise is equal to 1 meter of water increase in applied stress on the confined zone. The effect is the same as an equivalent decline in lower-zone artesian head caused by pumping. However, seepage stress increase caused by water-table rise is largely offset by concurrent changes in two other stress components.

One component results from change in buoyant support of the granular skeleton of alluvial or colluvial deposits within the depth interval of water-table change. For unsaturated materials

$$\gamma_d = (1-n)(G)\gamma_w \qquad (4.12)$$

where γ_d is the dry unit weight of deposits above the water table, n is the mean porosity, G is the mean specific gravity of the mineral grains, and γ_w is the unit weight of water. The situation for the submerged unit weight of deposits, γ' is

$$\gamma' = (1-n)(G-1)\beta_w \qquad (4.13)$$

Assume a porosity of 0.40 for the sand bed and a specific gravity of 2.7 for the sand grains. Applied stress due to the weight of deposits above the water table is equal to the weight of 1.6 m of water per meter of aquifer thickness. Applied stress due to the weight of deposits that are submerged below the water table is equivalent to 1.0 m of water per meter of

thickness. Thus, each meter of water-table rise will decrease the grain-to-grain stress by an amount equal to 0.6 m of water.

We also need to consider the weight of water contained in the deposits above the water table. Capillary water clinging to the sand grains is part of the weight (applied stress) tending to compact the underlying deposits. Change from a saturated to an unsaturated condition causes the stress condition of the retained water to change from neutral to an applied stress. Water is a neutral stress where its weight is supported by underlying pore water. Most Quaternary deposits have a specific retention of approximately 20% by volume – a moisture content of about 12% of the weight of the solids in such deposits. Because we are expressing stresses in terms of meters of water, we can assign a value of 0.2 of the volume of the deposit for the water of specific retention. During the water-table decline, as each meter of deposits becomes unsaturated, an increase in applied stress equivalent to 0.2 m of water will be applied to all the underlying deposits. The effect of a water-table rise will be to decrease the applied stress by 0.2 m of water per meter of water table rise. As in the case of buoyancy of the granular skeleton, the water of specific retention will also tend to offset the seepage-stress change applied to the lower, confined zones of ground water as a result of change in the position of the water table. The combined effect offsets most of the 1.0 m of seepage-stress change; 0.6 m + 0.2 m = 0.8 m.

The change in applied stress within a confined aquifer system due to changes in potentiometric surfaces can be summarized in equation 4.14. p_a is the applied stress expressed in meters of water, h_c is the head in the confined aquifer system, h_u is the head in the overlying unconfined aquifer, and γ_s is the average specific yield expressed as a decimal fraction in the interval of water-table fluctuation.

$$\Delta p_a = -(\Delta h_c - \Delta h_u \gamma_s) \qquad (4.14)$$

Changes in hydrodynamic stresses applied to beds in basin-fill alluvium, or hillslope colluvium, become effective – changing the grain-to-grain stresses – only as rapidly as the diffusivity of the bedded deposit permits decay of excess pore pressures to zero.

4.2.1.2 Rain and Hillslope Stability

Changes in ground-water levels greatly affect landslide susceptibility. Saturating the toes of unstable hillslopes substantially decreases the resisting force component of equation 4.4. Water-table rise resulting from construction of dams can be expected to promote failures on the adjacent hillslopes of newly filled reservoirs (Gillon, 1992).

4.2.1.2.1 Italy

A classic example of the effects of such water table rise – and of seepage forces induced by rain and changing reservoir lake levels– is the Vaiont, Italy landslide disaster of 9 October 1963 (Kiersch, 1964, 1988; Hendron and Patton, 1985, 1987). During the night, a rockslide 2 km long, 1.6 km wide, and 150–200 m thick, quickly slid down the south wall of Vaiont Canyon in the Dolomite Alps of northern Italy. The 270×10^6 m^3 landslide completely filled the deep reservoir for 2 km upstream from the dam. The landslide accelerated to speeds ~20 m/sec (100 km/hour) creating giant waves that rose more than 250 m above the level of the reservoir. The world's highest concrete-arch dam (261 m) survived being overtopped by a 245 m high wave, but 2000 people asleep in downstream villages were killed. Even far downvalley the flood wave was 70 m high. This calamity should have been foreseen, at least with today's diverse instrumentation and knowledge of ground-water hydrodynamics. Ever since the Vaiont disaster, geologists and engineers study the hillslopes adjacent to a proposed reservoir as well as the dam-site foundation.

The Vaiont case is summarized here because several years of water-level and hillslope-movement data provide partial insight about a changing hillslope environment. We know the times and amounts of water-table rise related to reservoir levels, but piezometers were not installed to measure potential for seepage forces generated by rain soaking into karstic limestone high on the mountainside.

The 1963 event was a reactivation of a Holocene landslide, and the failure plane was clay layer(s) in marl and limestone whose bedding paralleled the steep hillside (Müller, 1964; Semenza and Ghirotti, 2000; Rinaldo and Ghirotti, 2005). Slickensides on calcium montmorillonite clay layers record the Holocene landslide event. Instead of solid rock, the slope above the failure plane was a ground-water reservoir of porous rock rubble.

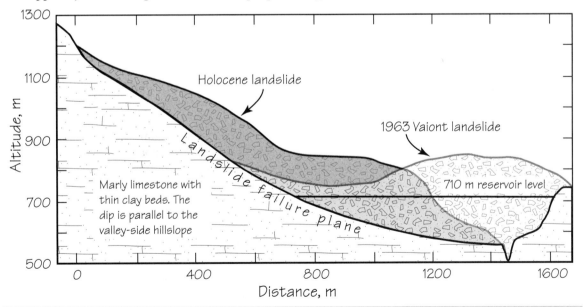

Figure 4.9 Cross section of the Vaiont Valley in the Dolomite Alps of northern Italy, outlining two deep-seated landslides. From figure 7 of Nonveiller (1987).

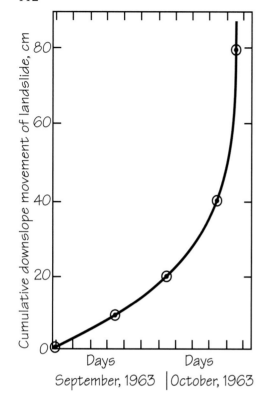

Cumulative downslope movement of landslide, cm

Days
September, 1963 | October, 1963

Figure 4.10 Exponentially increasing rate of downslope movement of the Vaiont landslide. From Kiersch (1964).

The lake level was at an altitude of 710 m when the Vaiont slide occurred. This large water-table rise had saturated the rubble at distances of as much as 400 m into the hillside (Fig. 4.9). Note how tectonically induced fluvial erosion during the Holocene notched an inner gorge into a glaciated valley floor to create a seemingly splendid site for a concrete-arch dam. Gorge cutting reduced the lateral support for the adjacent footslopes since the mid-Holocene. The stage was set for humans to shift the mountainside across the stability threshold.

The threshold had been crossed long before the cataclysmic phase, but nobody recognized the impending disaster. Bench-mark surveys made during the spring and summer months of 1963 revealed hillslopes that were slipping at a rate of 1 cm/week, a rate that increased to 1 cm/day in September. After a spate of heavy rains the creep rate became 20 and then 40 cm/day. By early October it was 80 cm/day. The exponentially increasing nature of this increase (Fig. 4.10) should have prompted immediate evacuation. Failure is inevitable after the stability threshold has been crossed (Schumm and Chorley, 1964), but the timing of many landslides is vague until it abruptly occurs. At Vaiont, such exponentially increasing rates were accompanied by opening of new fissures, with popping noises and snapping of tree roots in much of the mountainside.

The concluding rapid acceleration of the Vaiont landslide to 20 m/sec does not enter into this summary. It may have been the result of crack propagation in brittle rocks (Petley et al., 2002), low strength clay (Hendron and Patton, 1987; Kilburn and Petley, 2003), thermal heating of water or clay layers that would change pore pressures (Belloni and Stefani, 1987; Nonveiller, 1992; Veveakis, et al., 2007), or elimination of friction by rock fragmentation (McSaveney and Davies, 2005). The focus here is on precursor hydrogeologic changes.

The Vaiont landslide announced its presence shortly after the initial filling of the reservoir. A mass of hillslope detritus 500 m above the valley floor, and below an ominous 2 km wide M-shaped crack, began to move downslope at 10–35 mm/day (Fig. 4.11C). The stability threshold had been crossed as the factor of safety dropped below 1.0. On 4 November a 700,000-m³ landslide fell into the lake. This was a loss of footslope buttress material helping to restrain slippage of adjacent higher slopes.

The technical staff noted that movement of the main slide virtually ceased when the reservoir level was lowered. Note the synchronous peaks and lows of graphs A and C in Figure 4.11 in 1960 and 1962. It was assumed that future landslide movement rates could be controlled by changing the reservoir level. Unfortunately, lowering the reservoir level after mid-1963 did not halt hillside mass movements.

The revised game plan by the technical staff was to let slow landsliding gradually fill in part of the 160,000,000 m³ reservoir while a bypass tunnel was excavated in 1961 through the bedrock on the opposite of the valley. This tunnel would convey water from the upstream part of the reservoir and

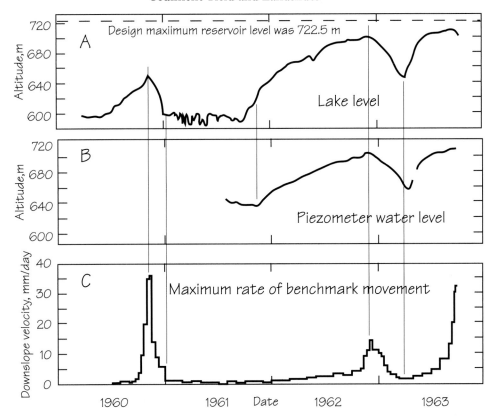

Figure 4.11 Graphs of factors affecting rate of downslope movement of the Vaiont landslide from March 1960 to 9 October, 1963. From figure 8 of Belloni and Stefani (1987).
A. Manipulations of the reservoir water level in an attempt to control landslide movement rate.
B. Water-level rise and fall in piezometer number 1.
C. Maximum downslope movement of benchmarks installed in the landslide.

allow the hydroelectric facility to operate after the slide moved into the reservoir. That was successful. Vaiont continues to generate electric power.

Filling of the reservoir in 1962 raised the ground-water table in the adjacent mountainside to an altitude of 710 m, the same as the lake level. Inflow of water into the hillside produced a seepage force that tended to hold the mountainside in place. Lowering of lake level would have an opposite effect. Each 10 meters of head change would have increased or decreased the seepage force by 1 kg/cm^2. Such lake-level drawdowns were done slowly to minimize "hydrostatic thrusting on joints" (Müller, 1987).

Once the water table had attained a stable and horizontal level, the lack of head differentials meant that no seepage stresses existed. Kiersch (1964) noted "Actual collapse was triggered by a rise in subsurface water level from bank infiltration with increased hydrostatic uplift". The substantial rise in the water table had reduced the grain-to-grain stress, the resisting force of equation 4.4. It did this by buoyantly supporting part of the weight of the intergranular skeleton of porous hillslope materials and by eliminating the weight of the capillary water in the formerly unsaturated zone. This amounted to an equivalent of 0.8 m of water decrease in resisting force for each meter of water-table rise – 0.8 (70) m

= 56 m of head, or 5 kg/cm² decrease in resisting force. Saturation of hillslope materials also locally changed capillary-water strength. These processes affected much of the footslope rubble above the failure plane (Fig. 4.9). Reversing the process increased resisting forces sufficiently to virtually stop bench-mark movements in 1961, and again in mid-1963. But then the slide stayed far to the unstable side of the factor of safety threshold.

The rapid increase of downslope landslide movement in September and October of 1963 coincided with the highest lake level of 710 m, and with prolonged heavy rains that introduced more hydrodynamic changes. About 200 mm of rain fell in only 8 days just before the final collapse. Karst solution tunnels in limestone high on the mountain would have funneled water downhill. The ever-widening fissures on the failing hillslope would have caught and trapped rainfall runoff. Clay layers may have been important locally in confining ground-water head. Increase of downslope directed seepage forces would have been equal to the loss of head along the flow path. Infiltration of rain also caused a rise in the hillslope ground-water table. This eliminated part of the capillary-water-strength component – a further decrease of hillside resisting strength.

The Vaiont story illustrates the several ways in which water may affect sensitivity of hillslopes to mass movements. We now turn to examples of how prolonged rain initiates debris flows.

4.2.1.2.2 California

The above discussions emphasize nontectonic controls of mass movements. Keep in mind that diverse, frequent mass movements are typical of tectonically active landscapes. The weather at a site, seasonal climatic setting, and the actions of humans all are important. Large, long rainstorms are especially important.

Most of the tectonically active California Coast and Transverse Ranges have a climate that favors multiple, fairly synchronous, landslide events. It is strongly seasonal semiarid and moderately seasonal thermic to strongly seasonal subhumid and moderately seasonal mesic (Appendix A). Both the strongly seasonal nature and the highly variable annual amounts of annual precipitation are important factors leading to winters with many landslides. The strongly seasonal climate, with rain concentrated in cool winter months, greatly increases the geomorphic effectiveness of the rainy season. Rainfall-per-month amounts are double to triple those of a similar, but non-seasonal, climate. Precipitation amounts not only vary greatly from season to season in California, but also from year to year (Bull, 1964).

The occasional particularly wet winter is a time of exceptional geomorphic work – including mass movements. A normally semiarid climate then can temporarily develop shallow ground-water conditions similar to those of a weakly seasonal, humid to extremely humid climate. But surficial materials quickly wash and slip away in the humid realm, whereas they accumulate in the headwaters tips of drainage networks, (the zero-order drainage basins of Dietrich et al., 1987) of California in the dry spells between wet winters. Occasional debris-flow events flush the accumulated detritus downvalley.

Other key variables of this interesting geomorphic setting should be kept in mind. These include a plant community that tolerates extreme annual droughts, and clayey (montmorillonitic) soil profiles that develop deep cracks during the hot, dry summers typical of this Mediterranean type of climate. These vertical fissures intercept overland flow at the onset of the rainy season.

Clayey colluvium has strong capillary tensile forces that prevail until soil-moisture deficits are reduced by rainfall-induced recharge. Increasing the thickness of the saturated zone during rainstorms enhances the likelihood of landslide initiation by decreasing capillary-moisture strength, and increasing seepage stresses.

Both intensity and duration of rainfall affect the generation of debris flows. Campbell (1975) identified the threshold moisture conditions as a 6.4 mm/hr rainfall intensity where the total seasonal antecedent rainfall had already attained 250 mm in the Santa Monica Mountains of southern California. He concluded that debris flows are most likely to occur where groundwater drains out of the colluvium on steep hillslopes. Such a rise in the ground-water table indicates the presence of seepage stresses and loss of capillary strength. Caine (1980) defined a limiting threshold where groundwater inflow exceeds outflow, thus promoting seepage forces.

$$I = 14.8D^{-0.39} \qquad (4.15)$$

where I is rainfall intensity in mm/hr and D is duration of rainfall in hours. Such conditions may also contribute to increased runoff.

How frequent are such times of many, widespread mass movements? They are concentrated mainly into those few exceptionally rainy winters that occur several times each century. These extraordinary winters have a critical combination of intensity and duration of rain (Cannon and Ellen, 1985; Cannon, 1988).

The storm of 3–5 January, 1982 is an excellent example. It triggered 18,000 debris flows in the 10 counties of the greater San Francisco Bay region, caused much damage, and many deaths. No wonder Californians need to be alert to landslide hazards! But such events rarely result from a single storm. Both the timing of landslides in this region and the threshold rainfall needed to initiate debris flows have been studied. Modeling by Iverson (1997, 2000) emphasizes response times to rainfall events.

The threshold rainfall intensity needed to trigger debris flows is different for a 6-hour than for a 24-hour rainstorm (Fig. 4.12, Wilson and Jayco, 1997). Normalized intensity data were plotted against

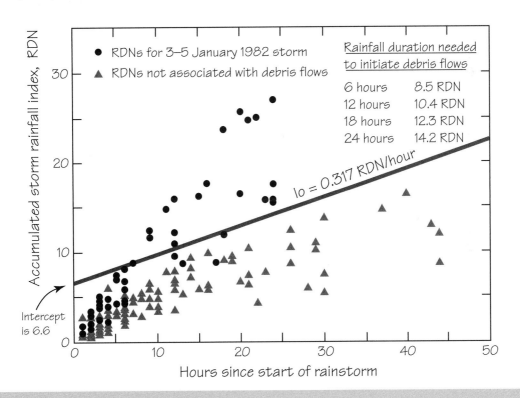

Figure 4.12 Rainfall thresholds needed to initiate debris-flows. Accumulated rainfall at many gages in the San Francisco Bay region, California were normalized by dividing the storm rainfall by the mean annual precipitation (MAP) of each station. The analysis includes orographic influences that influence rainfall frequency, the mean annual number of days with measurable rainfall (#RDs is the average number of rainy days). The plot is for normalized rainy days, RDN = MAP/#RDs. From Wilson and Jayko (1997) and Cannon (1988, figure 4.3).

time intervals for two classes of data – January, 1982 storm data, and data from storms that failed to produce debris flows. The line separating the domains approximates the rainfall threshold for debris-flow initiation in the region and was the basis of a landslide warning system from 1986 to 1995. It takes a large rainstorm to cause a debris flow, much more than the average rainy day total rainfall. Six *rainy-day normals* (RDN is the ratio between the mean annual precipitation and the average number of rainy days) will not do it even if that much rain falls in 6 hours. The threshold line of Figure 4.12 indicates that more intense rain – 8.5 RDNs – is needed for that time span. Lesser intensities are required if a storm persists for a longer time span. Seventeen RDNs spread out over a time span of 24 hours is more than sufficient to trigger debris flows, together with water runoff.

Nilsen et al. (1976a,b) report that mass movements in the San Francisco Bay region are triggered during storms where an additional 150–200 mm of rain falls after an initial 250–300 mm of rain earlier in a winter rainy season. About 60% of the total winter-season rainfall had transpired before landslide activity accelerated in dramatic fashion (Fig. 4.13 A, B). The 1972–73 rainy season began in typically abrupt fashion in mid-October and by mid-December more than 300 mm of rain had fallen. Only two landslides had occurred even though rainfall intensity had been high at times. Rainfall averaged 125 mm/month during this interval of minimal landslide activity. Intense storms returned after a dry spell in early January, but many landslides occurred this time. Five landslides occurred in the 3 months after the rainy season ended.

Nilsen's analysis nicely describes substantial response times of landsliding to rainfall. Low aquifer transmissivity maintained high ground-water levels and potential seepage forces, which kept the factor of safety threshold (equation 4.4) near or below a value of 1.0. A piezometric record of changes in hillslope ground-water levels would be valuable additional data. Although significant from a hazards perspective, only a minuscule proportion of the hillslopes failed during the 1972–73 rainy season. Hillslopes fail where concentration, instead of diffusion, of overland flow occurs.

Let us summarize these studies in the context of my hillslope ground-water model. Residents of landslide-prone coastal California know that the onset of heavy winter rains rarely triggers widespread landslides. Landslide movement occurs later in the rainy season during times of intense rainfall. Locations such as concave hillslope hollows are slowly shifted closer to failure-threshold conditions as the strong, dry hillslopes of summer become the progressively weaker hillslopes of winter. Prolonged winter rains favor gradual infiltration of water into hillslopes. Both rock type and tectonic setting influence infiltration rates. Brittle rocks produce fragmented, more permeable, colluvium than mudstone and weakly cemented sandstone.

The winter rainy season changes both the numerator and the denominator of equation 4.4 as ground water accumulates above low permeability bedrock and the water table rises. The added weight of increased soil moisture increases driving forces. Resisting forces decrease as grain-to-grain stresses are reduced by water-table rise that increases buoyant support for the grains. The downhill sloping layer of partially confined groundwater develops head differentials. Flow through confining beds results in seepage forces that further augment driving forces (Haefeli, 1948).

Elimination of capillary forces as surficial materials are submerged best explains why many mass movements occur during downpours of rain on hillslopes that are already close to a stability threshold. Saturation of surficial colluvium is more likely later in a rainy season, and occurs when other hydrologic factors have changed sufficiently to move parts of hillslopes close to stability thresholds. This final action in the evolving process of driving-force increases and resisting-force decreases shifts part of a hillslope across the equation 4.4 stability threshold. The slope fails. Clayey, saturated landslides accelerate when movement decreases their viscosity. This self-enhancing feedback mechanism transforms slumps into fast moving debris flows.

Ultimately, the propensity for landsliding in much of coastal California is the result of being a tectonically active landscape. Soft, crumbly rocks are raised to potentially hazardous heights. Hillsides are more likely to be underlain by soft, clayey rock types

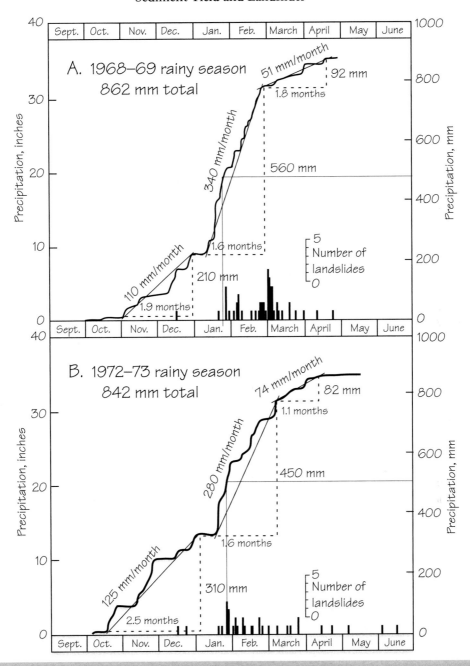

Figure 4.13 Relation of landslide activity in the San Francisco Bay region, California to seasonal rainfall accumulation. The steeper the slope of cumulative precipitation, the more intense the rainfall. From figure 8 of Nilsen et al. (1976).
A. 1968–69.
B. 1972–73.

in tectonically active landscapes than in inactive terrains where such materials are found primarily in valley floors. Valley floors are deepened by tectonically active downcutting and footslopes are steepened.

Such factors set the stage for occasional periods of widespread slumping of hillslopes and generation of debris flows. The damaging hillslope failures of the San Francisco Bay region are indeed notable. But the frequency and magnitude of landsliding is much less than in faster rising, extremely humid, mountains (Korup, 2005a,b; Dadson et al., 2004; Sarkar et al., 2006).

4.2.2 Landslides of Tectonically Active Regions

Weather events and human impacts are important direct causes of mass movements, but a strong case can be made that rock uplift is the important background factor. Tectonically induced increases of relief and orographic precipitation, plus fracturing of once solid rock are the ultimate controls of most mass movements.

Coseismic mass movements are common on unstable slopes created by the interactions of rapid uplift with fluvial and glacial processes (Morton, 1975; Plafker and Ericksen, 1978; Matthews, 1979; Adams, 1980; Keefer, and Tannaci, 1981; Pearce and, O'Loughlin, 1985; Keefer, 1999, 2000, 2002; Bull, 2007, Chapter 6).

Diverse rock types, and human disturbances of the landscape are present in the Appalachian Mountain region of the eastern United States and rain causes flooding when hurricanes come ashore. Mass movements do occur (Wieczorek et al., 2006), but at long-term frequencies far below those in Taiwan, New Zealand, the Himalayas, coastal Alaska, and the California–Oregon Coast Ranges. Landslide-prone rock types occur in valley floors instead of on ridgecrests in the Appalachian Mountains (Hack, 1965, 1973b) because of long-term denudation in the tectonically quiescent environment of this passive plate boundary margin.

Appalachian sandstones are even older than the Torlesse rocks of New Zealand but ridgecrests are much different. Folded strata sweep gracefully through a valley-and-ridge landscape (Fig. 4.14). This deformation occurred long ago in the ductile depth zone. Joint and fractures of the sandstone are so minimal that large streams such as the Shenandoah River cross where resistant sandstone is the thinnest. Water gaps along the ridgecrest record stream piracy during long-term drainage-net evolution. These sites have a joint spacing of 0.4 m as compared to the more general 0.8 m average spacing of joints (Thompson, 1949, p. 58). Silica recementation of joint planes is so advanced that joint abundance may not be indicative of relative rock mass strength. For a modern study of these passive margin landscapes – including where to go and what to read – I refer you to the discerning opinions of many authors in Pazzaglia et al. (2006).

These Appalachian strata were folded by the Paleozoic Taconic, Acadian, and Allegheny orogenies. It was a more regional uplift during the Cenozoic that raised this passive-margin landscape into the brittle depth zone of the crust. Old joints and fault planes have been cemented with silica, and production of new planes of rock weakness has been minimal.

In marked contrast, the greywacke sandstones and argillites in New Zealand have not been in the ductile depth zone. They have been constantly battered. "The Torlesse has spent nearly 100 My in the brittle zone during at least four periods of identifiable deformation" (John Bradshaw, University of Canterbury, written communication, 20 August, 2007).

The resulting ridgecrest processes and form could hardly be more different than the unbroken resistant sandstone crests of passive-margin landscape evolution (Fig. 4.14). The unraveling process shown in Figure 4.15 may be thought of as being either rapid stream-channel headcutting, and/or as successive collapses of oversteepened slopes that continues to the drainage divide. Two consequent processes occur now that this unraveling episode is complete. Patches of new vegetation are starting to cloak the steep bedrock surface. Rills and stream channels are starting to incise the broad swath of exposed rock created by the unraveling.

These rocks crumble and wash away so easily that immense amounts of sediment are yielded to streams. You might find the following opinion rather intriguing.

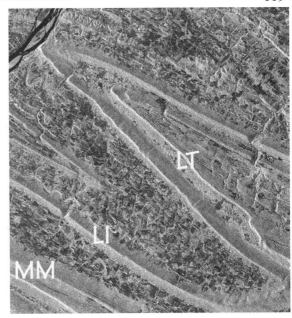

Figure 4.14 Space radar image showing how ridgecrest rock mass strength controls the topography of the Appalachian Mountains south of Sunbury, Pennsylvania, USA. The area shown is approximately 30 km by 31 km and North is towards the upper right. Resistant quartzitic sandstone forms the smooth texture of the forested ridgecrests above the patchwork of valley farms. The prominent ridges in the center-right are Little Mountain, LT, above the single ridge of Line Mountain, LI, and Mahantango Mountain, MM, is at the lower left. The Susquehanna River is at the upper left. The image center is at 40.85° N and 76.79° W. This is a grayscale version of NASA-JPL Space Shuttle image 1818 (SIR-C/X-SAR) of 6 October 1994. For a color version go to http://veimages.gsfc.nasa.gov/559/PIA01306_md.jpg

Uplift rate and rock mass strength are the ultimate causes of mass movements. Rock type regulates the type of mass movement and uplift raises susceptible earth materials to steep, unstable positions in a landscape. Climatic events and human impacts merely serve as triggering mechanisms for timing of mass-movement events. The crumbly rocks of the Seaward Kaikoura Range favor only shallow slumping landslide processes (Figs. 4.2, 4.15, 4.17B, 4.19). Rock mass strength is higher near the Alpine fault, 400 km to the southwest, where Torlesse sandstone and argillite have been metamorphosed to phyllite

Figure 4.15 A forested 400 m high, 44°, hillslope in the Seaward Kaikoura Range of New Zealand. The rock type is fractured, crumbly Torlesse sandstone. A stream that changed to a gully with a headcut started rapid hillslope unraveling that continued all the way to the ridgecrest (above and to the right of my fellow hiker). Other gullies have been healed by renewed plant cover (the pair of gullies at the left have young trees).

Figure 4.16 Highly unstable hillslopes underlain by sheared greywacke sandstone in highly seasonal, humid northern California. New scarps, N, are developing as an old landslide is reactivated. Lateral erosion induced base-level fall is the primary reason for the instability of this seacoast. Note the truncated alluvial fans, TF.

and schist. Huge, deep-seated landslides are an important aspect of the amazingly high sediment yield of the wet, northern flank of the Southern Alps (Hovius et al., 1997; Korup, 2004, 2005a; Korup and Crozier; 2002, Korup et al., 2004).

Lateral erosion induced base-level fall also is an important cause of landslides. The Mendocino seacoast of California shown in Figure 4.16 is being raised about 1 m/ky (Merritts and Bull, 1989). The combined effects of rocks that have been through the fault-slip mill and lateral erosion induced base-level fall have created a highly unstable and landslide-prone landscape. The buttress of footslope materials that once held this mountainside in place has been removed. Coastline retreat has initiated stream-channel entrenchment and gullies continue to expand rapidly. The resulting large sediment yields provide gravel that is deposited as debris-flow beds in the alluvial fans, only to be truncated by the waves and currents.

Lateral erosion by streams prevails over vertical erosion in tectonically stable valley floors. This tends to maintain steep midslopes, and to a lesser extent, the crestslopes on the adjacent hillslopes. These are the most likely sites for future landslides (Fig. 3.26B, 3.28).

Uplift of the Diablo Range foothill belt has raised clay-rich rocks to positions where they now underlie moderately steep slopes. The result is many landslides, earthflows, and debris flows in mountains with a strongly seasonal arid to semiarid, thermic to mesic climate. The presumably cooler and/or wetter full-glacial climates of the Late Pleistocene were conducive to increased landslide abundance in California (Stout, 1977, 1992; Ehlig and Ehlert, 1978, 1979; Yamanoi, 1979).

Diverse mass movements are common on hillslopes of rising mountains (Sharpe, 1938; Ritter et al., 1995). Slumps (Figs. 4.17) are characterized by abrupt downslope and rotational movement of blocks of rock and soil. An earthflow (Fig. 4.18) of wet mudstone moves ~0.2 to 4 m/yr, and is supplied by slumps at a headscarp that migrates upslope. Airborne laser altimetry (LiDAR) is an excellent tool for studying pulses of earthflow movement (McKean and Roering, 2004).

Debris flows and mudflows are intermediate between landslides and water floods in that they are mass movements that flow down stream channels (Blackwelder, 1928). Flow rates range from less than 1 to more than 100,000 m/hr (Costa, 1984). They are non-Newtonian fluids (Bingham substances according to Johnson, 1970) and are especially common in tectonically active and volcanic regions. Nontectonic factors that also promote debris flows are abundant water (intense rainfall) over short periods of time at

Figure 4.17 Examples of slumps.
A. Downslope movement along an upward-curving surface in mudstone and shale has rotated the head of a slump, H, backward and has created a characteristic bulge, B, at the toe of this landslide. Location is adjacent to a stream channel in the Ciervo Hills, Diablo Range foothills, California.

Figure 4.17 Examples of slumps.
B . Downslope movement along an upward-curving surface in brittle, fractured sandstone has rotated this slump as indicated by the tilted trees. Location is adjacent to Jordan Stream in the Seaward Kaikoura Range of New Zealand.

Figure 4.17 Examples of slumps.
C. Active, A, and old, O, landslides occupy most of a small drainage basin underlain by diatomaceous shale, Tumey Hills, Diablo Range foothills, central California. This tectonic activity class 1 drainage basin illustrates the potential for landslides in an arid region that has a mean annual precipitation of only about 180 mm.

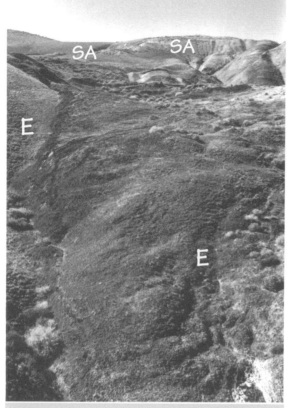

Figure 4.18 This earthflow in mudstone and shale of the Tumey Hills, Diablo Range foothills, California, is essentially inactive during drought years of less than 70 mm rain, but moves several meters during and after wet winters with more than 400 mm of rain.
E is edge where shearing occurs between the earthflow and the stable hillslope.
SA is broad concave source area that collects runoff, which is channeled into fissures.

irregular intervals, steep slopes with cracks and fissures that favor quick entry of runoff water, and a source material for the mud matrix (Cannon, 2001).

Uplift induced stream-channel downcutting has resulted in steeper adjacent footslopes in the Diablo Range. Cracks parallel to the slope contours of a 40–45° mudstone slope are indicative of the head of a landslide. About 0.1 to 1 m of surficial weathered mudstone can shift downslope as a debris slide during prolonged, intense winter rains. Deep

wetting and failure of less weathered bedrock causes a block to fail – a slump. Erosional base-level fall along the stream channel initiates upslope propagation of mass movements. Slumps are common near stream channels that are being affected by tectonically induced downcutting (Fig. 4.17C). Both processes extend the tectonically steepened portion of the hillside farther upslope and into the watershed headwaters.

Debris-slide processes occur in noncohesive materials such as alluvium or highly fractured rocks. Slopes in the humid, mesic Seaward Kaikoura Range of New Zealand are good examples.

Hillslopes unravel as headscarps migrate upslope from degrading stream channels in response to continued uplift of the Seaward Kaikoura Range. Tectonically induced stream-channel downcutting starts this process. Common sites for initiation of slumps are the especially steep footslopes on the outsides of river bends (Fig. 4.19A). The gradient of this braided reach of Jordan Stream is very steep, averaging 0.17. This results in sufficient stream power to flush away enormous amounts of recent landslide-produced detritus and further incise the underlying bedrock. Footslope slumps typically occur during or after periods of above-normal rainfall when hillslope seepage forces are largest.

Debris slides are caused by rainstorms, and earthquakes. Vegetation, including trees, offers little protection. Progressive debris sliding unravels hillslopes as 3 to 8 m of soil and colluvium are stripped to bedrock (Fig. 4.19B). Spatial decay of the effects of stream-channel downcutting applies to hills as well as to streams. Mass-movement removal of surficial materials decreases markedly in the upslope direction, partly as a function of change in style of landsliding. Recent footslope slumps have lowered the hillslope by about 10 to 20 m. Debris sliding has lowered the surface 8 m in the midslope and only 1 m near the ridgecrest. The resulting topography has convex footslopes and roughly planar midslopes and crestslopes.

Hillslope responses to uplift rates are partly a function of the work capacity of geomorphic processes. Debris-slide processes might match a stream-channel downcutting rate of 1 m/ky. It appears that a more potent geomorphic process is needed

Figure 4.19 Vertical view of debris slides and rock-fall deposits in fractured greywacke sandstone along Jordan Steam, Seaward Kaikoura Range, New Zealand.
A. Vertical aerial photo showing diverse range of hillslope debris-slide processes and resulting landscapes. **A,** Recently initiated slumps on the outsides of bends of downcutting trunk stream channel. **B,** Debris slides have removed the forest and colluvium, and exposed bedrock and progressive failure has extended to the watershed divide. Maximum denudation rates occur where footslopes slump into the actively downcutting trunk stream channel.
C, Dendritic patterns of chutes contribute to accumulations of talus and debris slopes that mimic a stream-channel network. **D,** Steep, small watershed where active debris slide and rockfall activity keeps hillslopes bare of bushes and trees. Watershed relief is 700 m. Photo courtesy of Alastair Wright, Environment Canterbury, New Zealand.

to keep pace with the current 4 m/ky. So, slumps instead of debris slides are the dominant footslope mass-movement process in this climatic and lithologic setting. However, denudation rates clearly are not spatially uniform, increasing from being paltry at the ridgecrest to profound on the footslopes.

Landslide detritus tends to dam streamflow. Steep longitudinal profiles of channels such as Jordan Stream are conducive for quick removal of slumps and mass movements.

Horizontal scree beds above the talus cone (Fig. 4.19B) probably were graded to a higher valley floor, and may date to the last full-glacial episode of climate-change induced valley-floor aggradation. Subsequent Holocene degradation has cut down through this aggradation event valley fill and then into bedrock. Such aggradation events temporarily prevent degradation of footslope bedrock while ridgecrest degradation continues. The midslope of this hill has a more uniform style of degradation over time spans of 100 ky because it is not affected by valley-floor aggradation events of 10 to > 50 m.

Aerial-photograph comparisons reveal the rapidity with which forested or tussock-grassland landscape is transformed in the Southern Alps. Progressive unraveling can shift a headscarp a kilometer upslope in only 20 to 50 years. Progressive hillslope collapse converts some densely vegetated slopes into broad sheets of talus that continue to receive rock-fall blocks from the upslope retreating headscarp. Such slopes are planar, being at the angle of repose, but the underlying bedrock has a convex form where talus thickens downslope. Reversal of the process may occur after the process has reached the upper boundary of a hillslope – the ridgecrest divide. Then, markedly reduced flux of detritus permits re-establishment of plants. They promote an opposite self-enhancing feedback mechanism leading to growth of new forests (Fig. 4.15).

Type and frequency of mass movements is also a matter of hillslope scale, geologic structure, and converging and diverging flow lines of diffusive processes. Faults and joints determine many of the locations of streams of talus that pour down chutes to cones and block fields. A dendritic example is shown at the two locations labeled C in Figure 4.19A.

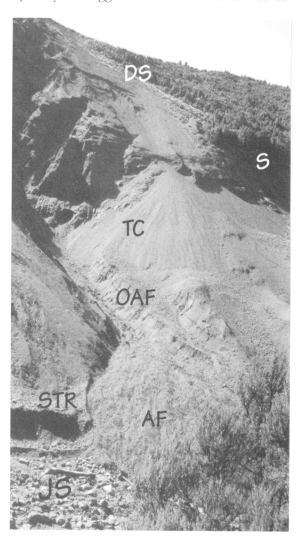

Figure 4.19 Views of debris slides and rock-fall deposits in fractured greywacke sandstone in the north fork of Jordan Steam, Seaward Kaikoura Range, New Zealand.

B. Ground view of hillslope to the right of location **B** on Figure 4.19A. **DS**, debris-slide area where intensity of landslide erosion decreases from midslope to crestslope parts of hillslope. **S**, scarp of a slump, the dominant footslope process. **TC**, talus cone derived from debris sliding and located where space is made available for deposition by the base-level fall created by slumping footslopes. **OAF**, incised surface of old alluvial fan that was graded to the higher prior fill stream terrace of Jordan Stream, **STR**. **JS**, steep active channel of Jordan Stream. **AF**, active alluvial fan that is graded to the present trunk stream channel.

Exceptionally large rainstorms temporarily raise the ground-water table in these exceptionally permeable materials. This may cause part of the accumulated mass of blocks to fail as a debris flow that surges downstream leaving a pair of impressive levees. Daily rainfalls of 250–400 mm have occurred four times since 1985 in the vicinity of Jordan Stream.

The ratio of fluvial to mass-movement degradation increases with drainage area. Mass movements are still the dominant process rapidly denuding the hillslopes at location D in Figure 4.19A. But here the dendritic pattern of the landscape also promotes streamflow that flushes some of the rocky detritus downstream, instead of allowing it to accumulate. Note the progressive increase of dendritic characteristics from least in area B, more noticeable in area C, and best in area D. The dynamic and ever changing hillslopes of the Seaward Kaikoura Range are becoming progressively longer and steeper in response to ~5–6 m/ky of rock uplift. They change from vegetated, soil mantled hillsides to bare crags and cliffs on which forests are less likely to be re-established between intervals of mass-movement activity.

Mass-movement processes are capable of shifting huge volumes of earth materials into stream channels. Disruption of vegetation and soil by mass movements greatly increases local hillslope sediment yields. Sediment yields in fluvial systems with active landslides comes mainly from sites of mass movements (Hovius et al., 1997; Korup, 2005a). Increase in width of the active stream channel downstream from a landslide source of sediment underscores the importance of such landslide sources of sediment.

The influence of active uplift on landslide distribution is illustrated by the semiarid Arroyo Ciervo watershed in the Diablo Range foothills (Fig. 4.20). This drainage basin is underlain by 60% diatomaceous shale, 8% mudstone and clay, and 32% soft sandstone and sand. Clayey rocks occur throughout the basin but landslides are not common near the mouth or headwaters. Mean annual precipitation increases with altitude from about 110 to 360 mm. The monocline at the basin mouth is so young that

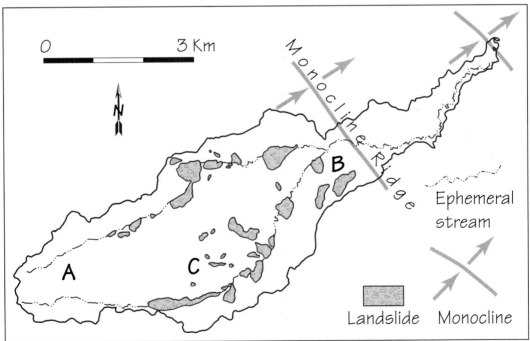

Figure 4.20 Distribution of landslides upstream from a tectonically active monocline in the Arroyo Ciervo drainage basin, eastern Diablo Range of California.

uplift has yet to create much local relief. Two range-bounding thrust faults may be active (Bull, 2007, figure 4.3D). Steep slopes and 300 m high hills are common in the shale and sandstone upstream from tectonically active Monocline Ridge. Accelerated valley downcutting between basin-position coordinates of 0.24 and 0.47 has caused many landslides, which occupy 8% of the total watershed area. Accelerated channel downcutting at location A is in the initial stages.

Arroyo Ciervo provides a nice example of how uplift interacts with fluvial processes to determine locations of landslides. Tectonically induced downcutting caused by Late Quaternary uplift of Monocline Ridge has extended to the headwaters streams that flow into the right-branch trunk stream channel (location C on Fig. 4.20). These mass movements consist of present-day footslope slumps. In contrast, landslides in that part of the basin immediately upstream from Monocline Ridge (location B) occur higher on midslopes or on footslopes adjacent to small tributary streams. The location B slides are a mixture of ages; some are only a few decades old, but others are so old that the hummocky topography typical of landslides has become quite subdued. This areal distribution of landslides illustrates the substantial reaction time for increased incidence of landslides as the effects of a tectonic perturbation – still ongoing – are transmitted by the stream subsystem to progressively more distant hillslopes.

A threshold hillslope steepness may be needed for abundant debris-flow production. Watersheds that are underlain by clayey rock, but have mean slopes of less than 0.2, may have high sediment yields but rarely produce debris flows. An example is the Diablo Range foothill belt in California. The threshold for common debris-flow production during my 1956 to 1986 study period appears to be at a mean slope of about 0.3 for basins underlain by mudstone and shale. Basins with lesser slopes may generate small debris flows from locally steep slopes but such mudflows are small and rarely reach the piedmont alluvial fans. Fans from gently sloping basins are composed chiefly of water-laid deposits. Debris flows are a more common output of drainage basins with mean slopes in excess of 0.3 and the abundance of debris flows in the stratigraphic record is greatest for fans derived from basins with mean slopes of 0.4 to 0.5.

Prolonged rapid uplift of the west side of the Panamint Range (Fig. 4.21A) has resulted in exceedingly steep hillsides whose highly rilled slopes are underlain by soft rocks that favor production of debris flows. Vegetation is sparse on the arid lower slopes, so intense rainfall from summer monsoon thunderstorms strips away the surficial weathered argillaceous rocks. Plutonic and metamorphic rocks furnish abundant large boulders that are transported by debris flows to the piedmonts of these tectonically active landscapes (foreground of Fig. 4.21A and Figure 4.21B). Tectonically induced downcutting maintains narrow valley floors in the mountains that confine debris flows. Such landscapes favor depths of flow needed to keep viscous debris flows moving as a series of surges.

Mass movements become ever more important as uplift changes low hills into lofty mountains. Sediment yields increase dramatically, not only because water flows over steeper slopes, but also because mass movements supply detritus to steep streams with large unit stream power. The way in which landslides in only 5% of a watershed can supply most of the sediment flux is so dramatic that it led to the concept of "partial-area contribution" of sediment yield (Ragan, 1968; Betson and Marius, 1969; Dunne and Black, 1970; Emmett, 1970; Dunne et al., 1975; Scoging, 1982; Poesen, 1984; Juracek, 1999). Streams entrain sediment as they undermine landslides that have moved into a stream channel. This process is much more effective in supplying detritus to streams than gradual downslope creep of hillslope soil and colluvium. Streams in rising mountains efficiently convey landslide detritus to downstream reaches and to the sea. Rising mountains with soft rocks are well known for having exceptionally large sediment yields (Kelsey, 1980, 1985, 1987; Carver, 1987; Nolan et al., 1987).

4.3 Summary

Large watershed sediment yields result from tectonic deformation that shatters strong rocks, and uplift that creates high, steep mountains that receive more rain and snow. Debris flows and rock falls are

Figure 4.21 Conditions conducive for production and deposition of debris flows in landscapes with a class 1 tectonic activity classification.
A. West side of the Panamint Range at Surprise Canyon, where a combination of steep slopes, clayey rocks, sparse plant cover, and narrow, steep valley floors favor production and conveyance of debris flows to the large alluvial fan in the foreground. Drainage-basin relief is 2800 m. Width of the mountain front in the foreground is 5 km. The fan is an area of tectonic base-level fall and thus is well suited for preservation of debris flows in the stratigraphic record.

Figure 4.21 Conditions conducive for production and deposition of debris flows in landscapes with a class 1 tectonic activity classification.
B. 1.3 to 2 m thick debris flow on the alluvial fan of Sparkplug Canyon, west side of the White Mountains, southeastern California. Highly viscous material oozed to a stop around fragile bushes in the foreground. Only a few meters away round protuberances mark the locations of 1 to 3 m mud-covered boulders transported by the same flow. Rock hammer for scale.

common, and deep-seated landslides occur where rocks are sufficiently cohesive. Sediment yields from steep mountains with soft rocks are especially impressive at latitudes conducive for frequent large rainfalls. Two representative climatic regions were examined in this chapter – humid to extremely humid New Zealand and the highly seasonal semiarid eastern margin of the Diablo Range in California.

Amounts of alluvial-fan area per unit drainage-basin area were used as a sediment yield index for the Diablo Range. Regressions of sediment-yield index and mean basin slope describe increase of sediment yield with steeper source areas. Doubling watershed mean slope increases the sediment-yield index by fourfold. About 1.8 times as much sediment is yielded from mudstone basins as from basins underlain by sandstone. The abundance of debris

flows is greatest for alluvial fans derived from basins with mean slopes of 0.4 to 0.5. Landslides in only 5–10% of a watershed can be a dominant source of sediment.

Rock type regulates the type of mass movement and uplift raises susceptible earth materials to steep, unstable positions in tectonically active landscapes. Climatic events and human impacts merely serve as triggering mechanisms that determine the times of mass-movement events.

The propensity for landsliding in New Zealand, and much of coastal California, is the result of being tectonically active. Soft, crumbly rocks are raised to potentially hazardous heights and landscape relief is increased. Hillsides are underlain by soft, clayey or highly fractured rock types whereas such materials are found primarily in valley floors in tectonically inactive settings. Valley floors are incised by steep stream and adjacent footslopes are steepened. Such factors set the stage for occasional periods of widespread slumping of hillslopes and of debris flows. Self-enhancing feedback mechanisms allow clayey slumps to become fast moving debris flows. Although impressive, the frequency and magnitude of landsliding in California is much less than in faster rising, extremely humid, mountains such as the Southern Alps of New Zealand, Taiwan (Kamai, et al, 2000) and the foothill belt of the Himalayas.

Californians believe that rainstorms cause destructive landslides and debris flows. Geologists know that earthquakes, steep slopes, clayey rocks, and geologic structures that parallel hillslopes increase mass movement hazards. The onset of heavy winter rains rarely triggers widespread hillslope failures. About 60% of the total winter-season rainfall occurs before accelerated landslide activity begins, and then only in truly unusually wet winters.

California has a strongly seasonal climate. Changes in soil-moisture conditions shift hillslopes closer to failure thresholds as the strong, dry hillslopes of summer become the progressively weaker hillslopes of winter. Rain concentrated mainly during the cool winter months greatly increases the geomorphic effectiveness of the rainy season. Even semiarid hillslopes can temporarily develop shallow ground-water tables during the occasional particularly wet winter. Seasonal ground-water table changes move hillslopes closer to failure thresholds, increasing the likelihood of mass movements.

A seasonal water-table rise increases the thickness of the saturated zone and enhances initiation of landslides. The added weight increases driving forces on the upper sections of hillslopes. Larger head differentials increases the potential for downslope directed seepage forces. Colluvium and clayey hillslope soils have strong capillary tensile forces that prevail until soil-moisture deficits are reduced by infiltrating rain. Elimination of water clinging to grains as deposits become saturated decreases the weight of water of specific-retention. Capillary moisture strength is eliminated where water tables rise to the land surface.

Footslopes are also affected. A water-table rise provides buoyant support to each grain that is submerged, thereby decreasing its cumulative weight on all underlying connected grains. Such footslope materials no longer are effective as a buttress holding the hillslope in place. This ground-water induced change applies to the common problem of landslides caused by filling of reservoirs upstream from newly constructed dams.

Intuitively, there might be a limit to which a hillslope can be steepened by tectonically induced downcutting of the valley floor to which it is linked. One suspects that landsliding becomes the dominant process shaping such steep hillslopes, which leads to the assumption of a steady-state maximum slope steepness that is determined largely by rock mass strength (Carson and Petley, 1970; Carson, 1976; Dietrich et al., 1992, 1993; Schmidt and Montgomery, 1995; Densmore et al., 1997; Hovius et al., 1998; Shroder, 1998; Whipple et al., 1999; Montgomery, 2001; Montgomery and Brandon, 2002; Burbank et al., 2003; Gabet et al., 2004; Korup, 2008). The next chapter examines this and other aspects of steady-state landscape models proposed by these and many other geomorphologists.

Chapter 5

A Debate About Steady State

5.1 A Century of Conceptual Models

It is so tidy just to conceptualize landscapes as interacting suites of landforms comprising the fluvial systems of mountain ranges. Increase of rock-uplift rate increases hillslope relief and steepness. The result is an increased erosion rate. Such responses between geomorphic variables is just as appealing to computer modelers as it was a century ago to men on horseback exploring the American West.

G.K. Gilbert (1909) and John Hack (1960) believed that landscapes tend to achieve a state of dynamic equilibrium: "When the ratio of erosive action as dependent on declivities (slope) becomes equal to the ratio of resistance as dependent on rock character, there is an equality of action" (Gilbert, 1877, p. 100). John Hack admired Gilbert's model: "The principle of dynamic equilibrium states that when in equilibrium a landscape may be considered a part of an open system in a steady state of balance in which every slope and every form is adjusted to every other" (Hack, 1965, p. 5). Steady-state watersheds have unchanging landscape configurations (Gilbert, 1877; Davis, 1898, 1899; Hack, 1960, 1965;

Roering et al., 1999; Willett et al., 2001; Whipple and Meade, 2004; Miller et al., 2007; Stolar et al., 2007).

Studies that incorporate the premise of steady state now use digital elevation models, numerical modeling, terrestrial cosmogenic nuclides, thermochronometers, and thermobarometers. But this appealing simplification of the real world remains loosely defined and unproven.

Burbank and Anderson say it best (2001, p. 145). They maintain that a steady-state landscape can be envisioned where denudation maintains hillslope

This dark flat-topped hill is a remnant of a marine terrace that has been raised above timberline in the Seaward kaikoura Range of the Southern Alps of New Zealand. Fluvial erosion has carved deep valleys but has yet to eradicate remnants of coastal landforms preserved locally on the ridgecrests. This tells us that denudation rates cannot be uniform even after >200,000 years.

Tectonically Active Landscapes, 1st edition. By W. B. Bull. Published 2009 by Blackwell Publishing, ISBN 978-1-4051-9012-1

length and relief through time. Their definition focuses our attention on the boundary landforms of hillslope subsystems. Steady state prevails only when erosion lowers ridgecrests and valley floors at the same rate.

Note my emphasis on the scale used by these geomorphologists. Like Gilbert and Hack, Chapter 5 mainly discusses drainage basins and their component hillslopes because fluvial systems are the basic elements of mountain landscapes. This choice places an emphasis on larger scales and longer time spans than stream-channel landforms, and on smaller spaces and shorter time spans than tectonic orogen processes such as crustal flux and erosion along a plate boundary. Of course we compare drainage-basin models with landscape adjustments that may be occurring at smaller and larger scales – stream channels and drainage basins. This chapter assesses the likelihood of a watershed-size landform achieving an unchanging configuration.

Landforms must achieve steady state before entire drainage basins. Some do, but for others steady state is impossible. Both early and present-day workers emphasize the importance of independent variables such as climate, rock uplift, and bedrock durability in this debate.

Attainment of landform equilibrium or landscape steady state may be a matter of having sufficient time to adjust to changes in the variables of a fluvial system. My choice of four independent variables (Fig. 5.1) may not rise to the top of your list. I emphasize the role of change instead of physical characteristics. Consider revising your numerical model to acknowledge these basic controls. Let us mention the importance of each.

Perturbations and adjustments of landforms cannot be transmitted elsewhere in a fluvial system unless continuity is present. See Bull (2007, section 2.5.2) for diverse examples of lack of fluvial-system continuity. A pulse of channel downcutting migrates upstream until it is stopped by the simple hydraulic jump of a waterfall. Rocky outcrops rising above a soil-mantled hillslope are marching to a different tune – they are little subsystems in the overall larger picture. Tors may rise ever higher above a hillslope that is being degraded at a spatially uniform rate. Such lack of fluvial-system continuity is one result of spatially variable rock mass strength.

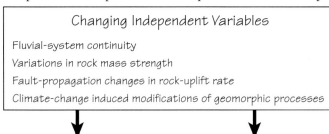

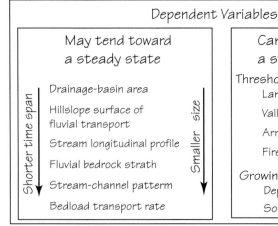

Figure 5.1   Flow chart summarizing the influences of several independent variables of fluvial systems on dependent landscape elements that may, or cannot, tend toward a steady-state configuration.

The Gilbert–Hack steady-state model accounts for spatial variations in rock mass strength by progressive changes in steepness of hillslopes. Disequilibrium rules when hard and soft rocks of rapidly rising hills have the same slope (Fig. 3.8A). Equilibrium is more likely when hard rocks occur on steep slopes, and with faster erosion rates, the soft rocks become ever more gently sloping (Fig. 3.8B). Steady state would exist where rates of hillslope denudation are the same, as steep slopes (underlain by hard rocks) erode at the same rate as gentle slopes (underlain by soft rocks). But there is a limit to the effectiveness of increasing the rate of hillslope erosion by making it steeper. Relatively harder rocks that become cliffs may become even more resistant. Exposed massive granitic rocks persist as expanding and rising outcrops because they no longer weather as fast as when they were in a moist subsoil environment (Wahrhaftig, 1965).

Spatial patterns of local rock uplift may change with the passage of time even where mountain-building forces maintain a generally uniform rock-uplift rate over long time spans. Each rupture of earth materials propagates the length of a master fault and a variety of secondary faults, such as those shown in Figure 4.1. Landscape response to folding above a blind thrust fault should change after the fault-tip propagates to the land surface. Landforms change in response to changing patterns of tectonic deformation. Are the rates of landscape change rapid enough to adjust to changes created by active faulting and folding? How slow must the rates of tectonic deformation be in order for landscape change to keep pace? How much time is needed after orogenic displacements cease for hillslopes to degrade uniformly and stream channels to maintain a constant pattern?

We also need to assess magnitudes and styles of geomorphic processes induced by major swings of climate in and out of ice ages during the Late Quaternary. Landscape responses are diverse and vary greatly in different climatic and lithologic settings. Such complications will add tricky doses of spice to your modeling recipe. Hooke (2003) considers asymptotic approaches to attainment of steady state likely in unglaciated terrain with soft rocks that erode quickly during frequent floods. U-shaped valleys with massive, hard rocks that were created by valley glaciers may require unrealistically long time spans to approximate steady state.

The sample of dependent variables listed in Figure 5.1 addresses two classes of processes and landforms. Some landscape elements may tend towards an unchanging configuration with the passage of time. The shorter the time span, and the smaller and more transitory the element, the more likely that steady state can be demonstrated. Intermediate-size elements such as soil-mantled footslopes created by flowing water are likely to achieve a balance between slope steepness and local flux of sediment where the base level of the adjacent valley floor is constant. Assumed attainment of steady state, but not yet proven, is all we can say about midslopes and drainage divides.

Response-time adjustments of midslopes varies greatly. Roering et al. (2005) note that "Short, steep hillslopes that erode via nonlinear slope-dependent processes can rapidly (~40 ky) adjust their morphology to climatic or tectonic perturbations". The landscape shown in Figure 4.2B would seem to fit this model because this style of hillslope morphology is typical of the Seaward Kaikoura Range. Adjustment that creates a typical morphology, and spatially uniform denudation rates, is possible. But does hillslope relief continue to change with the passage of time? Such change in the altitudinal positions of valley floors and ridgecrests would be indicative of persistent nonsteady-state conditions.

Equilibrium is most likely where powerful processes, such as streamflow, shift small volumes of materials over short time spans. The timing of strath formation for the Charwell River of New Zealand underscores this point. Lateral beveling of bedrock by the river can create the equilibrium landform of a strath. But this happens only when tectonic and climatic controls on unit stream power and bedload transport rate allow a reach of the stream to achieve, and maintain the base level of erosion (Fig. 5.2).

Times of strath formation along the Charwell River were modulated by Late Quaternary climatic changes in a tectonically dynamic setting. The valley floor downstream from the active Hope fault was either aggrading or was catching up to new base levels of erosion most of the time.

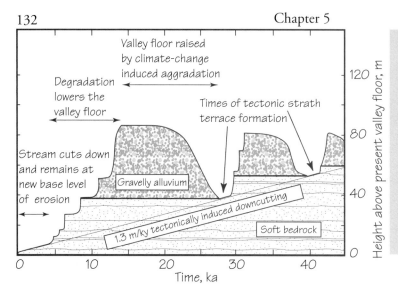

Figure 5.2 Changes in the streambed altitude of the Charwell River, New Zealand reflect the combined influence of tectonic and climatic controls during the past 45 ky. Tectonic strath terraces are created only during brief time spans that follow climate-change modulated episodes of tectonically induced downcutting. Simplified from figure 5.24 of Bull (1991).

Each valley-floor aggradation event forced the stream to catch up to a new base level of erosion. The stream had to degrade through the most recently deposited valley fill, and then through a thickness of bedrock equal to the amount of rock uplift since the last time the stream attained type 1 dynamic equilibrium. The Charwell River barely had enough time to bevel a new tectonic strath after attaining the base level of erosion, before the onset of the next aggradation event. Strath formation only occurs here during interglacial times of reduced bedload transport rate and can be interrupted by brief earthquake-induced episodes of aggradation (Section 2.4).

The key point is that stream channel steady-state conditions were achieved occasionally under conditions of uniform uplift rate of 1.3 m/ky and major climate-change induced fluctuations in bedload transport rate and unit stream power. Nearby larger streams had the same climate-change perturbations, but their much greater stream power has allowed them to achieve steady state quickly, and maintain it longer, perhaps 40–80% of the time.

Longitudinal profiles of streams may also achieve steady-state configurations. The gradient-index parameter of Hack (1973a,b, 1982) is nice for assessing if a reach of a stream is at steady state. See Bull (2007, fig. 2.25) for Charwell River examples.

Larger landforms – valley cross sections and basin areas – may respond to climate-change perturbations very slowly. These landscape elements are huge when compared to valley-floor landforms. Geomorphic processes change them slowly. Such long time spans are needed that fluctuations in the independent variables may be integrated into long-term averages. But we have yet to demonstrate that a tendency for a time-independent landscape has actually occurred at specific field locations. Nonsteady state processes, such as landslides, limit choice of potential field study areas. Only the computer models with constrained boundaries suggest steady state.

Threshold-based landscape change is incompatible with the steady-state concept. An over-steepened hillslope either is completely stable (not moving) or an incipient irreversible mass-movement has begun. The threshold described by equation 4.4 has been crossed. Likewise, a pulse of valley-floor deposition or a period of arroyo cutting only occurs when a geomorphic-process threshold is crossed.

Slopes dominated by threshold processes cannot achieve a time-independent configuration. Landform evolution is away from steady state. Toppling of a rock monolith is one example. The 1 mm crack that marks the initial separation of the monolith from a massive cliff initiates a non-reversible process – a stability threshold has been crossed.

Deep-seated landslides disrupt fluvial characteristics of landscapes, such as the drainage network, by shifting huge volumes of rock downslope. Such destruction and disruption of stream channels is still another source of lack of continuity in fluvial

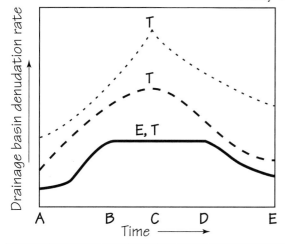

Figure 5.3 Hypothetical threshold-equilibrium plots depicting changes in sediment yield after initiation of uplift at time A. T is a threshold. E is a time span of equilibrium conditions. The top plot shows an abrupt threshold between the domains of increasing and decreasing sediment yield after abrupt cessation of mountain-range tectonic uplift at time C. The middle plot portrays a more gradual change, but the threshold is still a point in time. The lower plot has a period of equilibrium conditions, E, between times B and D where flux of sediment from the watershed remains constant. This time span is also a threshold because it separates domains of increasing and decreasing sediment yield.

Landforms that become progressively larger or smaller are situations of change – not steady state. A delta, sand dune, glacial moraine, or alluvial fan would not exist unless deposition had increased the size of the deposit.

But the interrelations between adjacent deposits may be constant. Stratigraphic boundaries between piedmont deposits derived from different source areas may remain constant (vertical instead of overlapping) with the passage of time. An example already noted (Section 1.2.3) is the vertical stratigraphic contact between adjacent alluvial fans. This was described by Hooke (1968) as steady-state rates of fan accumulation. Once again, using a smaller scale makes it easier to assume steady state. At the large scale, fans grow in volume.

Consider threshold-equilibrium plots depicting several of many possible trends in sediment yield from a watershed in rising mountains that abruptly become tectonically inactive when mountain-building forces shift to a different fault zone (Fig. 5.3). The combination of progressively steeper hillsides and increasing orographic precipitation (Norris et al., 1990; Beaumont et al., 1992; Willett, 1999; Burbank and Anderson, 2001) favors exponential increases of sediment yield between times A and C in the upper plot. Tectonic uplift is assumed to cease abruptly in this hypothetical example, whereupon flux of sediment decreases exponentially. The slope between C and E is not as steep as between A and C because isostatic rock uplift continues after tectonic uplift ceases.

The lower plot is different in that sediment yield remains constant between times B and D, and thus depicts a balance between tectonic and geomorphic variables such that flux of sediment does not change with the passage of time. In the example shown here, the steady-state condition continues past the time of cessation of tectonic uplift into the time span where rock uplift is purely the result of isostatic adjustments resulting from erosional unloading of the crust.

This format of pre-steady-state, steady-state, and post-steady-state conditions can also be applied to small spaces and short times such as the characteristics of a short reach of a powerful stream. For each geomorphic setting, one needs to consider

systems. Mass movements are not a steady-state process and such hillslopes cannot return to a fluvial-erosion steady state if renewed slip occurs. You may not agree with this perspective.

Some geomorphologists use a model where steady-state is achieved by just using a much longer time span. For example, landslides may be a non-steady state process, so define a model where mass movements are small (shallow debris slides and small slumps) and use a time span of 1 My. This should create and maintain an unchanging landscape. Here, by using the Burbank and Anderson (2001) definition, I will show that steady state cannot exist even with such extreme assumptions.

landscape evolution in terms of rates of rock uplift, rock mass strength, weathering and erosional processes, and response times to tectonic and climatic perturbations. Geomorphologists vary greatly in their opinions as to the viable length of equilibrium time spans (from B to D in Figure 5.3), or if equilibrium is ever actually achieved.

Mountainous landscapes may be regarded as forever changing, or as remaining essentially unchanged during a time span that is sufficiently long to allow attainment of a dynamic equilibrium between rock uplift and the geomorphic processes that tend to lower land-surface altitude. Consider three possible scenarios.

1) Uplift accelerates stream-channel downcutting, which steepens hillslopes sufficiently to accelerate denudation by mass movements and flowing water so that erosional lowering of the landscape approximates rock uplift rates – the G.K. Gilbert–John Hack model of dynamic equilibrium.

2) The concentrated power of streams downcuts faster into valley floors than diffusive erosional processes can lower the intervening ridgecrests. Streams in this case have extremely short response times to rock uplift as compared to their adjacent ridgecrests. Relief and slope length increase proportionately to uplift rate at a pace that is largely a function of rock mass strength and annual unit stream power. Some authors class such landscapes as "pre-steady-state mountains" (Abbott et al., 1997; Hovius et al., 1998; Burbank et al., 1999). This term connotes eventual attainment of a topographic steady state.

3) Consider trunk streams that respond to uplift by degrading their valley floors, and then remain at their base levels of erosion for long time spans. Large streams quickly attain and maintain type 1 dynamic equilibrium longitudinal profiles even during continued rapid uplift of a mountain range. They do not downcut further after uplift has ceased because they are already graded to a stable base level. Erosion then lowers ridgecrests towards stationary valley floors. Steady state is never approximated in this scenario of constantly changing landscapes.

Steady state seems more plausible if we move up to orogen-scale landscapes (Pazzaglia and Knuepfer, 2001). A rising crustal block consists of mountains whose slopes become progressively steeper as relief increases. Denudation rates increase exponentially with increase of hillslope steepness (Strahler, 1950, 1957; Ahnert, 1970), and ever-steeper hillslopes accelerate hillslope degradation by landslides (Hovius et al., 1998; Burbank and Anderson, 2001). Steady-state is where the overall flux of sediment yielded from an orogen balances the rock uplift rate. This is the flux steady state of Willett and Brandon (2002). Local areas in mountains may have denudation rates that depart from the norm over short time spans, so we purposely use sufficiently long time spans to smooth out such oscillations.

Erosion rates cannot counterbalance fast uplift rates in extremely arid mountains. But orogen-scale steady state seems plausible for landscapes in extremely humid climatic settings such as Taiwan and New Zealand. The odds improve if a model uses time spans longer than 100 ky to average out the effects of Late Quaternary climatic oscillations. Let us think of this large-scale long-term orogenic flux balance as being a geophysical steady state.

Geophysical-scale attainment of topographic steady state for orogens is conceptually tidy. If "mountain building has been sustained for tens of millions of years, then it may be almost inevitable that the range is in steady state" (Burbank and Anderson, 2001, p. 220). Hooke (2003) concludes that asymptotic approaches to steady state are more likely in humid, lofty mountain ranges with erodible rocks. The requisite time spans to attain approximate landscape steady state are just too long for low hills of hard rock in dry climates. Unfortunately, time and size scales limit the ways in which to test the validity of this most interesting orogen-scale concept.

Workers have turned to low-temperature thermochronometers, such as fission-track dating using apatite and zircon (Fitzgerald and Gleadow, 1988; Kamp et al., 1989; Kamp and Tippett, 1993; Stuwe et al., 1994; Gunnel, 2000; Gleadow et al., 2002) and U–Th/He dating of apatite (Farley, 2000; House et al., 2001). Thermochronometer-based constant erosion rates can be interpreted as representative of orogen-scale steady state (Pazzaglia and Brandon, 2001; Gleadow and Brown, 2000; Batt et al., 2001). Fission-track dating and isotopic analyses are important inputs into many exhumation studies, and will continue to provide vital information about the tectonic evolution of orogens over time spans longer than 1 Ma.

Such coarse approximations of geomorphic-process rates lead to only the most general of topographic conclusions. For example, in the Southern Alps both the presence of residual uplands (Figure 6.5 presented later in Chapter 6) and remnants of flights of marine terraces reveal a recent doubling of rock-uplift rates. This increase occurred at about 135 ka (Bull and Cooper, 1986; Bishop, 1991), but is too recent to be detected by thermochronometers. Indeed thermochronometers are so distantly related to topography that Willett and Brandon (2002) prefer to use the term "exhumation steady state". When used in that context, thermochronometers have much to offer active tectonics investigations.

Given the importance of the debate about steady state, it makes sense to investigate landscape evolution with numerical models. See Tucker and Bras (1998) for a summary of diverse models. Modeling emphasizes how flowing water creates drainage nets, but addition of diffusive hillslope-process terms by Clem Chase allowed development of realistic convex ridgecrests and drainage nets. Many numerical models appear to validate the supposition that landscapes can attain steady-state drainage nets and hillslopes (Koons, 1989; Willgoose et al., 1991; Chase, 1992; Kramer and Marder, 1992; Leheny and Nagel, 1993; Howard, 1994; Stüwe et al., 1994). Although the computer models mimic some aspects of landscapes quite well, some less obvious situations may depart from real-world mountain ranges.

The numerical models simplify one essential aspect – boundary conditions are assumed to remain constant with the passage of time. Hasbargen and Paola (2000) state "Constant boundary conditions inevitably result in stable networks". It appears that numerical models are set up with strong self-arresting feedback mechanisms between stream and hillslope processes. Hasbargen and Paola believe that this "dampens perturbations and drives landscapes to stable forms over longer time scales".

A century of endorsements for time-independent landscape evolution has many geomorphologists believing that mountainous landscapes do indeed attain unchanging topographic characteristics. The Gilbert–Hack hypothesis seems so logical.

Arthur Strahler's (1957) study of valley-side slope angles promotes the idea that systems that adjust their morphology so as to attain a time-independent shape. But have we really shown that hillslopes presumed to be in steady state have a spatially uniform and constant rate of degradation?

Humid soil-mantled hillslopes of the Oregon Coast Ranges seem ideal for quantitative tests of steady-state. Sediment-yield studies indicate erosion rates of 0.05–0.3 m/ky (Beschta, 1978; Reneau and Dietrich, 1991; Heimsath et al., 2001), which is the same as the estimated rock uplift rate (Kelsey et al., 1996) and nearby stream channel tectonically induced downcutting rates (Personius, 1995; Almond et al., 2007). This apparent balance between rock uplift and erosion is suggestive of a tendency toward, or presence of, a steady-state topography (Reneau and Dietrich, 1991; Roering et al., 1999, 2001, 2005; Montgomery, 2001).

Heimsath et al. (1997, 2001) used cosmogenic radionuclides to estimate long-term rates of rock weathering and watershed erosion in this Oregon study area (Fig. 3.29). The results are 0.1 ± 0.03 mm/yr, the same as the Reneau and Dietrich estimate, and although only half the estimated minimum uplift rate, are regarded as suggestive of approximate equilibrium. Models, such as those of Roering et al. (1999) assume a spatially uniform erosion rate and define steady state as when each point in the Oregon study watershed erodes at the same rate.

The potential conceptual pitfall is that it is easy to assume that spatially uniform hillslope erosion rates prove the existence of a static, unchanging landscape. Have the watershed boundaries changed in either numerical models or in real world settings? We need to show that the relative positions of drainage divides and trunk stream channels remain constant in order to conclude that steady state has been achieved.

Note that I initially focus on whether or not the steady-state concept is applicable for individual fluvial systems. I lay aside evaluations of flux, thermal, and exhumational steady states (Willett and Brandon, 2002). As a geomorphologist I prefer to emphasize the more meaningful topographic steady state. By first examining the watersheds that drain a large mountain range, I use a spatial scale that is intermediate between entire mountain ranges and individual landforms within specific fluvial systems. A 10,000 km^2 mountain range is merely an aggregate of many drainage basins. One can assume that regional topographic steady state prevails if the component watersheds have achieved steady-state conditions.

It is dangerous to assume the converse, which is just what appears to have happened. Studies that conclude that steady state has been achieved at the orogen scale, run the risk of being wrong if it is also assumed that steady state also pertains to watershed topography. Demonstration of a balance between denudation and uplift in the geophysical sense – flux, temperature, and exhumation – for a region may have little bearing on topographic evolution in a watershed. Too many earth scientists have taken crustal-dynamics equilibrium to also infer topographic equilibrium.

Whether or not the component landforms of fluvial systems approximate topographic steady-state conditions is considered in the ensuing discussions as we progress from small- to large-scale landscape elements.

1) Longitudinal profiles and other characteristics of streams do indeed tend toward and then fluctuate about a steady-state condition. Straths are maintained as a type 1 dynamic equilibrium landform while degrading at the same rate as rock uplift (Fig. 5.2). The present interglacial climate is a time of strath cutting for many streams. Gradient-index evaluations of long reaches of ongoing strath cutting nicely demonstrate attainment of equilibrium conditions at the longitudinal-profile scale of landforms (Bull, 2007, section 2.6.1).

2) At a larger scale we need to discern if hillslope subsystems maintain dynamic equilibrium concurrently with their adjacent valley floors. Do watersheds ever achieve a synchronous condition where degradation rates, slope length and relief remain constant with the passage of time?

3) Then I appraise the evolution of landscapes of large mountain ranges with different tectonic histories.

I follow the advice of Schumm and Lichty (1965) who reasoned that application of a geomorphic model should be done in the context of a defined space and time. Although I focus mainly on changes of individual hillslopes and watersheds during the past 100 ky, most of my conclusions are just as applicable to changes of mountain-range topography during the past 10 My.

Let us start our inquiry with an example of how terrestrial cosmogenic nuclides can be used to assess possible attainment of local hillslope steady-state conditions.

5.2 Hillslope Degradation

Analysis based on terrestrial cosmogenic nuclides analyses (Gosse and Phillips, 2001; Bierman and Caffee, 2001, 2002) allowed Monaghan et al. (1992) and McKean et al. (1993) to quantify rates of production of soil and downslope creep of a clayey ridgecrest. The study area is in rolling hills northeast of Mt. Diablo (see Figure 2.13A). Present mean annual precipitation is 445 mm and a strongly seasonal Mediterranean climate concentrates rainfall in the winter months to nourish a seasonal, continuous grass cover. Lithology, vegetation, and climate favor diffusive creep of hillslope materials instead of surficial rill wash. This allows use of ^{10}Be as a chronometer to study bedrock-to-soil conversion rates (Fig. 5.4). McKean et al. used concentrations of ^{10}Be (half-life of 1.5 My) in samples collected from a line of three test pits to estimate conversion of soft Eocene shale to a montmorillonite-rich soil at a rate of 0.031 g/cm^2/yr^{-1}. Creep moves clayey soil containing the ^{10}Be down the slope of this 50 m long section of ridgecrest.

Does this hillcrest landform approximate steady-state conditions? The creeping soil mantle is about 135 cm thick, so the estimated downslope transit time (residence time of ^{10}Be in the soil) is about 12 ky. This is much shorter than the ^{10}Be residence time of 100 ky in stable soils not subject to creep (Pavich et al., 1985; Pavich, 1989).

Unruh and Sawyer's (1997) estimates of maximum rates of uplift of active anticlines in the foothills southwest of Mt. Diablo (Fig. 2.13A) range from 0.6 to 1.4 m/ky. So the 0.22 m/ky rate of uplift assumed by Monaghan et al. (1992) and by McKean et al. (1993) seems reasonable for this less active area.

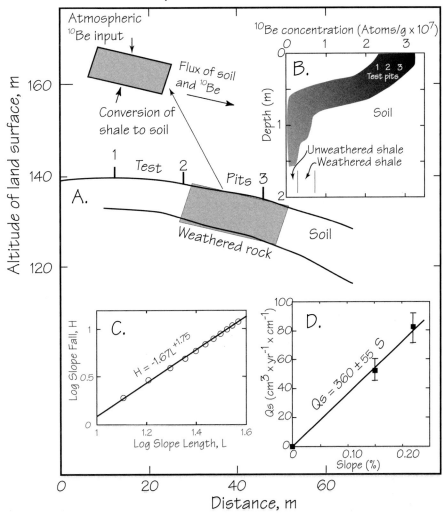

Figure 5.4 Use of terrestrial cosmogenic nuclides to evaluate soil production and downslope flux rates on a ridgecrest underlain by Eocene shale 10 km northeast of Mt. Diablo, California. This figure uses data and graphics from three figures in McKean et al. (1993).
A. Topographic profile of the ridgecrest showing the locations of three test pits from which [10]Be samples were collected. The inset figures summarize the [10]Be soil mass balance model.
B. An overall profile of [10]Be concentration in test pits 1, 2, and 3. The trend of the broad swath of results reflects an increase in [10]Be concentration with depth and also downslope, and is based on duplicate laboratory analyses made on 16 soil samples.
C. This power function regression describes the convexity of this ridgecrest for the distance shown in Figure 5.4A. The tight fit results from using data from a map with a 1 meter contour interval and the exceptionally uniform bedrock, soil, and plant characteristics of this carefully selected study site.
D. Linear regression of percent slope and soil-creep flux rate. The error bars indicate a 15% uncertainty estimate of the experiment's result.

A mass-balance model was used in this ideal ridgecrest study area to test a G.K. Gilbert hypothesis. Soil flux increases downslope at a rate that is proportional to slope steepness and when in dynamic equilibrium the ridgecrest morphology does not change with the passage of time. Is steady state present in this small part of a rising soil-mantled landscape that has an ideal combination of rock type, climate, and geomorphic processes?

Eocene shale weathers to highly plastic clayey soil without noticeable changes in pedogenic characteristics over the 50-meter transect. Soil-profile thickness increases only slightly downslope (Fig. 5.4A). Changes in the concentration of ^{10}Be with depth and down the slope allows assessment of rates of soil-creep flux and the rate at which shale is converted to soil over a time span is estimated to be about 3,500 years.

^{10}Be concentration decreases with depth (Fig. 5.4B). Unweathered shale contains virtually no ^{10}Be but has minor amounts of ^{10}Be where soil has fallen into cracks. The results from each test pit clearly show that ^{10}Be cations have an atmospheric source and are quickly adsorbed onto clay minerals in the soil. Other factors that contribute to higher concentrations of ^{10}Be near the surface include a sufficiently high soil field capacity to easily absorb the occasional 80 mm of rain that may fall during 5 days in a wet winter. One would also expect a decrease with depth in the intensity of soil mixing caused by bioturbation and downslope creep. Soil-flux rates increase downslope, from test pit 1 to 2, to 3, producing a band instead of single line. The fairly broad swath shown in Figure 5.4B is a product of my combining the results from the three test pits.

Uniform soil composition and moisture conditions constrain spatial variations in either the coefficient of diffusion or the rate of conversion of shale into soil. Such uniformity in controlling variables also produces a smooth surficial topography as suggested by the r^2 of 0.999 for the regression of slope fall versus slope length (Fig. 5.4C). An exponent of $+1.75$ for the equation 3.7 power function describes a similarly strong ridgecrest convexity as noted for other tectonically active parts of the Diablo Range. This supports McKean's premise of recent uplift of the study site.

The soil-creep flux rate does indeed increase with slope steepness. The coefficient of 360 in the Figure 5.4D equation is an order of magnitude larger than the mean value of 45 for 34 other hillslopes in California, Oregon, and Washington. Rapid soil flux here may be ascribed mainly to the relatively much greater weakness of these montmorillonite-rich soils when wet, and possibly to large shrink–swell characteristics in the strongly seasonal climate.

Use of the mass-balance model to quantify transport rates of this soil and its atmospherically derived cosmogenic ^{10}Be indicate a systematic behavior of interrelated geomorphic processes. The model predicts a linear increase in creep flux with increasing slope gradient. It does not necessarily prove the Gilbert hypothesis of ridgecrest convexity as being an unchanging state of dynamic equilibrium. The McKean et al. (1993) ^{10}Be data indicate a small downslope increase in the diffusion coefficient and a downslope decrease in bedrock weathering rate.

Two interpretations are possible. Only a 15% difference in ^{10}Be concentration would give results supporting uniform diffusion and weathering rates, and 15% is within the estimated error of this technique. This allows for the possibility of steady-state ridgecrest conditions.

Steady-state conditions are possible for selected soil-mantled ridgecrests and midslopes if it can be shown that these are unchanging surfaces of detrital transport.

Alternatively, non-uniform systematic processes reflect continuing changes in ridgecrest morphology that are in part a function of Late Quaternary rock uplift. Valley floors would be lowered more rapidly than ridgecrests in a nonsteady-state fluvial system. The resulting increase in footslope steepness would propagate up the hillsides. This response to ongoing rock uplift would account for:
1) The observed increase in diffusivity coefficient,
2) The downslope decrease in the rate of production of soil by weathering of the underlying shale,
3) The strong ridgecrest convexity indicative of Late Quaternary increases of hillslope height and steepness (Section 3.4.1).

Steady state may never be proved for entire watersheds in the Coast Ranges of California and Oregon. Even if uplift rate has been constant, we know that Quaternary climate has repeatedly and abruptly changed amounts and styles of precipitation, plant communities, burrowing rodent populations. The rheological characteristics of the soil and rock-weathering and slope-denudation rates would also change. The solution is to model using sufficiently long time spans to average out such oscillations. The 3,500 years interval studied at the Mt. Diablo site is so short that uplift effects can not be separated from possible continuing impacts from the Pleistocene-Holocene climatic change.

Impressive quantitative studies form the basis for the present bias for virtually ubiquitous steady-state hillslopes in the region. Modeling of slope degradation in the Oregon Coast Ranges (Roering et al., 1999) shows a clustering around their estimated constant rate. "However, modeled erosion rates for 30% of our study area differ by more than 50% of the assumed constant rate, suggesting that the equilibrium assumption may be only approximately met" (Roering et al., 1999, p. 868). This is hardly convincing in view that their selection of study watersheds concentrated on the more likely approximations of equilibrium landscapes. Their superb topographic maps showing the spatial distribution of hillslope curvature and gradient for a small watershed (Roering et al., 1999, plate 1) are remarkably similar to the characteristics of my example of a tectonically active Diablo Range watershed (Vigorous Vale of Figure 3.27A, B). Both study areas strike me as being situations where uplift rates exceed denudation rates and watershed relief is increasing.

We should expect midslope denudation rates to be approximately uniform over short time spans in drainage basins where rock uplift, climatic, and lithologic factors are constant. Numerical values describing such spatial uniformity of erosion will change gradually as relief and hillslope steepness change. Such changes are universal at the drainage-basin scale if one accepts the axiom that relief and length of hillslopes are never constant throughout a given fluvial system. Ridgecrests and valley floors comprise such a small percentage of most watersheds that modelers could mistakenly assume that spatially constant erosion rates prevail for an entire drainage basin. This never occurs in the context of more useful longer time spans.

Never ending landscape change is the rule, not the exception. The contrasts depicted in Figure 3.32 are profound. Tectonic base-level fall produces a progressive increase of hillslope steepness from the headwaters to the basin mouth. Nonsteady-state hillslope erosion results in two domains after uplift ceases. Attainment of stream channel type 1 dynamic equilibrium at the basin mouth migrates upstream creating a stable base level to which the hillslopes are graded. Although headwaters reaches may continue to show a downvalley increase in hillslope mean slope (segment A–B), the rest of the basin reverses trend and hillslopes become progressively gentler (segment B–C of Figure 3.32). The most realistic model of landscape evolution is for conditions represented by segment D–E to persist as long as valley floor is being downcut as streams attempt to match uplift rates. Then reversal occurs to create a BC segment, and point B migrates up the A–B–C plot at an exponentially decreasing rate.

My model for explaining spatial variations of hillslope steepness in tectonically active and inactive drainage basins includes much more than changing relief between ridgecrests and valley floors. Entire hillslopes continue to evolve over time spans of 10 ky to more than 1000 ky in the Vigorous Vale and Dormant Hollow study areas in the Diablo Range of California (Section 3.7). Although the basin-position coordinate at which a stream achieves the base level of erosion migrates upstream, it never reaches the headwaters.

Attainment of valley-floor equilibrium conditions moves as a wave up the trunk valleys of the watershed. The hillslopes adjacent to the sequence of stable valley floors also adjust their shapes, but at much slower rates than the valley floors. The plots of convex–straight transitions for Dormant Hollow (Figs. 3.30, 3.31) would be horizontal if all slopes were eroding at the same rate – steady state. Instead, the sloping trends reveal 1) the rate at which the valley floors achieve a stable base level, and 2) continuing changes in hillslope morphology at all basin-position coordinates.

I conclude that steady-state hillslope development remains unproven, even for ideal watershed study sites. Many erosional landforms do not even tend toward a steady state. They, and depositional (growing) landforms, are better modeled and studied in the field using the allometric-change model (Bull, 1975, 1991, section 1.9), which models increase or decrease of landform size.

Future studies of landforms and landscapes should place less emphasis on whether or not steady state is likely. Investigations should include the following essential tasks.

1) Assess the relations between stream channels and the hillslope and larger landscape elements in the fluvial system selected for study. Waterfalls, cliffs, and valley-floor locations of continuing deposition (such as debris fans) may represent unwelcome noise in a study of an otherwise smoothly functioning system.

2) Morphometric analyses of characteristic hillslope shapes in different parts of the drainage basin as defined by hydraulic coordinates. Data for such work may be based on field surveys, contour maps, laser altimetry, and digital images. Characteristic watershed landforms should include average form as well as spatial variations with increasing distance downstream from the headwaters divide.

3) Evaluation of the drainage-net characteristics. Density and order of streams should provide planimetric data to be used in analyses of adjacent hillsides. Stream-channel characteristics, estimated present annual unit stream power, and flood-discharge power, are useful for judging the ability of the trunk stream channel to degrade at a rate equal to that of ongoing rock uplift. Erosional power per unit area and rock mass strength (Selby, 1982a,b; Augustinus, 1992, 1995) determine rates of change.

4) Future studies should make more use of weathering and slope-stability information contained in soil profiles. Soils data allow assessment of possible spatial and temporal variations in sediment-flux rates, and provide information about ranges of ages of Pleistocene and Holocene landforms. A complete hillslope study should include a field survey of soil-profile characteristics, relative ages, and style of soils genesis. Most mountain ranges have a wide range of soil ages.

Distinguishing between Pleistocene and Holocene soils is easy because climates were much different. The result is obvious differences in the geochemistry and genesis of soil profiles. The common occurrence of Middle Pleistocene and late Holocene soils in adjacent areas of a soil-mantled hilly landscape is clearly a case of nonsteady rates of degradation (Bull, 1977). Geomorphologists should seek signatures of surficial-erosion history contained in the soil profiles of ridgecrests, midslopes, and footslopes.

5) Use the soils-inventory information to devise studies using cosmogenic isotopes to better define rates of soil production, surficial erosion rates, and rates and processes of downhill diffusive movement of regolith and overlying soils. For example, see the study by Phillips et al. (2003).

6) Numerical modeling will benefit from all the preceding steps, which identify the more important unresolved problems. Inclusion of more essential assumptions will result in more realistic modeling results. Just including ridgecrest and valley-floor hillslope boundaries that change with time will add an imperative first step.

Maybe we can answer questions such as:

1) Has the stream channel and associated valley floor at the mouth of the watershed achieved the base level of erosion?

2) How far upstream has this base-level control progressed, and how do spatial shifts in the base level represented by equilibrium reaches of the trunk stream channel affect adjacent footslopes?

3) Which parts of the watershed fluvial system adjust quickly and definitively to stable base-level controls and associated feedback mechanisms, and which parts of landscapes are little affected by stable base levels?

4) What are the rates of change and hillslope morphologies adjacent to stable reaches as compared to upstream reaches that are still actively downcutting? What are the rates of lateral erosion induced base-level fall responsible for valley-floor widening in reaches of stable base level, and how does this type of base-level fall influence adjacent footslopes and midslopes? We need to compare and contrast these much different styles – tectonically induced downcutting and lateral erosion induced base-level fall.

6) Are crestslope morphologies different adjacent to stable and downcutting reaches of the trunk stream channel?

7) What are the reaction and response times of drainage divides to changes in midslopes?

Numerical modeling has much to offer, especially when based on complete and diverse field data, and realistic assumptions. The above challenges are not easy, they will require creative and talented earth scientists.

Mountain ranges will never achieve steady state if individual drainage basins are forever changing. The great contrasts in styles of change for individual watersheds such as Vigorous Vale and Dormant Hollow apply to entire mountain ranges. Mountain-range landscapes consist of suites of drainage basins such as these two representative watersheds.

We should further consider the role of active landsliding to better understand how the obvious increase of landslide abundance in rising mountains influences spatial variations of erosion rates. Let us also continue to examine mountain-range evolution over time spans of millions of years.

5.3 Erosion of Mountain Ranges

5.3.1 Southern Alps

Possible attainment of topographic steady state should also be assessed on the scale of large drainage basins or entire mountain ranges, especially where long-term climatic and lithologic characteristics are uniform. Two vastly different mountain ranges are considered here, the Southern Alps of New Zealand and the Sierra Nevada of California.

The Southern Alps of New Zealand are underlain by fractured greywacke sandstone, which has been metamorphosed to schist near the plate bounding Alpine fault on the northwest side of the range and in the Otago district northwest of Dunedin (Fig. 1.9). The humid to extremely humid (2,000 to 15,000 mm mean annual precipitation), mesic to frigid, weakly seasonal climate (Appendix A) maintains powerful rivers in this rapidly rising, young mountain range. Downstream reaches of rivers easily keep pace with uplift rates by quickly eroding to their base levels of erosion after each surface rupture event.

Uplift began only 3 to 5 Ma (Berryman et al., 1992; Batt et al., 2000) and has accelerated with time. Increase of uplift rate perhaps may be linked to shifts in the pole of rotation of the Pacific plate

that increased convergence with the Australian–India plate. Uplift rates for the central section of the Southern Alps are high, being 5 to 8 m/ky (Suggate, 1968, Cooper and Bishop, 1979; Bull and Cooper, 1986; Simpson et al., 1994; Yetton and Nobes, 1998). Most of the uplift of the Seaward Kaikoura Range has occurred since 1 Ma.

The Southern Alps, even on the drier eastern flank, has impressive areas of landslides (Fig. 5.5). Exposures of landslide thickness in road cuts and stream channels in the Crown Range indicate that footslopes have greater thicknesses of landslide deposits than midslopes. This terrain may be as nonsteady state as Jordan Stream hillslopes in the Seaward kaikoura Range (Fig. 4.19).

A key question – does an extremely humid climate favor attainment of steady-state conditions for the Southern Alps? Keller and Pinter (1996, p. 311) would say yes: "The best cases for real world equilibrium mountain ranges are those with very high rates of uplift and precipitation. The Southern Alps … ".

John Adams (1980, 1985) cited the prevalence of sharp, spiky peaks and steep mountainsides as indicative of attainment of steady state for the Southern Alps. His appealing argument seems

Figure 5.5 Mass movement processes dominate this landscape where a humid climate, weathered biotite schist, and seismic shaking favor hillslope degradation by mass movements. Car on highway for scale. Mt. Scott in the Crown Range, South Island, New Zealand.

142   Chapter 5

quite reasonable, so much so that Adam's hypothesis has been adopted by many recent workers for other mountain ranges. Hillslopes are thought to become progressively steeper and therefore more unstable, and I agree with that (Fig. 4.19). But does this accelerate denudation rates everywhere in the component drainage basins of a mountain range? Does landsliding promote spatially uniform denudation rates so that degradation of ridgecrests equals that of the valley floors?

Field evidence at many sites in the Southern Alps clearly shows the opposite is happening. Degradation rates on ridgecrests, midslopes, and footslopes are diverging instead of converging on an areally uniform erosion rate. The Figure 4.19 example has two orders of magnitude variation of surficial erosion.

Equilibrium landforms are present where broad valley floors underlain by straths show that the base level of erosion has been achieved. Such rivers cut down at rates that match uplift rates because exceptionally large annual unit stream power is capable of doing more work than transporting huge amounts of bedload supplied by the many landslides.

Tree-ring analyses of cedar trees growing on a flight of stream terraces along the Karangarua River reveal rapid formation of a new terrace after

Figure 5.6 A stream has incised this valley floor to create landslide-prone hillsides. This 0.3 km² watershed post-dates the ~100 ka marine terrace whose shore-platform remnants are buried under loess on the watershed divides. Kaikoura Peninsula, South Island, New Zealand.

each recent surface-rupture event along the range-bounding Alpine fault (Yetton, 1998). Each coseismic-uplift event causes landslides that promote brief aggradation of the valley floors, followed by renewed stream-channel downcutting, which completes the process of creating a low fill terrace (Bull, 2007, figure 6.18).

Ridgecrest denudation rates in the Southern Alps are equally impressive but for the opposite reason. They are much, much slower. Flat-topped peaks and ridgecrests, and uplands of gently sloping hilly terrain (Fig. 6.5, presented later), have remnants of marine shorelines with sparsely scattered beach pebbles (Cooper and Bishop, 1979; Bull and Cooper, 1986, 1988). Strandlines even occur on the main divide crossing of the Routeburn Track, and near Arthurs Pass 335 km to the northeast (Bull, 1984). These widespread strandline remnants occupy only a minute portion of the Southern Alps landscape, but at many localities they clearly allude to ridgecrest degradation rates that are insignificant. Net degradation is zero where uplifted marine-terrace shore platforms, with quartz beach pebbles are present.

Uplifted marine terraces show that degradation rates have been spatially and temporally variable in the Southern Alps during the past 300 ky. Ridgecrests with beach pebbles are sites where surface uplift is equal to the sum of tectonic and isostatic rock uplift. My preferred landscape-evolution model is not steady state. Instead degradation rates become progressively less with increasing distance away from the 4 to 8 m/ky maximum rates in the valley floors. Abrupt transition to slower erosion rates would be the preferred model upon reaching the level of the inherited stable terrain of former uplands with their remnants of former shorelines (Section 6.2). Two examples illustrate this point.

The first example is hardly alpine, being the highest marine terrace on Kaikoura Peninsula. A small watershed of only ~0.3 km² has been cut into the broad tread of a planar terrace tread that has been raised 105 m and tilted slightly. A detailed study of this terrace flight (Ota et al., 1996) included analysis of the marine fauna preserved just above the shore platform of the terrace shown in Figure 5.6. They concluded that a distinctly cold-water fauna did not date to the time of the last major interglacial warm waters at about 125 ka. Their age estimate of roughly 100 ka included luminescence analysis. This age estimate was used to assign ages for four younger marine terraces on the premise that sea-level high-stands were synchronous with those of the much studied flights of global marine terraces on New Guinea. Persons who prefer to include two additional younger terraces on Kaikoura Peninsula would have an older age estimate for this highest terrace, perhaps about 180 ka. In any case, study of post-marine terrace drainage basins provide a way to estimate rates of hillslope and valley-floor degradation.

Only nontectonic changes in landform altitude are considered in Figure 5.6. The ridgecrests of the small watershed have been rising because there has been net accumulation of eolian loess on these gently sloping, slightly convex ridgecrests. Maximum denudation rates occur along the valley floor, which has downcut sufficiently to create unstable adjacent footslopes. The result is many landslides. Obviously, steady-state conditions do not exist here. Instead, mean hillslope degradation rates for the past 100 ky increase progressively from the ridgecrest minimum to the valley floor maximum. This is obvious, but do such coastal landforms represent landscape evolution high in the Southern Alps of New Zealand?

Rugged hillslopes of the Southern Alps (Fig. 5.7) rise above treeline only 18 km north of the watershed of Figure 5.6. The Seaward Kaikoura Range is rising much faster than Kaikoura Peninsula, but marine-terrace remnants are present on some ridgecrests. An example is shown in the banner photo on the first page of this chapter. Evidence for these flat ridgecrests being of marine origin include remnants of shore platforms and degraded sea cliffs, and sparse beach pebbles. Even the modern shore platforms are largely swept clean of beach gravel (Bull, 2007, figs. 2.7, 2.9). Beach pebbles are not common on the marine-terrace remnants of the Seaward Kaikoura Range because there is much less quartz here than on the schistose western flank of the Southern Alps. Weathering removes other lithologies quickly in this humid climate.

The flat top of Mt. Stace in the foreground of Figure 5.7A provides a dramatic contrast with the general character of the rugged mountains in the background of the view. Another remnant of the same marine terrace 1.5 km to the east is just as striking (Fig. 5.7B) and beach pebbles were collected on the Hill 1168 marine-terrace remnant (Fig. 5.7C). These two marine-terrace remnants have been raised to the same altitude, perhaps because this former shoreline is parallel to the nearby active Kowhai thrust fault.

One has to be careful that such pebbles are not crop stones from one of the 11 species of moa. These flightless ostrich-like birds ranged in height from 1 to 4 m and recently became extinct. The pebbles shown in Figure 5.7C have the shape and impact marks of being thrashed on a beach. Yes, moa could have picked up such stones. Gentle grinding in a moa's crop to break down coarse vegetation imparts a definitive sheen to the surfaces, like that of rocks tumbled in a rock polisher. With time, the impact marks on crop stones become less obvious but the fractures caused by being caught between larger rocks in a streambed or on a beach remain. Hapuku river pebbles are subangular, not rounded like the beach pebbles shown in Figure 5.7C.

Uplift rates for this part of the Seaward Kaikoura Range appear to be about 5 m/ky (Wellman, 1979; Bull, 1984, 1985). Sufficient marine-terrace remnants are present here to allow altitudinal spacing analysis that matches the spacing of terrace heights with the same spacing at another South Pacific island where each marine terrace can be dated – New Guinea (procedure is discussed later in Section 8.3.1). This terrace appears to have been formed at the 215 ka sea-level highstand.

Ridgecrest degradation since ~215 ka has been minimal, much like the Kaikoura Peninsula example of Figure 5.6. But here the valley floor of the downcutting stream is 650 m lower than at the time of a continuous shoreline between Mt. Stace and Hill 1168 (Fig. 5.7D). Hillslope degradation is much less than valley-floor degradation because unit area power of flowing water to erode bedrock, and shift landslide debris, is much smaller than near the trunk stream channel of the Hapuku River watershed. It is wishful thinking to conclude that such landscapes represent steady state over the long time spans represented by this view high in the Southern Alps.

Ridgecrest degradation will never match tectonically-induced stream-channel downcutting in

Figure 5.7 Remnants of marine terraces preserved on hilltops above treeline in the Seaward Kaikoura Range of the Southern Alps, New Zealand.
A. Flat summit of Mt. Stace, altitude 1167 m.

Figure 5.7 Remnants of marine terraces preserved on hilltops above treeline in the Seaward Kaikoura Range of the Southern Alps, New Zealand.
B. View towards the north of the flat summit of Hill 1168. A closer view of the outcrop below this shore platform is shown in Figure 4.2A.

Figure 5.7 Remnants of marine terraces preserved on hilltops above treeline in the Seaward Kaikoura Range of the Southern Alps, New Zealand.
C. Beach pebbles collected from the summit of Hill 1168. Scale on the tape is in mm.

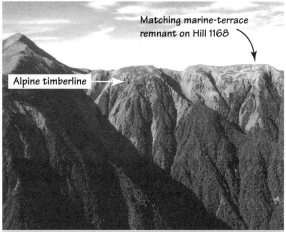

Figure 5.7 Remnants of marine terraces preserved on hilltops above treeline in the Seaward Kaikoura Range of the Southern Alps, New Zealand.
D. View from Mt. Stace towards Hill 1168 to the east. The Hapuku River is far below this view, 650 m lower than the skylined ridge.

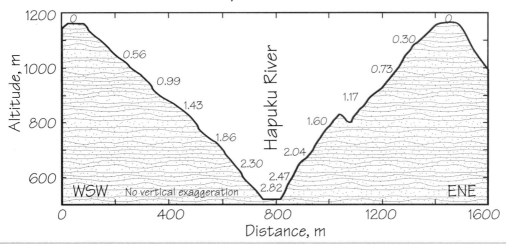

Figure 5.8   Mean rates of hillslope denudation along a cross-valley section between Mt. Stace (at distance of 50 m) and Hill 1168 (at distance of 1450 m). The valley floor of the Hapuku River is the fastest eroding part of this rising landscape. Denudation rates, in m/ky, decrease with distance from the active channel and net erosion on both ridgecrests since ~215 ka is zero.

these rising mountains (Fig. 5.8). Even where erosion from converging hillslopes has eliminated a planar (marine-terrace remnant) ridgecrest, the rate of ridgecrest lowering only changes from zero to the very slow rate of the adjacent crestslopes. Watershed relief becomes greater and hillslopes steeper in the Southern Alps of New Zealand – this model applies to rising mountains in general.

A larger river would have cut down closer to sea level and the maximum degradation rate would exceed 4 m/ky. This fairly small stream is close to its headwaters so it cannot keep pace with the 5 m/ky uplift rate. Strath terraces do not occur here. The apparently broad valley floor suggested by Figure 5.8 reflects the highly braided stream-channel pattern needed to shift enormous quantities of boulders downstream. Straths are common in downstream reaches of the Hapuku River, after large tributaries have greatly increased unit stream power.

Tectonically induced downcutting of trunk stream channels steepens the adjacent footslopes in the Southern Alps, making them progressively more unstable. Denudation of ridgecrests lags so far behind valley floors that ridgecrest spreading occurs. Ridgecrest spreading landforms – ***ridge rents***

– unequivocally indicate continuing nonsteady-state conditions. Rapid stream-channel downcutting creates such unstable hillsides that upslope-facing scarps are formed. Slowly degrading ridgecrests spread when lateral support is reduced beyond a threshold value. Both Mt. Stace and Hill 1168 have ridge rents. See Section 6.2 for a complete discussion of this subject.

The slopes of the Seaward Kaikoura Range tend to be uniform (Figs. 4.2B, 5.8) instead of having convex crestslopes and concave footslopes. This suggests that shallow mass-movement processes promote collapse of the highly fractured greywacke sandstone and argillite. Such uniformly steep topographic slopes suggest that mass movements here are mainly earthquake triggered (Densmore and Hovius, 2000). The steepness of these uniformly sloping valley sides might be a function of how wet the climate is (Carson, 1976; Gabet et al.; 2004).

A threshold-slope model seems to apply here, where erosion is less likely to steepen large areas of hillslope beyond a critical inclination. Threshold slopes (Carson and Petley, 1970) have been reported in many places, including the rising mountains of the Andes (Safran et al., 2005), Himalayas (Burbank et al., 1996a), the Tibetan Plateau (Whipple et al., 1999), and the Olympic Mountains in the USA state of Washington (Montgomery, 2001).

Mass movements in the Seaward Kaikoura Range are not along deep-seated failure planes. The only obvious slump in greywacke is near Barretts Hut (fig. 6.7 in Bull, 2007). Rockfalls are ubiquitous and shallow soil and debris slides are easy to find. This bedrock washes and slides away in little bits instead of as large, thick masses. Frequent large rainfalls cause erosion events that promote landsliding, which results in a high sediment yield.

Study watersheds 400 km to the southwest are much different. Metamorphism of fractured greywacke to phyllite and low-grade schist has increased rock mass strength sufficiently that mountains fail as deep-seated landslides (Korup, 2006).

Rapid incision by the Hapuku River steepens adjacent hillslopes until they fail by shallow landsliding. Steep hillsides expand and now have reached the point where only isolated remnants of marine terraces remain. In a study of the Himalayas, Burbank et al. (1996a) concluded that such stream-channel downcutting creates hillslopes that have reached a threshold steepness determined by their rock mass strength (see equation 4.5 discussion). It would seem that slope height and relief are limited by rock mass strength (Schmidt and Montgomery, 1996; Montgomery and Brandon, 2002).

Korup (2008) used the regional similarity of rock mass strength of the Torlesse greywacke to study threshold slopes in New Zealand mountains with different climates and uplift rates. I summarize his results, using the tangent of slope steepness (tan β) to compare three of his six study areas, which include the Seaward Kaikoura Range. The west-central Southern Alps site has a mean annual precipitation of 6 to 15 m, a rock uplift rate of 5 to 8 m/ky, and a density of recent landslides of 8.3 $\pm$ 4.7 per km². The Seaward Kaikoura Range has a mean annual precipitation of 2 to 5 m, a rock uplift rate of 4 to 6 m/ky, and a density of recent landslides of 8.9 $\pm$ 2.5 per km². The rugged Raukumaras are near East Cape in the North Island. They have a mean annual precipitation of 1.5 to 4 m, a rock uplift rate of 1 to 2 m/ky, and a density of recent landslides of 8.0 $\pm$ 4.1 per km².

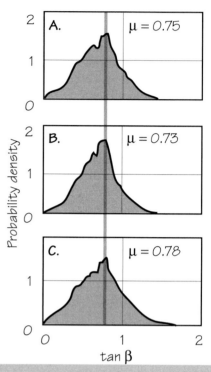

Figure 5.9 Probability density plots of slope-tangent values for New Zealand mountain ranges underlain by fractured greywacke sandstone. Thick vertical grey line is the mean tan β value for all six mountain ranges in figure 1 of Korup (2008).

A. Raukumara Range   37° 44' S, 177° 52' E.
B. Seaward Kaikoura Range 42° 15' S, 173° 38' E.
C. Western Southern Alps 43° 7' S, 170° 54' E.

The probability density plots of Figure 5.9 describe similar patterns of slope steepness. The means of slope steepness are virtually the same, mean tan β ranges from 0.73 to 0.75. The vertical gray line shows the modal averages for all six study areas and describes a regional threshold slope. Attainment of threshold slopes does not imply steady-state hillslopes, as is clearly demonstrated in the previous discussions of the Seaward Kaikoura Range in this chapter. Constant rock mass strength of these study areas appears to be the key controlling variable, overriding local differences in climate and uplift rate.

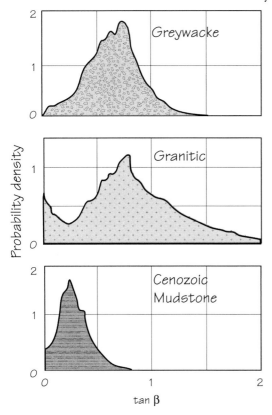

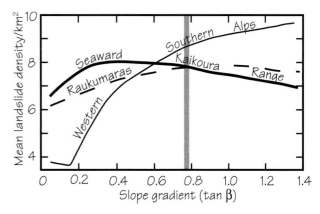

maximum at ~0.45, much lower than the ~1.1 maximum for the western Southern Alps study area. The Raukumara Range has an intermediate relation of slope–landslide density.

One might expect a higher frequency of landslides with increasing slope steepness in the Seaward Kaikoura Range. Korup notes that these steeper slopes have little remaining potential for soil and debris slides. A decrease in landslide abundance would also result if steeper slopes were associated with larger landslides.

The younger, upper parts of the Hapuku hillslopes (Fig 5.8) expanded laterally since ~200 ky and the gorge became ever deeper. Channel downcutting rates have been less than uplift rates. This nonsteady-state landscape changed at a threshold slope determined by the rock mass strength of greywacke and argillite.

So why does valley-floor downcutting fail to cause a self-arresting feedback mechanism leading to steady-state drainage basin landscapes? We need to take into account the many orders of magnitude difference in rates of denudational processes on different parts of the landscape. Response times to uplift may be so long that downcuttng has ceased in valley floors by the time that ridgecrest denudation rates fully adjust to tectonic base-level fall.

Different rock types have a variety of probability density plots (Fig. 5.10) with differing ranges and peak values. When compared to greywacke, mountains underlain by granitic and gneissic rocks have a broad, dispersed pattern of slope steepness. Weak, massive rock types have less dispersion of slope-steepness data. Soft mudstone and sandstone of Cenozoic age has the sharpest peak, and a much lower threshold-slope mode. Rock mass strength controls the equation 4.5 internal angle of friction and landsliding threshold slopes.

The density of recent landslides varies with slope gradient, but the patterns vary greatly between greywacke study areas; three are shown in Figure 5.11. The Seaward Kaikoura Range has a broad

I conclude that the Southern Alps presently are far removed from, and will never achieve, steady-state conditions. Threshold slopes are ubiquitous and may record large areas of uniform erosion rate. Steady state is not present because relief and slope length continue to increase in response to rock uplift. Marine-terrace remnants, ridgecrest-spreading landforms, and characteristics of footslope and valley-floor landforms convincingly underscore this conclusion. Nevertheless, assertions of probable steady state continue to be made for mountain ranges such as the Southern Alps.

One would expect a reversal of the Southern Alps style of differential landscape denudation after cessation or reduction of uplift. The reverse case for tectonically inactive terrains would be for ridgecrests to be lowered faster than the adjacent valley floors, which would maintain their position in the landscape by remaining at the base level of erosion. Neither landscape-evolution model ever approximates a steady state.

5.3.2 Sierra Nevada and Appalachian Mountains

We shift our attention from the young Southern Alps of New Zealand to two old mountain ranges in North America – the Sierra Nevada of California, and the Appalachian Mountains of the eastern United States. Our objective here is to see if Cenozoic-length time spans are long enough for attainment of landscape steady state. The Appalachians have been subject to continuing isostatic uplift in response to gradual erosion of this passive margin. The Sierra Nevada were persistently dormant until they were raised during a sudden uplift pulse at ~4 Ma that was caused by delamination of the batholithic root. For drainage basins in both mountain ranges, we would like to see if hillslope length and height have remained constant. The Sierra Nevada background presented in Section 1.2 is used here and especially in Chapter 8, which is devoted to the subject of the effects of upwelling asthenosphere on landscape evolution.

The imposing Sierra Nevada is much different than the younger Southern Alps and eastern margin of the California Coast Ranges. Mesozoic intrusion of batholiths of the California arc (Dickinson and Snyder, 1978) created mountains as lofty as the Andes of South America. Prolonged denudation created a terrain of low relief by 10 Ma (Lindgren, 1911; Christensen, 1966; Huber, 1981; House et al., 1998). About 350 km of the present range crest is higher than 3,000 m, due in part to rejuvenated uplift of a residual mountain range of modest relief in the north and substantial paleorelief in the south (Wakabayashi and Sawyer, 2001). Studies of landscape evolution of the Sierra Nevada are complicated by Plio-Pleistocene uplift and glaciation. Here, we are interested mainly in how the fluvial systems were functioning before these perturbations. Large rivers draining the western slope during this tectonically inactive long time span remained at their base levels of erosion until ~4 Ma.

My simple model for this 50 My time span of tectonic inactivity would be slow ridgecrest degradation that would gradually decrease relief while large rivers remained at their base levels of erosion. Both the huge area and long time span promote crustal isostatic adjustments. Minor valley floor downcutting would continue in response to isostatic uplift resulting from overall denudation of the range. Uplands and ridgecrests would be the landforms of maximum denudation. They would be progressively lowered towards the relatively stable valley floors that gradually became wider. Rates of landscape change may have been minimal in the foothills as compared to the higher relief and steeper topography closer to the range crest. This passive style changed abruptly with renewed uplift at ~4 Ma. Fortunately, a geochemical test reveals where erosion was fastest.

Thermobarometers are able to evaluate long-term trends in landscape evolution. A U–Th/He isotope thermobarometry analysis by House et al. (1998) assesses mineral closure cooling temperatures as a way of estimating rates of long-term denudation of the Sierra Nevada. Their study evaluates 70 million years of erosion of the Sierra Nevada, including the effects of late Cenozoic rejuvenation of the mountain range. Their study (Figure 5.12) reveals a twofold variation in apparent ages that suggests overall rates of Cenozoic degradation of ridgecrests were twice those of the intervening valleys. These apparent ages would all be the same if the Sierra Nevada landscape was characterized by steady-state

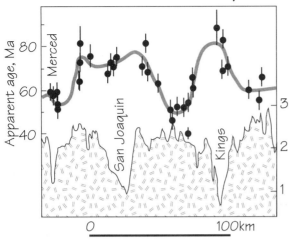

Figure 5.12   (U–Th)/He ages along a range-parallel transect at 2,000 m altitude in the Sierra Nevada, California. Topographic profile, with large vertical exaggeration, along the transect shows the locations of major river canyons. Replicate analyses of samples of inclusion-free apatite from granitic plutons with mean ages ranging from 44 to 85 My. General trend through the sample points is shown by broad line. From figure 3 of House et al. (1998).

denudation rates during the Cenozoic – spatially uniform rates of exhumation. I presume that post-4-Ma tectonically induced canyon incision would diminish this contrast. Ridgecrests may have been lowered ten times faster than the valley floors prior to tectonic rejuvenation of the mountain range.

This example of nonsteady-state denudation has relative rates of lowering of ridgecrests and valley floors opposite those of the Southern Alps. I presume that a Southern Alps style of landscape evolution prevailed in the late Mesozoic. Then ridgecrest denudation became dominant as Sierra Nevada rivers approached their long-term base levels of erosion. The style of landscape evolution since 4 Ma has been similar to the Southern Alps as ridgecrests became ever higher above adjacent rapidly incising inner gorges (Section 8.2).

Such nonsteady-state denudation may continue for more than 50 My, especially in the headwaters of streams. The next example is a sequence of stream captures that increased drainage-basin area for a dominant stream in the Valley and Ridge Province of the Appalachian Mountains in the eastern United States. Stream capture in the Santa Catalina Mountains of Arizona (Section 2.3.1) is another example of nonsteady-state evolution in a tectonically inactive landscape.

The characteristic topography of the Appalachian region during the Cenozoic is the result of nonsteady-state erosion of linear belts of rock. Resistant ridgecrests have tended to become progressively higher than adjacent valleys during the late Cenozoic. The Shenandoah River evolved into a dominant stream because it joins the Potomac River just upstream from the resistant sandstone strata of Blue Ridge (Judson and Kauffman, 1990). The powerful Potomac River was able to cut down through the resistant sandstones of Blue Ridge.

The erosional base-level fall migrated along the softer carbonate rocks beneath the Shenandoah River and promoted extension of the steepened headwaters divide. One after another, small streams that crossed, but lacked the unit stream power to cut down through Blue Ridge, were captured by the advancing headwaters of the Shenandoah River (Judson and Kauffman, 1990).   The abandoned stream channels of the captured streams remain as wind gaps notched into the sandstone ridgecrest. *Wind gaps* (abandoned stream channels notched into a ridgecrest) were used as the key evidence for the sequence of stream captures. Each wind gap marks where additional flow was diverted into the ever more powerful Shenandoah River. Faster denudation of the relatively less resistant valley-floor rocks increased local relief so the wind gaps are now far above the present Shenandoah River. Stream capture in a tectonically passive setting results in nonsteady-state landscape evolution.

Accelerated downcutting by the Potomac River (Bierman et al. 2004) implies departure from a previous base level of erosion. Have parts of the Appalachian Mountains been raised in the recent geologic past? Does continuing isostatic rebound of the landscape favor non-steady state landscape evolution? Also, a Potomac River headcut may have migrated upstream as a result of a nontectonic base-level fall. One possibility would be upstream headcut migration after the river dropped into a submarine canyon during a sea-level lowstand (Bull, 2007, p. 67). The causative perturbation is now below sea level, but the subaerial base-level fall continues to migrate upstream.

As in the case of the Diablo Range streams in California (Fig. 3.8), lithologic control of erosion rates has a profound affect on landscape evolution. Differential erosion rates create situations of base-level fall and stream capture. Steady state is never attained in such situations.

By considering diverse data from landscape elements ranging in size from watersheds to large mountain ranges it is obvious that geomorphic steady state is never achieved. Spatially uniform degradation of the midslopes of hills is possible where lithology is uniform. Reaches of streambeds can achieve dynamic equilibrium, and some concave footslopes may be equilibrium surfaces of detrital transport. But geomorphic steady state should continue to be defined in terms of entire hillslopes, and of fluvial systems. Most mountainous landscapes are better modeled as situations of changing dimensions and shapes. Allometric-change models may be better suited for such studies than models that assume steady state.

5.4 NonSteady-State Erosion of Fluvial Systems

Let us further examine how landscapes are forever changing. The positions of the two hillslope boundaries – ridgecrests and stream channels – change with the passage of time in the young Southern Alps and Diablo Range foothill belt, as well as in the old Appalachians and Sierra Nevada. Heights of ridgecrests above adjacent valley floors are either increasing or decreasing in all drainage basins.

How does such nonsteady-state behavior affect geomorphic processes and landscape evolution? A book should be written on just this topic. I will illustrate my nonsteady-state approach here by

assessing where sediment is coming from in tectonically active and inactive fluvial systems. The data input consists of constrained estimates – no matter, because the intent here is just to better understand how uplift rate affects how fluvial systems operate.

I will assess spatial variation of sediment yield in tectonically active and inactive drainage basins. Two watersheds, Vigorous Vale and Dormant Hollow (Section 3.7), in the Diablo Range of central California are well suited for this comparison.
1) Which parts of these markedly different landscapes yield most of the sediment to their respective trunk streams?
2) Are some elements of the landscape eroding faster because of tectonic uplift?
3) Are some landscape elements yielding about the same amount of sediment regardless of uplift rate?

Vigorous Vale is mostly convex hillslopes, whereas Dormant Hollow is mostly concave footslopes. My analysis uses the areas of different hillslope topographic elements in each watershed, together with estimates of denudation rates. Tectonic controls on spatial variations in relative denudation rate are summarized in Figure 5.13. My impression is that the overall denudation rate for the Vigorous Vale watershed is 1.0 m/ky – roughly two-thirds the rock uplift rate. For tectonically inactive Dormant Hollow the estimated denudation rate is 0.4 m/ky. The analysis results would not change if these crude approximations were changed. It is more important that you know how I estimated the denudation rates for hillslope elements in different tectonic settings. Chapter 3 hillslope morphology models are useful here.

Uplift affects ridgecrest morphology by making summits more convex and steeper. Unfortunately we know little about the time for a ridgecrest to react and respond to a stream-channel downcutting event, and how much of that perturbation is eventually transmitted to the ridgecrest. We do know that some ridgecrests in the Southern Alps have hardly changed, despite being in a rapidly rising mountain range with impressive rates of stream-channel downcutting (Figs. 5.7, 5.8).

I take a less extreme position for Vigorous Vale ridgecrests; eroding slowly but lagging far behind concurrent slope changes on the midslopes

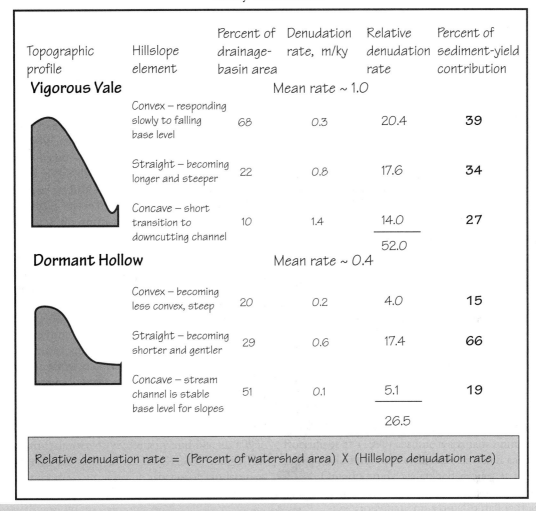

Topographic profile	Hillslope element	Percent of drainage-basin area	Denudation rate, m/ky	Relative denudation rate	Percent of sediment-yield contribution
Vigorous Vale			Mean rate ~ 1.0		
	Convex – responding slowly to falling base level	68	0.3	20.4	**39**
	Straight – becoming longer and steeper	22	0.8	17.6	**34**
	Concave – short transition to downcutting channel	10	1.4	14.0 ____ 52.0	**27**
Dormant Hollow			Mean rate ~ 0.4		
	Convex – becoming less convex, steep	20	0.2	4.0	**15**
	Straight – becoming shorter and gentler	29	0.6	17.4	**66**
	Concave – stream channel is stable base level for slopes	51	0.1	5.1 ____ 26.5	**19**

Relative denudation rate = (Percent of watershed area) X (Hillslope denudation rate)

Figure 5.13 Influence of the distribution of convex, straight, and concave hillslopes on denudation rates in two Diablo Range, California watersheds with similar climate and rock types, but with contrasting uplift rates. Values of relative erosion rate for a hillslope element is equal to the sediment flux from that type of hillslope multiplied by its proportion of total watershed area.

and footslopes where denudation is much faster. My impression is that the ridgecrests are being denuded rather slowly at about 0.3 m/ky, which means that local relief is increasing with the passage of time because the valley floors are being lowered at a much faster rate. The opposite situation prevails in Dormant Hollow where tectonic inactivity should favor ridgecrests that are less convex and not as steep. This should be conducive for slow denudation. My estimate for ridgecrest denudation is only

slightly lower though, 0.2 m/ky, because midslopes have remained steep.

Midslopes in Vigorous Vale become both steeper and longer as continuing uplift increases local watershed relief. Midslopes in Dormant Hollow either decrease slightly in steepness or remain the same because of continuing lateral erosion induced base-level fall. Substantial erosion of these steep slopes should mean that this hillslope element is a major source of sediment. My midslope denudation rate estimates are 0.8 m/ky for Vigorous Vale and 0.6 m/ky for Dormant Hollow.

The largest contrast in erosion rates between tectonically active and inactive drainage basins is along the trunk stream channels and the adjacent footslopes. Rapid downcutting in Vigorous Vale does not favor development of concave footslopes. Footslopes subject to continuing base-level fall amount to little more than junctions between the valley floor and adjacent straight footslope. Even if the tectonic setting favored broader, concave footslopes, they would be steeper than in a tectonically inactive setting (Section 3.5.1.1). My impression is a footslope denudation rate of about 1.4 m/ky, which will be maintained if uplift continues at 1.5 m/ky.

Dormant Hollow represents the opposite condition, with most of the drainage basin having been at the base level of erosion for more than 1 My. Footslopes graded to a stable base level are essentially surfaces of transportation, but I suspect that some erosion is occurring near the junction with the midslopes. So I assign an average footslope denudation rate of 0.1 m/ky. Landscape change is so slow along headwaters tributary streams that they have yet to achieve the base level of erosion. This means that a small part of Dormant Hollow behaves in a manner more like that of Vigorous Vale.

Sediment yield is a function of both spatial variation in erosion rates on different types of hillslopes, and on the percentages of a watershed that are composed of convex, straight, and concave slopes. The low denudation rate for convex slopes in Vigorous Vale is offset by the fact that two-thirds of the watershed consists of such slopes. Footslopes and stream channel may be eroding very fast but comprise only a small proportion of the fluvial system. This tectonically active drainage basin has roughly similar relative denudation rates for the convex, straight, and concave hillslope areas.

In contrast, sediment yield from tectonically inactive Dormant Hollow is mainly from midslopes, whose relative denudation rate is about four times that of either the convex or concave slope areas. Thinking about landscape evolution in a non-steady-state manner reveals the profound difference between the behavior of the tectonically active and inactive terrains.

Summary

John Hack (1965, p. 5) admired Gilbert's (1877) model: "The principle of dynamic equilibrium states that when in equilibrium a landscape may be considered a part of an open system in a steady state of balance in which every slope and every form is adjusted to every other". The Gilbert–Hack steady-state model accounts for spatial variations in rock mass strength by progressive changes in steepness of hillslopes. Burbank and Anderson (2001, p. 145) envision such steady-state landscapes as where denudation lowers ridgecrests and valley floors at the same rate. Dynamic equilibrium is so appealing, and mathematical models that assume the constraint of steady state are much easier to format. Each of us needs to consider how universal such attainment of steady state might be.

Some landforms are created by processes that promote a tendency towards an unchanging configuration with the passage of time. The longitudinal profile of a bedrock strath, once attained, retains a sufficiently steep valley-floor surface to convey the average flux of bedload. As such it maintains a constant position in a rising landscape, and remains at the same altitude in tectonically inactive settings. Equilibrium straths in tectonically active settings degrade at a rate equal to the rate of uplift – a nice example of steady state. Rivers with large unit stream power are more likely to achieve this steady state. Soil-mantled footslopes created by flowing water may also achieve a balance between slope steepness and concavity, where the local flux of sediment and the base level of the adjacent valley floor is constant. A sufficiently long time span and soft rocks make this more likely.

The variables of time and rock mass strength limit achievement of steady-state conditions for larger landforms, and for entire drainage basins. Are rates of landscape change sufficiently rapid to adjust as tips of active faults continue to propagate? Steady-state models of fluvial systems need long enough time spans to even out the large Late Quaternary variations in climate (Whipple, 2001). Steady-state mountains are even more unlikely where erosional processes alternate between fluvial and glacial, or between fluvial and eolian. Steady state also seems unlikely where horizontal tectonic motion is affecting the macro-geomorphic form of mountain belts (Willett et al., 2001).

There is a limit to the effectiveness of increasing the rate of hillslope erosion by making it steeper. Relatively harder rocks often become cliffs whose dryness makes them even more resistant to weathering.

Slopes dominated by threshold processes cannot achieve a time-independent configuration. Landslides are a common example where hillslope materials are either stable (lacking failure planes) or are moving downslope. They can never be a steady-state landform. Hillslopes are driven still further

from a fluvial-erosion steady state if renewed slip occurs. Such destruction and disruption of stream channels, and sources of sediment, creates a lack of continuity in fluvial systems.

Chapter 5 introduced Part III with a distinct geomorphic flavor. The next three chapters are much different. Chapter 6 looks at how erosion influences styles of displacement on geologic structures. Chapter 7 examines how propagation of active faults affects geomorphic processes and landforms. Chapter 8 has a geophysical flavor – landscape evolution as affected by upwelling of asthenosphere.

Chapter 6

Erosion and Tectonics

Early work in tectonic geomorphology focused on how geomorphic processes and landforms responded to tectonic base-level changes. Such work included obvious tectonic landforms such as triangular facets and fault scarps. Then signatures that reveal rates of tectonic deformation were deciphered in the hills and streams of fluvial landscapes. This productive subject, as summarized by Merritts and Ellis (1994), is the main focus of this book.

Brilliant recent work gives tectonic geomorphology additional dimensions. Erosion that creates landforms can affect styles of faulting and folding. Drainage nets might change as faults propagate. Computer modeling and geophysical studies have opened fascinating, diverse opportunities. Chapter 6 illustrates the importance of removal of crustal mass on landscape at scales ranging in size from a ridgecrest to crustal plates.

The approach here is to use geologic structures of progressively larger size and time spans of 10 ky to 50 My. Exfoliation joints is the first topic, then ridgecrest spreading. Erosion also controls fault zone partitioning. At the continental scale, erosion

> Erosion of rising mountains creates fractures and ridgecrest rifts. Such unloading may also affect range-bounding faults and lithospheric upwelling. This view is of rapidly eroding Uwerau and Manakau in the Seaward Kaikoura Range of New Zealand.

Tectonically Active Landscapes, 1st edition. By W. B. Bull. Published 2009 by Blackwell Publishing, ISBN 978-1-4051-9012-1

Figure 6.1 Exfoliation joints in granitic rocks of a glaciated valley, South Fork of the Kings River, Sierra Nevada, California. A. Widely spaced joints parallel the hillslope where exposed at the mouth of a tributary stream valley.
B. Smooth exfoliation-joint surface, **S**, of this granitic hillslope was exposed by mass movements that deposited the talus blocks in the foreground.

induced by long-term plate collision affects the influx of crustal material into the Himalayan orogen.

6.1 Exfoliation Joints

G.K. Gilbert (1904) proposed the idea – controversial at that time! – that Pleistocene glacial erosion was responsible for the formation of large exfoliation joints and sheets that parallel the surfaces of granitic domes and valley sides of the Sierra Nevada of California. Pleistocene glaciers created U-shaped valleys whose shape implies removal of footslope rock that supports hillslopes. Melting of the ice removed more of the lateral support for these cliffs of massive batholithic rocks.

     The structural response after retreat of the valley glaciers was development of tension fractures normal to the minimum compressive stress (Holzhausen, 1989). Prominent sheeting that parallels cliffy hillslopes (Fig. 6.1) played an important role in the style, frequency, and magnitude of mass-movement processes during the Holocene. Electronic surveying and rock mechanics analyses by Bahat et al. (1999) indicate that composite fractures result from merging of fan-shaped cracks and plumes only a few meters across. The stress that caused initial exfoliation of Half Dome in Yosemite National Park may have been approximately 0.1 to 0.5 MPa. Cracks propagate parallel to the cliff face from such nucleation zones. Longitudinal splitting and buckling processes result in exfoliation cracks that extend for hundreds of meters (Fig. 6.1B). This process reduces hillslope stability resulting in mass movements ranging from rockfalls to rock avalanches. More rocks fall during earthquakes in glacially scoured valleys, such as Yosemite, than in the glaciated range crest (Bull, 2007, figs. 6.37, 6.38, 6.48A)

6.2 Ridgecrest Spreading

Gravity induced rock-failure processes, and the resulting geologic structures and landforms, are quite different where rocks lack the cohesive strength of massive batholithic rocks. Rapid deepening of valleys eroded into crumbly, fractured rocks by fluvial or glacial processes also reduces lateral support. This may cause ridgecrests to collapse, forming uphill fac-

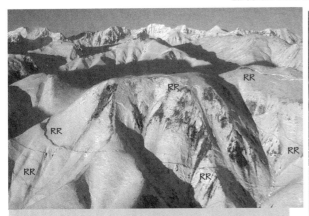

Figure 6.3   Double-crested ridge at right center and foreground. Two antislope scarps, **AS**, in the grassy area at upper left; Aspen Highlands Ski Area, Colorado, USA. Ridgecrest rock is sandstone and shale of the Maroon Formation. Photograph courtesy of McCalpin and Irvine (1995).

Figure 6.2   Prominent antislope scarps near Gillett Pass in the snow-capped Alaska Range were formed, or reactivated, during the 3 November 2002 magnitude Mw 7.9 Denali earthquake (Eberhart-Phillips et al., 2003). Rock types include fine-grained schistose sedimentary rocks and volcanic rocks of Devonian and older age that are sufficiently fractured that this rugged landscape lacks bold outcrops. This 1980 m high mountain is between the main trace and a small splay of the Denali fault. These ridge rents, **RR**, are repositories for debris from rock and snow avalanches. Hillslope-fall ratio of largest antislope scarp at left side of view is 0.37. Photograph courtesy of Peter Haeussler, U.S. Geological Survey.

ing scarps. These are nontectonic and are called *ridge rents*, *antislope scarps* and *sackungen*. They are more indicative of ridgecrest spreading than sagging, as might be construed by the term "sackungen".

Tectonic movements can also reduce lateral support of hillslopes. Migration of a transpressive duplex along the Hope Fault of New Zealand converted wrench-fault produced thrust faults into normal faults (Eusden et al., 2000). Local compression changed to extension and ridge rents formed as a result of tectonically induced ridgecrest-spreading processes.

Such erosion-induced collapse structures, being a mass-movement threshold process, are another indication of nonsteady-state landscapes.

Ridge rents occur mainly in rising mountain ranges such as the Southern Alps of New Zealand, San Gabriel Mountains of southern California, and the Alaska Range. Earthquakes are a likely way to initiate ridgecrest spreading and continue subsequent propagation of the uphill facing scarps.

They also occur in deep, steep valleys produced by glacial erosion in fairly stable tectonic settings such as the Rocky Mountains, Scotland, and the European Alps. Nonsteady-state landscape evolution in these settings consists of glacial erosion that lowers valley floors faster than ridgecrests. Glaciation also reduces hillslope stability by changing valley cross sections from V-shaped to U-shaped.

Ridge rents (Figs. 6.2, 6.3, 6.4) may look like tectonic fault scarps. Single- and multiple-rupture event ridge-rent scarps in loose materials degrade in a manner that can be evaluated by diffusion-equation modeling. Materials eroded from reactivated scarps accumulate intermittently on footslopes and in adjacent hillside trenches. They can be studied using conventional trench-and-date stratigraphic procedures that evaluate earthquake recurrence intervals associated with tectonic scarps (McCalpin, 1996; Yeats et al., 1997; McCalpin and Hart, 2004).

Important differences between tectonic fault scarps and scarps created by gravitational ridgecrest

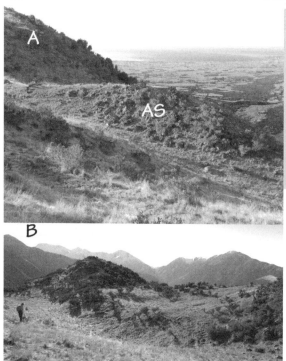

Figure 6.4   Ridge rents in fractured grey-wacke sandstone and argillite on Swyncombe Ridge, Seaward Kaikoura Range, New Zealand.
A. Antislope scarp, **AS**, photographed from the ridgecrest is positioned at hillslope-fall ratio of only 0.04.
B. Longitudinal trough in double-crested ridge-crest. Person for scale is at left side of trough, which extends diagonally to the lower-right.

spreading were recognized by Beck (1968). Ridge-rent scarps tend to parallel ridgecrests in a pattern that mimics a topographic fabric rather than a tec-tonic fault pattern. Sets of ridge rents lack net throw, which rules out the possibility that they are a distrib-uted surface expression of a single fault at depth. Beck also realized that the adjacent valley floors con-stitute a base level for rock-collapse processes. Ero-sion induced gravitational hillslope failures do not extend beneath valley floors.

Given these constraints, possible modes of failure may seem a bit odd when compared to tec-

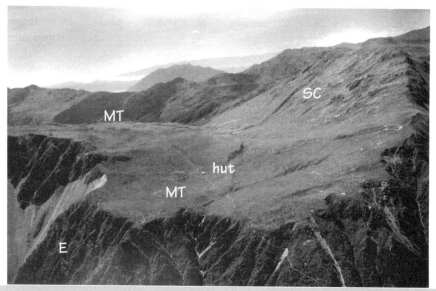

Figure 6.5 Topographic contrast between two ages of landscape. Rapid tectonically induced erosion, **E**, has reduced lateral support of ridgecrests. Low relief upland topography dates from a time of slower uplift rate when this was a seacoast. **MT** is marine terrace remnant. **SC** is a degraded sea cliff. Crest of the Kelly Range, central Southern Alps of New Zealand. Hut for scale.

tonic faulting. An important question to consider is "why should mountain-size rock masses fail by gravitational collapse of ridgecrest areas instead of by landslides ?".

Undercutting of valley sides by glaciers steepens footslopes sufficiently to create U-shaped valley cross sections. Such extreme removal of lateral support for the adjacent hillsides is conducive for ridge-rent formation, but only in settings with favorable rock types. Ridge rents are also common in unglaciated tectonically rising mountain ranges with steep-sided river valleys underlain by highly fractured sandstone and argillite (Fig. 6.4).

The Kelly Range study area of Beck (1968) is on the west flank of the Southern Alps near the junction of the Alpine and Hope faults. A mix of Late Quaternary glacial and fluvial valley erosion rapidly incised the deep valley of the Otira River below broad, low-relief, ridgecrest areas (Fig. 6.5). These gently sloping uplands of the Kelly Range are flanked by rugged slopes steeper than 45°. This is where lateral support for upland terrain has been removed during the Late Quaternary.

The Otira River watershed in the Southern Alps of New Zealand has a 2,000 m high headwaters divide with small glaciers only 11 km from the Figure 6.6 transect. Late Pleistocene valley glaciers were more than 600 m thick. Mean annual precipitation presently is 5,000 to 10,000 mm/yr in most of the watershed. Present lack of U-shaped footslopes is the result of rapid fluvial erosion and post-glacial mass movements. The wide, flat valley floor of the Otira River indicates the presence of sufficient annual unit stream power for the river to easily degrade into its bed at the same rate as this part of the Southern Alps are rising, about 5 m/ky. Excess stream power widens the valley floor and removes hillslope detritus. The resulting broad strath depicts a type 1 dynamic equilibrium stream channel.

Opposite the Kelly Range, much of Goat Hill slopes 40° to 45° and the crestslope has many large ridge rents. My choice of rock-type symbol in Figures 6.6 and 6.8 implies important lithologic controls on landscape evolution. Highly fractured greywacke sandstone and low-grade schist are not conducive for preservation of glacial U-shaped valleys, but greatly enhance prospects for ridgecrest

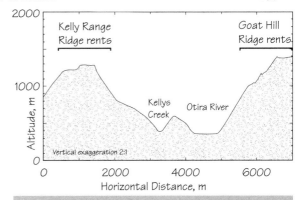

Figure 6.6 Topographic profile across the Otira valley between two ridgecrests that have many ridge rents. Transect is along a N 41° W line that passes through spot-altitude point 1391 m on the 1:50,000, Otira, 260-K33 Infomap of the New Zealand Department of Survey and Land Information.

spreading.

The small Kellys Creek watershed is tributary to the Otira River and lacks the wide equilibrium valley floor of the Otira River, even at a distance of only 1 km upstream from the Otira Valley base level. It has the characteristics of a type 2 dynamic equilibrium channel. Ice spilled over a low divide into this watershed during the Pleistocene but fluvial erosion has created most of the present landscape.

Beck mapped ridge rents in the Kelly Range (Fig. 6.7). He noted the highly fractured nature of the low-grade schistose metasediments. He correctly assumed that the recent glacial deepening of adjacent large valleys created the abrupt changes in relief and the steep valley side slopes conducive for formation of ridge rents. The main ridgecrest is about 800 m above Kellys and Seven Mile Creeks. Some of the uphill facing scarps are more than 0.5 km long with scarp heights of as much as 10 m. All scarps face the ridgecrest and nearly all of them are within 1 km of the ridgecrest.

Gravity driven hillslope scarp formation is more likely to occur where heights above the valley-floor base level are greatest. Gravitational driving forces increase progressively above a valley floor to a hillslope position where a failure threshold can be crossed at times of seismic shaking. So it is useful to note the hillslope-fall ratio, which ranges from 0 at the ridgecrest to 1.0 at the base of the footslope. Hillslope-fall ratios of <0.05 for the ridge rents shown in Figures 6.3 and 6.4 are very close to the ridgecrest. The lowest antislope scarp in the Kelly Range has a hillslope-fall ratio of 0.37. The most prominent uphill-facing scarp in the Alaska Range example (Fig. 6.2) has a hillslope-fall ratio of 0.37, but smaller scarps further down the mountainside have ratios >0.40.

Work in the Kelly Range since 1968 is enlightening. Beck (1968) noticed the break-in-slope to more gently sloping uplands and referred to this as a "pagoda-like" cross-section of the Kelly Range. Bull and Cooper (1986, 1988) described flights of raised marine terraces in this part of the Southern Alps. Both the Kelly Range and Goat Hill have nice examples of degraded sea cliffs and rounded quartz beach pebbles are fairly common on remnants of shore platforms. They concluded that the rolling uplands of the Kelly Range date to before 135 ka when the uplift rate was much slower. Uplifted remnants of coastal landforms indicate minimal ridgecrest erosion, which favors preservation of the ridge rents.

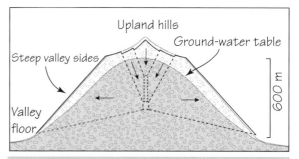

Figure 6.8 Generalized cross section of the Kelly Range showing one possible mechanism for formation of antislope scarps on this 800 m high ridge. Thin accumulations of valley-floor alluvium have been omitted, but a postulated ground-water table generating 600 m of head has been added. Hillslope-fall ratio of lowest scarp is 0.14. Modified from Figure 3a of Beck (1968).

The most likely structural configuration suggested by Beck for gravity-deformed ridgecrests is shown in Figure 6.8. The "keystone" block and adjacent slices move downward whenever the lower hillslopes shift horizontally. The figure implies discrete movements of blocks, but I like Beck's idea that distributed shear also may be an important concurrent process in crumbly, fractured rocks.

A good example of low rock mass strength in the Southern Alps was provided by the collapse of the uppermost 60 m of Mt. Cook, which resulted in a ~12 × 10^6 m³ rock avalanche in 1991 (Chinn et al., 1992; Owens, 1992). The landslide revealed the inside of the highest peak in the Southern Alps. It resembles a rough pile of layered bricks instead of

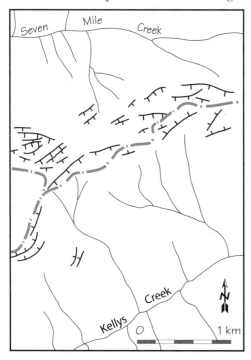

Figure 6.7 Map showing distribution of ridge rents in the Kelly Range, Southern Alps, New Zealand. The broad gray line shows the main ridgecrest divide and the black lines show locations of the antislope scarps and their downthrown sides. Hillslope-fall ratio of lowest antislope scarp is 0.37, but most are above 0.20. From part of figure 1 of Beck (1968).

Figure 6.9 View of the summit ridge of Mt. Adams. All of this cliffy slope collapsed on 6 October, 1999 generating a ~12 × 10⁶ m³ rock avalanche that dammed the Poerua River in Westland. The fractured nature of the bedrock shown in this exposure is typical of much of the Southern Alps. Photo courtesy of Tim Davies of the University of Canterbury.

susceptible terrains to cross a stability threshold. Admittedly, earthquakes are not common in places such as central Colorado (Fig. 6.3) and Scotland.

Ridge rents seem to be common in humid mountain ranges such as the Southern Alps, the Alaska range, and coastal ranges of the American northwest and Canada. They may be absent or rare in dry regions such as the Sinai Peninsula and the deserts of the American Southwest, and the foothill belt of the Diablo Range of central California.

I would add large seepage forces (see the Section 4.2.1.1 discussion) to the list of conditions conducive for lateral spreading of ridgecrests. Groundwater tables roughly parallel the land surface in humid terrains and downhill flowage creates seepage forces equal to the loss of hydraulic head. Ridgecrests that rise only 50 to 100 m above adjacent valley floors will have head losses generated by groundwater flow that are insufficient to promote ridgecrest collapse. However, rapid elevation of ridgecrests such as the Kelly Range and concurrent valley-floor downcutting may produce hydraulic head losses in excess of 500 m. These seepage forces would be directed down and out from the hillslope. A 600 m head loss exerts 60 kg/cm² downslope force. While this may be insufficient to overcome rock mass strength, it moves the mountainside closer to the hillslope stability threshold. Then seismic shaking may trigger gravitational collapses recorded by the ridge rents.

6.3 Erosional Controls of Fault Zone Partitioning

Styles of tectonic deformation along transpressive plate boundaries are functions of plate motion directions relative to the strike of the boundary. Thrust and strike-slip components may occur on a single fault, or on separate fault zones. Such partitioning occurs as one of two styles, serial or parallel (Fig. 1.10). Basal dips of a thrust fault complex become less as it expands out from a transpressional mountain front. This increases the likelihood of separating tectonic displacements into discrete strike-slip and thrust faults (McCaffre, 1992).

The Alpine Fault of New Zealand is amazingly linear in Figures 1.8, 1.9. Norris and Cooper (1995) found that between Haast and Hokitika it

solid rock. The scar left by the ~12 × 10⁶ m³ Mt. Adams rock avalanche (Hancox et al., 2005) revealed a similar low rock mass strength (Fig. 6.9).

Ridge-rent scarp morphologies suggest abrupt, rapid initial surface rupture. Neither field observations nor historical seismic records have documented aseismic episodes – random collapses – resulting in ridge rents. It seems reasonable to infer that earthquakes provide a mechanism that allows

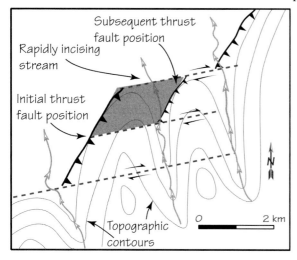

Figure 6.10 Diagrammatic map showing two types of erosional control on transpressional faulting along the central Alpine fault, New Zealand.
1) Rapid erosion of deep valleys favors creation of thrust-fault segments on northwest sides of ridges and strike-slip fault segments across the valley mouths.
2) Gray parallelogram is area where erosion through part of a thrust necessitates formation of new thrust and strike-slip faults along the east and south sides of the parallelogram. Figure 15 of Norris and Cooper (1995).

consists of alternating 2–10 km long sections. More northerly striking sections with oblique-thrust displacements alternate with more easterly striking sections with dominantly strike-slip displacements (Figs. 6.10, 6.11). The result is a serial fault partitioning.

Such separation into alternating zones of thrust and strike-slip faulting is the result of nonuniform denudation of a mountain front. Maximum crustal unloading occurs in the trunk valley floors where rivers and glaciers achieve maximum erosive potential. Alternation between thrust-dominated and slip-dominated short segments results from oblique shear within the Alpine fault zone. Fault-zone partitioning is modulated by the unloading caused by deep incision of valleys on the hanging-wall block. This causes serial partitioning that is characteristic of the central Alpine fault.

Serial partitioning also develops where erosion prevents further propagation of a low-angle thrust complex (Norris and Cooper, 1997). Down-

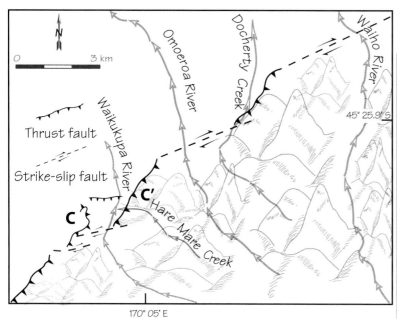

Figure 6.11 Serial partitioning of the Alpine fault of New Zealand between the towns of Fox Glacier in the southwest corner and Franz Josef in the northeast corner. Thrust-fault segments occur mainly on the northwest sides of ridges and strike-slip fault segments occur across the valley mouths. C–C' geologic section is shown in Figure 6.12. From figure 1 of Norris and Cooper (1995).

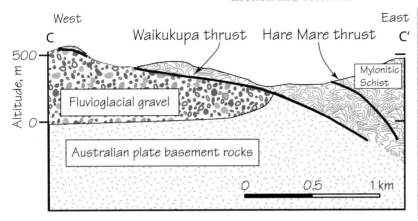

Figure 6.12 Geologic cross section parallel to thrusting direction along line of section shown in Figure 6.11. Waikukupa River has removed much of the rock wedge above the Waikukupa thrust fault since about 60 ka. From figure 3c of Norris and Cooper (1995).

cutting by New Zealand rivers has been sufficiently rapid to cause relocation of an active oblique thrust fault. Such downcutting behavior was described by Norris and Cooper (1995, 1997) at their Waikukupa River site where erosion regulates the near-surface characteristics of the Alpine fault (Fig. 6.12). Thrust faulting emplaced mylonite and cataclasite over fluvioglacial gravels between about 65 ka and 20 ka. Concurrent river erosion reduced the taper of the thrust wedge below its critical value (Davis et al., 1983). This resulted in a shift to a new range-bounding fault, and generated a new fault – the Hare Mare thrust – since about 20 ka.

At larger spatial and temporal scales, the Southern Alps are an example of where concentration of erosional activity along a transpressional plate boundary has altered structural behavior through the influences of rapid uplift and exhumation on crustal strength. The study by Koons et al. (2003) nicely illustrates how fluvial, glacial, and mass-movement processes influence crustal mechanical processes through erosional-thermal coupling. Very young zircon fission-track ages (~1 to 4 Ma) recorded by Tippett and Kamp (1993) attest to the magnitude of recent exhumation. The northeast trending Southern Alps are ideal for promoting large orographic rainfalls from frequent incoming storms from the west (Koons, 1990; Beaumont et al., 1992; Willett, 1999). Long-term uplift rates increased at about 135 ka from 5 to 8 m/1000 yr (Bull and Cooper, 1986). They are modeled here as 8 m/1000 yr, a rate sufficient to have caused thermal weakening of crustal materials, as hot rock was advected into the zone

of greatest exhumation thus weakening the upper brittle crust. This advection influenced the style of plate-boundary faulting.

Koons et al. (2003) note that mylonitic lineations and fault-plane striations have different orientations. Basal fault plane cataclasite striations represent near-surface shear orientations. They plunge about 22° toward 071°, but mylonitic stretching lineations created in the deeper ductile depth zone plunge at about 42° towards a bearing of 105°. The mylonitic signal lags behind the cataclasite by the time span for the deep mylonite to be exhumed. This disparity should be even greater for the Southern Alps, which has become less dominantly strike-slip within the past 5 Ma (Walcott, 1998) as a result of shift in the nearby tectonic plate pole of rotation.

This disparity is best modeled by separation into two shear zones in the lower crust as a function of temperature (Fig. 6.13). Lateral strain abandoned the vertical structure and shifted to the dipping structure. This occurred when the upper crust along the Alpine fault was weakened by exhumation, combining with the convergent strain to form a single oblique fault that accommodated the plate motion. The two components of strain remain separate in the lower crust. Strain along this part of the Australian–Pacific plate boundary has been partitioned into a near vertical lateral structure and a dipping convergent structure. The lateral ductile shear zone is not exhumed and is initially hidden. The convergence shear zone imposes its lineation textures on the near-surface materials. This explains why the ductile structures appear more convergent than

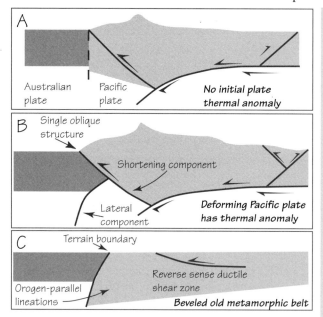

Figure 6.13 Diagrammatic cross sections of the Southern Alps showing long-term evolution of faulting. Australian indentor plate is little deformed by the transpressional collision. From figure 4 of Koons et al. (2003).
A. Initial lack of horizontal variation in crustal rheology produces two distinct structures.
B. Rapid exhumation of Pacific plate rocks results in thermally thinned and weakened crust where both strain components are accommodated by a single structure in the upper crust but diverging with depth.
C. This terrain boundary could be misleading in the distant future when eventual exhumation of deep crust would reveal two structures with apparently different strain histories.

would be predicted from the plate vectors. Eventual long-term post-orogenic exhumation of the lateral structure will appear more strike-slip than might be expected (Fig. 6.13C).

6.4 Consequences of Erosion Induced by Long-Term Plate Collision

Thermal weakening of crustal materials that is induced by erosion has been important for much lon-

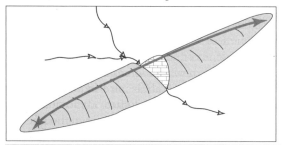

Figure 6.14 Self-enhancing feedback mechanism between fluvial erosional unloading of crust and active folding. Diagrammatic sketch of a river incised through the center of doubly plunging anticline. Erosional unloading of the crust enhances local tectonic deformation. Figure 1 of Simpson (2004).

ger time spans in the Himalayas than in the Southern Alps. The Indian indentor plate has been driving into the Asia plate for at least 45 My, and flow of crustal materials around the corners of the indentor plate has profoundly affected landscape evolution. The novel concept that fluvial and glacial erosion can remove sufficient crustal mass to influence mountain-building forces and even rock metamorphism is applicable on a scale from anticlines to large mountain ranges.

Simpson (2004) notes that rivers commonly have excavated gorges across the middle parts of doubly plunging anticlines (Fig. 6.14). He ascribes this to local unloading of the crust that results in maximum anticlinal flexure centered on a doubly plunging fold. Previously it was thought that these were antecedent streams that over long periods of time had eroded down through pre-existing folds rather than flowing of around them. I suspect that still might be the case for some incised folds. Two requirements must be met before Simpson's model is applicable. First, the fold must have greater mass than that easily supported by the elastic strength of the earth's crust. Second, the local crust should be actively deforming at the time of river incision.

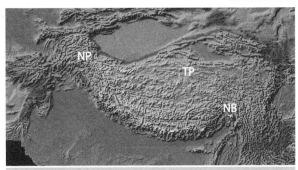

Figure 6.15 Composite satellite image of the Himalayan–Tibetan region. NP is Nanga Parbat and NB is Namche Barwa. TP is the Tibetan Plateau. Image provided courtesy of http://rsd.gsfc.nasa.gov/users/marit/projects/modis_refl/index.html).

He cites the actively deforming fold-and-thrust belt in Iran as an example. Oberlander (1985) noted for the larger streams of the Zagros orogen an "uncanny homing instinct of the larger streams, which find all the anticlines they can, while open structural spillways often gape on either side". If so, this situation is a self-enhancing feedback mechanism. Erosion that unloads the crust concentrates rock uplift in the vicinity of ongoing river excavation. My survey of orbital images of the Zagros region found few large rivers, and only the occasional notching of the center of an anticline.

Both timing and rate of erosion may influence the thermal (metamorphic) evolution of crust thickened by plate collisions (England and Thompson, 1984). This proposition was developed in the lofty Himalayan region of Asia where intense monsoon rainfalls contribute to rapid erosion. Exceptionally steep terrain also is responsible for the phenomenal sediment yields of southeast Asia watersheds (Holeman, 1968; Summerfield and Hulton, 1994).

Magnitudes and rates of crustal processes make the Himalayas ideal for study of feedback relationships between erosion and tectonism (Brozovic et al., 1997). The high-altitude eastern and western parts have reliefs of 4 to 7 km between valley floors and summits. Vigorous surface and tectonic processes promote rapid erosional exhumation of young crustal rocks. The Nanga Parbat and Namche Barwa areas (Fig. 6.15) are the focus of studies about the relationship between erosion and tectonics in the context of the India–Asia collision.

The initial impact of the rapidly moving (150 mm/yr) Indian plate with Asian continental crust at approximately 55 Ma did little more than start to close the Tethys Sea, terminate marine sedimentation by 25 Ma, and fold sediments. The duration and magnitude of this continental collision since 40 Ma has proceeded to change features ranging from atmospheric circulation (Raymo and Ruddiman, 1992) to rock metamorphism (Zeitler et al., 2001). About 2,000 km of northward movement of the Indian plate has compressed the crust into twice its normal thickness, allowing the Tibetan Plateau and adjacent mountains try to rise several kilometers by means of crustal buoyancy. This long-term tectonic plate interaction requires comparable right-lateral displacement towards China.

The tectonically driven indentor block – the Indian plate – has corners at its leading-edge, so adjacent crustal materials are shoved aside (Fig. 1.13). Both ends of the Himalayas are areas where orogenic forces turn sharply about a vertical axis. These two tectonic syntaxes developed fairly late in the collisional history and are areas of exceptionally rapid erosion, rock uplift, and rock metamorphism. In a general sense they can be regarded as large antiformal structures that attest to the large magnitude and continuing high rates of continental shortening. They are named after lofty peaks in their cores – Nanga Parbat in the west and Namche Barwa in the east (Wadia, 1931; Gansser, 1991).

Both syntaxes are areas where high-grade metamorphism has overprinted basement rocks just since the Miocene epoch. Zeitler et al. (2001) believe they "owe their origin to rapid exhumation by great orogen-scale rivers". Let us examine how rapid degradation of the Himalayas since 4 Ma has influenced creation of syntaxial massifs that are developed atop weak crust that is relatively hot, dry, and thin.

Data obtained from many global positioning satellite monitoring stations provides key information regarding present-day tectonics of the Tibetan Plateau (Q. Wang et al., 2001; Jade et al., 2004).

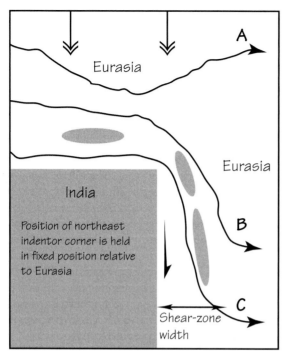

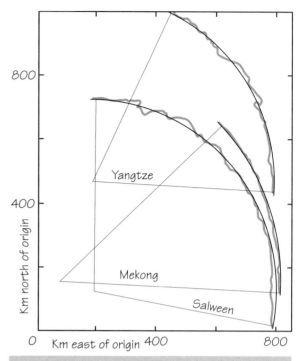

Figure 6.16 Sketch showing how movement of the corner of Indian indentor plate shears river watersheds. Tectonic deformation changes circles to ellipses and rotates them, and the position of a river shifts from A to B to C. From figure 6 of Hallet and Molnar (2001).

Figure 6.17 The Salween, Mekong, and upper Yangtze rivers in southeastern Tibet follow arcs of circles that are roughly concentric with the eastern syntaxis of the Indian plate indentor. Small circles on the earth surface remain circular on this plot of Lambert azimuthal projections. Plot origin is at 91° E and 25° N. Figure 8 of Hallet and Molnar (2001).

These data allow the assessment of whether the Tibetan Plateau (Fig. 6.15) deforms as rigid plates and blocks that float on the lower crust, or by continuous deformation of the entire lithosphere. They conclude that Tibet behaves more like a fluid than like a plate. Crustal material is moving eastward (Lavé et al., 1996) and flowing around the eastern end of the Himalayas. Of course some deformation of the brittle upper-zone crust occurs along fault zones between rigid blocks. Zhang et al. (2004) conclude that "deformation of a continuous medium best describes the present-day tectonics of the Tibetan plateau; ... crustal thickening dominates deformation on its eastern margin except near the eastern syntaxis, where rapid clockwise flow around the syntaxis, not rigid body movement occurs". ... "material within the plateau interior moves roughly eastward

with speeds that increase toward the east, and then flows southward around the eastern end of the Himalaya".

Hallet and Molnar (2001) recognized that such eastward directed deformation should gradually modify the shapes and orientations of large watersheds. Drainage basins near the syntaxis corner have become progressively narrower, which increases the probability of stream capture. Indeed, the ancestral Tsangpo-Irrawaddy River appears to have been captured by Brahmaputra River (Fig. 1.13) headward cutting promoted by the tectonics of the eastern syntaxis (Koons, 1995). The Hallet and Molnar conceptual model is illustrated in Figure 6.16. Three of

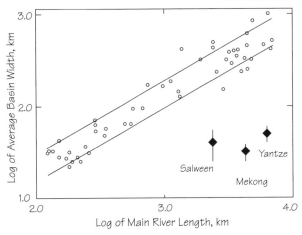

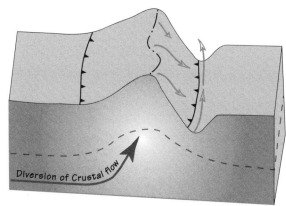

Figure 6.18   Relation of mean basin width to length of trunk stream channel for the 50 longest rivers in the world. Lines show 1 σ departure from linear regression. Figure 2B of Hallet and Molnar (2001). Magnitude of long-term tectonic deformation of the three rivers of Figure 6.16 is shown by the three points with the 1 σ error bars.

Figure 6.19   Sketch of mature tectonic aneurysm. Deepening of large river gorge reduces topographic loading, which initiates advective crustal flow causing mountains to rise and further increase the rate of erosion. This self-enhancing feedback mechanism steepens the crustal thermal gradient, and raises the brittle–ductile transition. Decompression melting may produce low-pressure, high-temperature metamorphic rocks. Figure 5 of Zeitler et al. (2001).

the largest rivers in the world flow through this region, but they are parallel to one another as they pass the eastern syntaxis. The three large rivers follow arcs that are approximately concentric to the location of the eastern syntaxis (Fig. 6.17). The resulting highly anomalous basin widths depart greatly from the norm in the Figure 6.18 regression. Orientations and spacings of these three major rivers illustrate the profound long-term influence of the India–Asia plate collision.

Young metamorphic rocks in both Himalayan syntaxes appear to have formed just yesterday in a geologic time sense (since 4 Ma). This latest development in the long history of crustal collision is a consequence of prolonged deep erosion by large rivers. A purely isostatic model would increase relief as large, powerful rivers removed mass centered on the valley floors. The Indus and Tsangpo Rivers (Fig. 1.13) have easily kept pace with the uplift and have excavated 4 to 7 km deep valleys as they turn south and cross the Himalayas. The resulting self-enhancing isostatic feedback mechanism (fig.

1.4, Bull, 2007) raised mountainous divides, further increasing orographic precipitation (Roe et al., 2003). However, the highest peaks in both syntaxes now appear to be centers of structural, geophysical, and petrological anomalies. They have rapid exhumation rates, and are the locations of very young high-temperature low-pressure metamorphic rocks.

Peter Koons (1998) explains this paradox with a second self-enhancing feedback mechanism that has become dominant. The valley incision feedback noted above results in local rheological weakening of the crust as erosion removes the stronger upper crust. Flow paths of crustal particles will be concentrated where the local geotherm is steepened from below by rapid uplift of hot rock. Such transfer of the crustal materials results in partial melting that further reduces the relative strength of crustal materials and further concentrates strain. A threshold is crossed where focusing of strain and rapid exhumation leads to metamorphic and structural overprinting of the crust. High-temperature lower crustal rocks promote development of large mountains perched atop hot, weak crust (Figure 6.19).

This rheological weakening favors development of young metamorphic rocks under the massifs of Nanga Parbat and Namche Barwa instead of the valley floors. The process is described as being a tectonic aneurysm by Zeitler et al. (2001) that centers both uplift and consequent erosion in rapidly rising summit areas (Figure 6.19). They note "Inherent in this model is the notion that feedback can amplify rather local geomorphic processes to the point where they can exert profound influence on the metamorphic and structural evolution of rocks at considerable depth". These fascinating perspectives are a rather large-scale and broad-brush introduction. Let us consider where and how rainfall is eroding these lofty mountains.

The locations and rates of tectonically induced erosion of the Himalayan Mountains are a function of summer monsoon rainfall. This important control on landscape evolution varies in both time and space. Recent work of Bodo Bookhagen and co-workers provides critical information as to how fluvial systems behave in this huge, lofty region. The emphasis here is on how topography influences the amounts and locations of rainfall generated by orographic lifting of monsoon rain clouds streaming in from the Bay of Bengal.

Their use of a watershed approach lets one note changes in controlling variables over a 2,600 km swath of drainage basins (Fig. 6.20) where majestic mountains rise above the plains of India. The Himalayas are very good at squeezing water from clouds. Their southern flank (the Siwaliks and the Lesser Himalayas) are truly soaked, while sufficient precipitation is left for the lofty Greater Himalayas. Still further inland, the vast Tibetan Plateau is semiarid to arid even though it receives 80% of its annual precipitation of ~250 mm from the summer monsoon.

The strength of a given summer monsoon is generated by processes in the Arabian Sea to the west of India, but the influx of moisture comes from the southeast through the Bay of Bengal to the east of India. Clemens and Prell (2003) used principal components analysis of suites of Arabian Ocean variables to decipher when the monsoon was weak or strong (Fig. 6.21).

Their climate proxies included variables such as regeneration of biological material in the intermediate and deep waters, post-depositional diagenesis, and the $\delta^{18}O$ record. Of course such variables are affected by factors other than climate, which is why they used a statistical stacking procedure of variables – each of which is partly controlled by Late Quaternary climatic changes.

The resulting picture of variable monsoon strength in Figure 6.21 suggests substantial variation that may be periodic. Although not a perfect match, it appears that Milankovitch orbital parameters of planet Earth partly determine monsoon strength. The yearly heating of the Asian Plateau and latent heat export from the southern subtropical Indian Ocean both appear to be maximized at times of obliquity maxima, which has a 41 ky cycle (Imbrie et al., 1984; Berger, 1988). A monsoon maximum, such as in the early Holocene (Fig. 6.21 and discussed later) was characterized by warm southern hemisphere summers followed by cold southern hemisphere winters. Such Milankovitch factors maximized latent heat export during the summer monsoon.

Most of the Himalayas get 80% of their annual precipitation from the summer monsoon. The corners – the Nanga Parbat and Namche Barwa tectonic syntaxes – get 40% of their annual precipitation during the winter. Along strike of the Himalayas and at altitudes of less than 500 m, there is an overall six-fold east-to-west decrease in amount of monsoon rain with increasing distance from the Bay of Bengal source. East to west precipitation amounts are uniformly distributed above 500 mm and this indicates a different rainfall generating process for the Bodo Bookhagen–Douglas Burbank model.

One might think that the higher, steeper mountains would get most of the rain. Instead, the first rise in the topography gets dumped on. A single rise in topography from 500 to 5,000 m altitude creates a single large, 1 to 5 m, peak of annual precipitation (Fig. 6.22A). These mountains are so efficient in wringing out moisture from the monsoon clouds that very little remains when the monsoon clouds rise to 4,000 m altitude.

Nepal has a pronounced step in the topography, the Lesser Himalayas rise up from the Gan-

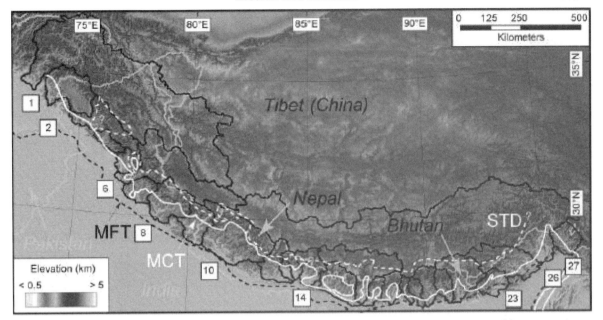

Figure 6.20 Adjacent drainage basins draining the mountain front of the Lesser Himalaya and Greater Himalaya. The large Indus River (drainage basin 1) swings around the western syntaxis of Nanga Parbat, and the large Tsangpo River (drainage basin 25) swings around the eastern syntaxis of Namche Barwa (Fig. 6.15). MFT is the main frontal thrust, dashed black line; MCT is the main central thrust , white solid line; and STD is the southern Tibetan detachment, white dashed line. This is a grayscale version of a color figure supplied courtesy of Bodo Bookhagen.

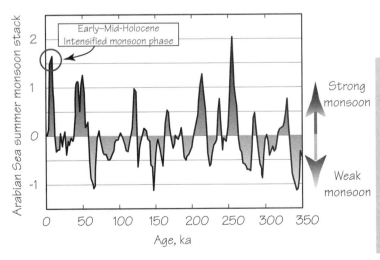

Figure 6.21 Oscillation between strong and weak monsoon rains During the Late Quaternary. Figure is from Clemens and Prell (2003) figure 4, who combined five monsoon proxies from the northern Arabian Sea to assess monsoon variability. This drafted version was furnished courtesy of Bodo Bookhagen.

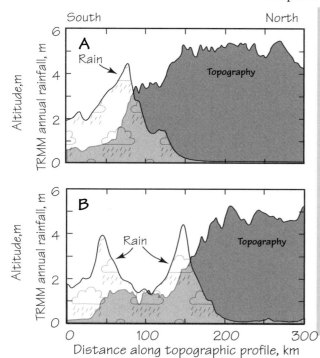

Figure 6.22 Topography and rainfall distribution resulting from orographic lifting of monsoon clouds up the south side of the Himalaya Mountains. Rainfall data are from satellite observations over 9 years with a resolution of 5 × 5 km (TRMM – Tropical Rainfall Measurement Mission). Drafted from figures supplied courtesy of Bodo Bookhagen.
A. Section across Bhutan showing a single broad peak of annual rainfall that coincides with a single high and steep mountain front.
B. Section across Nepal showing a pronounced double-peak in amount of annual rainfall caused by range-bounding and internal mountain fronts separated by 100 km.

ges River plains, and the Greater Himalayas rise up to magnificent peaks higher than 5,000 m. The result of this two-stage style of topographic control is a pronounced double peak in annual rainfall (Fig. 6.22B). Again intense monsoon rains are dropped upon encountering the initial first rise. Then precipitation decreases for the next 100 km before abruptly rising in a second peak that coincides with the second rise in the topography.

The high, extremely steep part of the Himalayas is only subhumid to semiarid, under the current climate as surveyed by a weather satellite (TRMM – Tropical Rainfall Measurement Mission). Minimal precipitation results in minimal erosion by streams, except during incursions of above-normal summer monsoon rainfall.

The style depicted for present conditions can undergo major changes, as is indicated by the occasional year of much stronger monsoon inflow. Rain penetrates much farther into the Himalayas, using the access provided by the larger valleys (Bookhagen et al., 2006a). Large amounts of rain and snow then can fall at the higher altitudes, and this is where maximum relief and hillslope steepness are present.

The resulting increase of landslides and stream power greatly increases sediment yields.

The major increase in monsoon rainfall in the early Holocene (Fig. 6.21) shifted the intensity of hillslope and stream-channel processes much farther back into the Himalayas. Then the Holocene climate became progressively drier.

As at present, precipitation amounts during the Holocene varied. Monsoonal droughts were as long as 1,000 years as noted in the $\delta^{18}O$ of stalagmites in a Chinese cave (Y.J. Wang et al., 2001; Dykoski et al., 2005) and in Tibetan Plateau lacustrine pollen (Gasse et al., 1991). Times of relative drought occurred at ~8.5, 6.4, 3.2, and 2.7 ka. These coincide with times of stream-channel downcutting in the Sutlej Valley watershed as is shown by dating of recently abandoned fill-terraces treads with [10]Be and [26]Al terrestrial cosmogenic nuclides (Bookhagen et al., 2006a).

The responses of the ~50,000 km² Sutlej Valley watershed (basin 6 on Figure 6.20) to Holocene climate variations were studied by Bookhagen et al. (2006a). Influxes of greater moisture triggered widespread landsliding on hillslopes, and created un-

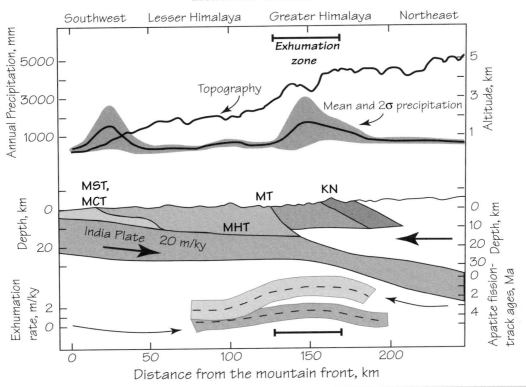
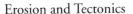

Figure 6.23 Coupling between surface processes and tectonism in a swath across the Sut-lej region, drainage basin 6 of Figure 6.20. Rugged topography and high rainfall erode the Greater Himalaya, causing exhumation of higher grade metamorphic rocks in a block between a thrust fault to the southwest, MT, and a normal fault to the northeast, KN. MHT is the main Himalayan thrust fault, MCT is the main central thrust fault, MBT is the main boundary thrust fault. From figure 4 of Thiede et al. (2004).

stable glacial moraines. The landscape was already close to mass movement stability thresholds. The synchronous large increase in stream power was more than capable of flushing these massive increases of sediment yield downstream. The steep stream-channel gradients in the interior of the Sutlej Valley watershed, together with the increase of annual stream power put the operation of this part of the fluvial system strongly on the degradational side of the threshold of critical power.

The threshold was crossed further downstream where valley floors are wide and gradients not as steep. The resulting fill terraces record such incursions of stronger monsoon rainfall. Deposition switched to incision of the fill-terrace treads when

monsoon rains retreated to lower altitudes as depicted in Figure 6.22. Annual stream power in the upper and middle reaches of the Sutlej Valley watershed decreased, but sediment yield and bedload transport rate decreased relatively more. Fill-terrace reaches were then on the degradational side of the threshold of critical power.

The flight of fill-terrace treads is preserved because of continued net tectonically induced downcutting by the Sutlej River. The highest Holocene fill-terrace tread is 120 m above the active channel.

A two-peaks style of monsoon rainfall is present in the Sutlej region (Fig. 6.23), and the Greater Himalaya peak occurs in rugged terrain with steep stream channels. Frequent landsliding accelerates hillslope erosion, and the increased sediment is swept downstream. Crustal loading is reduced. Rapid erosion is partially offset by isostatic uplift, which brings deeper, younger, higher grade metamorphic rocks to the surface of this tectonic aneurism – a process called *exhumation*.

Apatite fission-track analyses are an essential part of this scenario because they provide a way to estimate the ages when rocks cooled enough to preserve tracks produced by radioactive fission. Tracks are linear damage trails in the crystal lattice created by nuclear fission of trace [238]U nuclei (Wagner and Van den Haute, 1992; Dumitru, 2000). New tracks have similar track lengths but older tracks gradually anneal, and are erased if rock temperatures are hotter than ~110–150°C. Apatite fission-tracks can be preserved where exhumation brings rocks into cooler depth zones near the land surface. Below ~60°C, the apatite fission-tracks are virtually stable because the annealing process becomes very slow. Microscopic examination of apatite determines the abundance of etch pits, which is directly proportional to the time when the rock became cool enough to preserve the fission tracks. Note how the trend of fission-track ages becomes younger in the exhumation zone of Figure 6.23. Fission-track analyses provide estimates of the rate of crustal exhumation. For greater detail see figure 4 of Thiede et al. (2004).

The thickness of rock removed by erosion since ~3–5 Ma appears to be inversely proportional to estimates of specific stream power based on present areal variations of precipitation generated by TRMM satellite data (Fig. 6.24). Schmidt hammer measurements suggest a similar compressive rock strength of the Lesser and Greater Himalaya. The locations and heights of topographic relief control the distribution of rainfall (Fig. 6.22) and maximum stream-channel downcutting occurs where stream channels are steep and narrow. Erosional lowering of the Lesser Himalaya decreases local monsoon rainfall and promotes a shift of the moisture flux to the Greater Himalaya. Specific stream power can also be increased where rapid uplift on active thrust faults enhances local orographic precipitation.

Unit stream power typically decreases upstream in a watershed. The reach with maximum capability to do work – the maximum specific stream power reach of Bodo Bookhagen and his co-workers (Bookhagen, et. al., 2006b) – should be in a downstream reach where large flows of water can efficiently degrade active channels that are still fairly steep. The trunk channel reach with maximum steepness typically occurs 20 to 50 km upstream from maximum cross-valley relief (Fig. 6.25).

The location of the reach with maximum specific stream power is controlled by spatial dis-

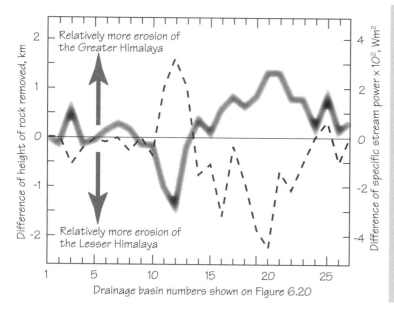

Figure 6.24 Variations in bedrock erosion (broad line) of the Greater and Lesser Himalaya as a function of location along 2,600 km of mountain front, from west to east. Eroded volumes are a minimum because much greater volumes of rock have been eroded than is represented by this relief-based plot. This erosion is relative to spatial variations in specific stream power (an index based on slope and discharge of a stream – shown by the dashed line). Figure supplied courtesy of Bodo Bookhagen.

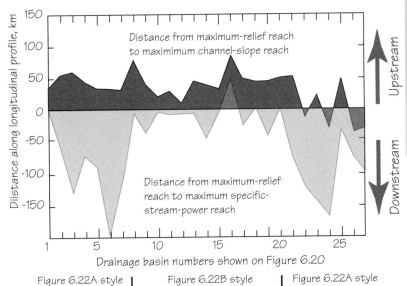

Figure 6.25 Spatial variations of locations of steepest channels and maximum specific stream power reach, relative to the location of the maximum cross-valley relief reach. Figure supplied courtesy of Bodo Bookhagen.

tribution of monsoon rainfall as well as by downstream increase of drainage-basin area. This reach is far downstream from the reach with maximum relief where the style of rainfall mimics that shown in Figure 6.22A. The two reaches roughly coincide where rainfall mimics the style shown in Figure 6.22B. The central Nepal watersheds would be an example. Stream discharge appears to be the dominant variable controlling the location of maximum stream power in the Himalaya.

Spatial or temporal variations in drainage-basin area directly control the magnitude of fluvial erosion. The potentially huge discharges of the Tsangpo River, which flows through and around the eastern tectonic syntax, and the mighty Indus River that flows around the western syntax are examples. These are basins 25 and 1 on Figure 6.20. Flood events are more than tenfold larger than for other watersheds studied by Bodo Bookhagen and Douglas Burbank. Their capacity for erosion results in tectonic aneurisms (Fig. 6.19). These are dominant fluvial systems (Section 2.3.1), so they presumably have become much larger since 5 Ma. Location is as crucial as basin size in the Himalaya, and both of these rivers extend far into the semiarid zone. Having semiarid headwaters diminishes their relative importance much of the time. That changes dur-

ing times of incursion of greater monsoon rainfall. That is when much of the geomorphic work of the Indus and Tsangpo Rivers is done.

6.5 Summary

Geomorphic approaches are now being included in geologic studies ranging from fractures to faults to collision zones (Summerfield, 1996, 2005; Church, 2005). Including the various impacts of tectonics on topography has resulted in greater breadth of geologic models.

Rock uplift causes rivers and glaciers to deepen their valleys, thereby decreasing lateral support of hillslopes. Low, gently sloping hills lack the driving forces to create long fracture planes. High, steep mountains are more likely to have a lateral support anomaly that exceeds the rock mass strength. Gravity induced rock-failures range from exfoliation joints in massive plutonic rocks to ridge rents in crumbly, fractured rocks.

Rapid, tectonically induced deepening of valley floors can result in sufficient crustal unloading to affect styles of surficial faulting. This caused serial partitioning of the oblique-thrust Alpine fault of New Zealand. Dominantly thrust displacements on the northwest sides of ridges alternate with dominantly strike-slip displacements across valleys.

Over time spans of millions of years, fluvial and glacial erosion can remove sufficient crustal mass to influence mountain-building forces and even rock metamorphism. Vigorous erosion and crustal-thickening processes since 40 Ma promote rapid erosion that exhumes young crustal rocks in the Himalayas of Asia. Rapid incision of deep valleys results in local rheological weakening as erosion removes the stronger upper-crust rocks. Rapid uplift of hot rock results in a tectonic aneurysm. Partial melting reduces strength of crustal materials and further concentrates strain. Such focusing of strain and rapid exhumation leads to large mountains of young metamorphic rock perched atop hot, weak crust. Examples are the syntax massifs of Nanga Parbat and Namche Barwa that mark the corners of the India indentor plate that is being shoved into Asia. The huge Tsangpo River that flows through and around the eastern tectonic syntax, and the mighty Indus River that flows around the western syntax can generate tenfold larger flood events than intermediate watersheds.

Topography influences erosion by Asian summer monsoon rainfalls. At present, the first topographic rise is so efficient in extracting moisture that much less remains when monsoon clouds rise to 4,000 m altitude. Rain penetrates much farther into the Himalayas when stronger monsoons flow up the larger valleys. Large amounts of rain and snow then can fall at the higher altitudes. A major increase in monsoon rainfall occurred in the early Holocene, resulting in more landslides and much larger sediment yields. This pulse of increased rainfall caused an aggradation event that is recorded by fill terraces along the Sutlej and other rivers.

Chapter 7

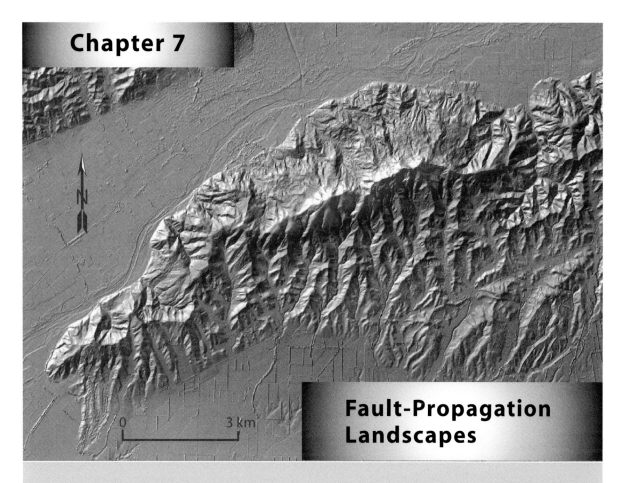

0   3 km

Fault-Propagation Landscapes

The South Mountain–Oak Ridge anticline rises above the deltaic Oxnard Plain of Southern California. The anticline is above a blind thrust fault and has grown 15 km towards the west since 0.5 Ma. Rates of reverse slip are ~5 m/ky (Huftile and Yeats, 1995). Lateral growth of the anticline is most likely the result of decreasing fault slip to the west from 5 to 1–2 m/ky instead of lateral propagation of the Oak Ridge reverse fault. Quaternary displacement is ~2.5 km.

Geomorphic analyses by Azor et al. (2002) show that the new north-side mountain front has a sinuosity of almost 1.0 (instead of 2). Ratios of width of valley floor to valley height decrease westward from 1.5 to 0.5. These V-shaped valleys also have high stream-gradient indices.

Image furnished courtesy of Edward Keller, and the Environmental Studies Program at the University of California at Santa Barbara.

Drainage nets change as faults propagate. This chapter continues the discussion about how erosion and deposition interact with landscape evolution by examining the geomorphic consequences of propagating geologic structures. First we examine how drainage nets change with lengthening of normal faults – slowly in the Nevada Basin and Range Province, and quickly in Greece. Then we discuss how landscapes change with active folding and thrust-faulting – slowly in New Zealand, and at

Tectonically Active Landscapes, 1st edition. By W. B. Bull. Published 2009 by Blackwell Publishing, ISBN 978-1-4051-9012-1

a quicker pace in California. Pure strike-slip faulting may not create base-level changes. So propagation of the northern tip of the Death Valley–Fish Lake transtensional fault zone is discussed here.

Faults evolve by horizontal and vertical propagation. Examination of the record of tectonic deformation provided by active and inactive geologic structures is hindered where rocks are hidden by deposition of younger sediments on the piedmont side of the fault and eroded from the landscape on the mountain-range side of the fault. Our interpretations of Quaternary tectonics are enhanced by geomorphic investigations that study how faulting has affected erosion of the drainage networks of streams, and the style and thickness of recent deposition in aggrading reaches of tectonically active reaches of fluvial systems.

Keller et al. (1998, 1999) found geomorphic criteria useful in analyses of propagation of thrust faults. James Jackson spotted useful tectonic and structural information by examining changes in how stream drainage networks evolve in response to tectonic deformation along fault zones whose surface-rupture length becomes progressively longer. His drainage evolution studies provide valuable insight regarding propagation of a normal fault in west-central Nevada, discussed in Section 7.1.1 (Jackson and Leeder, 1994) and for growth of fault-propagation folds in the Otago fold-and-thrust belt of New Zealand, discussed in Section 7.2.1 (Jackson et al., 1996; Bennett et al., 2006). Morewood and Roberts (1999, 2002) used both the streams of fluvial systems and marine terraces in their studies of landscape evolution associated with rapid normal faulting of the Gulf of Corinth rift in Greece.

An important conclusion of the Greece and New Zealand studies is that landforms record the initial generation of new faults as well as lengthening of faults by propagation of the fault tips. The common model adopted here – for both normal and thrust faults – is for rapid initial lengthening of short faults into an integrated single fault zone (Cowie and Roberts, 2001). This regional set of master faults have individual lengths that exceed the local seismogenic crustal thickness. This is followed by lengthening of the surface-rupture extent of each fault zone by episodic (earthquake-induced) propagation of the fault tips. Uplift becomes dominant so the uplift-rate/propagation-rate ratio increases during the evolution of either normal or thrust faults (Morewood and Roberts, 2002; Bennett et al., 2006). The Figure 7.1 model is applied for normal faulting in Greece, which is discussed in Section 7.1.2.

Normal and thrust faulting are also typical of releasing and restraining stepovers along strike-slip faults. This tectonic style commonly leads to propagation of oblique faulting on one side of a pull-apart basin, or on one side of a squeeze-up, and progressive diminishment of faulting on the other side. In such a process, new transverse structures and associated landscapes form, as the bend or stepover region migrates. This can result in progressive asymmetric development of a strike-slip duplex.

An example is the Hope fault of New Zealand. Eusden et al. (2000) describe a 13 km long and 1.3 km wide transpressional duplex structure that has migrated northeast. This leading portion of the duplex structure is rising on thrust faults while in the trailing southwest portion the formerly active duplex structures are now collapsing, undergoing a reversal of slip to become normal faults. The associated landscapes undergo equally profound changes.

Many other possibilities include progressive elongation of a pull-apart basin where deposits become progressively younger in the direction of basin expansion and stratigraphic thicknesses of shingled strata within a single basin can exceed 10,000 m (Crowell, 1982). See Wakabayashi et al. (2004) for a detailed summary of how releasing and restraining bends of strike-slip faults propagate.

7.1 Normal Faulting

7.1.1 Nevada Basin and Range Province

The range-bounding Pearce normal fault in west-central Nevada has propagated northward sufficiently fast that it has affected locations of Pliocene–Pleistocene stream-channel entrenchment and alluvial-fan deposition. The study area of Jackson and Leeder (1994) on the west flank of the Tobin Range was the site of the 1915 Dixie Valley magnitude Mw 7.2 earthquake (Wallace, 1977, 1984). Incremental increases of total vertical displacement on the Pearce–Tobin fault appear to be accompanied by increases in fault

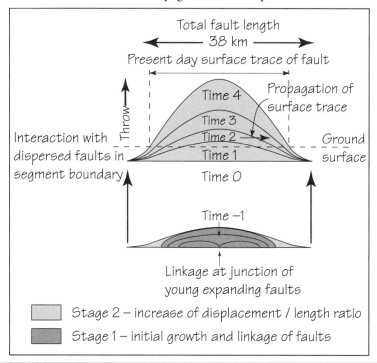

Figure 7.1 Two-stage fault growth model for the South Alkyonides normal fault, Greece. Stage 1 consists of rapid radial propagation and linkage of incipient small faults at time −1. A single large fault is formed at time 0. The style of fault displacement changes in Stage 2. Total fault length remains the same and the rate of lateral propagation of where the fault tip-line intersects the ground surface becomes progressively slower. Stage 2 characterizes the growth of the South Alkyonides fault over the past few hundred thousand years during which the length of the fault at depth has remained fixed. The vertical/horizontal ratio of tectonic displacements has increased. Formatted from figure 6 of Morewood and Roberts (2002).

length (Watterson, 1986), or by coalescence of fault segments (Dawers and Anders, 1995).

The Tobin Range piedmont is between two fault zones and is being tilted to the north (Fig. 7.2). The range-bounding Pearce and Tobin faults form a large én echelon step in the mountain front. Geomorphic changes include entrenchment of piedmont stream channels that shifted the apexes of active alluvial-fan deposition far downstream. The 1915 surface ruptures on the Tobin and Pearce faults overlap by about 2 km, but the Tobin fault did not have a surface rupture further south in 1915 (Wallace, 1984). Jackson and Leeder estimate that the Pearce fault propagated ~50 m northward during the 1915 event. It is unlikely that 30–50 m incremental

increases in fault length would be obvious to paleoseismologists after earthquake-generated slip. So we turn to longer term, geomorphic, evidence for cumulative displacement. Tectonic deformation can affect stream behavior even where the tip of a propagating fault has yet to rupture the piedmont surface.

Responses of the Tobin Range fluvial systems to Quaternary normal faulting support several interesting conclusions, which are nicely illustrated by figure 13 of Jackson and Leeder (1994). Streams X, Y, and Z are permanently incised into the Tobin piedmont, in contrast to minor, intermittent channel incision of the tectonically undisturbed piedmont streams farther north (Fig. 7.2). This tectonically induced downcutting clearly is the result of base-level

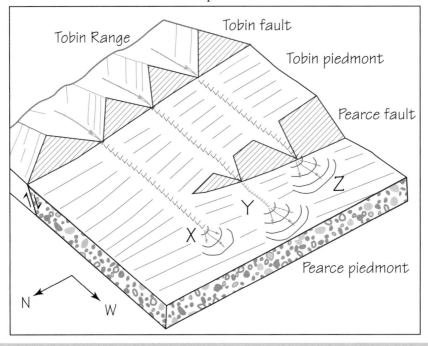

Figure 7.2 Sketch showing progression of stream and alluvial-fan evolution where the active Pearce and Tobin faults overlap in west-central Nevada, USA. Streams X, Y and Z are incised into the northward tilted part of the Tobin piedmont. Stream channel X is slightly incised into the Pearce piedmont and ends at a young alluvial fan. Stream channel Y now is being backfilled downstream from the Pearce fault and alluvial-fan deposition by stream Z is adjacent to the scarp of the active Pearce fault, presumably having buried the earlier stages in its evolution represented by entrenched channels X and Y. Figure 13 of Jackson and Leeder (1994).

falls created by episodic movements on the Pearce fault that have raised and tilted the block between the two faults. The morphology and geomorphic processes of the three streams reveal the sequence of landscape responses to propagation of the Pearce normal fault. All three streams end in young fans, but the stream channel in the adjacent upstream reach is different in each case.

Stream-channel entrenchment may be the result of either climatic or tectonic perturbations. Part of the narrowness of stream X downstream from the projected trace of the Pearce fault might be related to recent climate-change perturbations. If so, the terminal stream-channel reach would stay sufficiently far to the degradational side of the threshold of critical power to maintain modest channel entrenchment, despite the tendency of local uplift to

reverse the trend. Uplift has moved this reach closer to the threshold, but not across it to a depositional mode of operation.

Fan deposition is a local base-level rise, which might explain why the incised channel for stream Y has recently crossed the threshold of critical power. It has begun to aggrade between the Pearce fault and the fan apex. This is indicated by a much broader, flatter valley floor than for stream X. This process of fanhead trench backfilling will ultimately shift the fan apex upstream to the Pearce fault.

The fan Z apex location coincides with the trace of the active Pearce fault, and reflects input from two other variables. Base-level fall caused by movements on the Pearce fault is larger and older than for streams Y and X. The resulting greater entrenchment of the former piedmont reach between the Tobin and Pearce faults has increased the watershed sediment yield. The Tobin piedmont reach of stream Z, has been converted from an area of deposition to erosion. Propagation of the north tip of the Pearce fault is slow and intermittent but is clearly recorded by these types of changes in the fluvial systems along this part of the Tobin Range.

7.1.2 Greece

The Gulf of Corinth rift in central Greece is rapidly extending (10–12 m/ky). The south side of the gulf is bounded by a sequence of three north-dipping normal faults. West to east these are the Egion, Xylokastro, and South Alkyonides faults. Displacement rates have been estimated by geodesy, trenching of Holocene deposits, raised marine

terraces, and long term hanging-wall stratigraphic offsets. They show that the adjacent Xylokastro normal fault has a displacement rate several times greater than the South Alkyonides fault and possibly more than the Egion fault.

Late Quaternary landscape evolution along the South Alkyonides fault (Fig. 7.3A) is diverse. A throw rate of 2–3 m/ky is constrained by tilted marine terraces (Leeder et al., 1991; Armijo et al., 1996), trenching studies (Pantosti et al., 1996; Collier et al., 1998) and geodesy (Clarke et al., 1997). Focal mechanisms provide a fault-plane dip of $45 \pm 5°$ (Jackson et al., 1982) for a displacement rate 2.6–4.7 m/ky.

The strands of the South Alkyonides fault comprise a surface trace length of about 38 km (Morewood and Roberts, 1999). Observed slip vectors indicate prominent fault-normal stretching of a

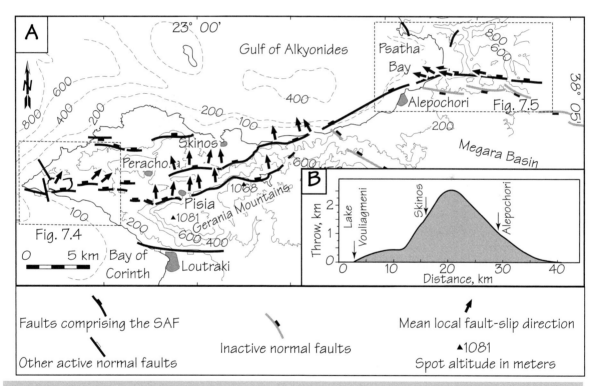

Figure 7.3 Features of the South Alkyonides fault (SAF), Greece. Map and profile from figure 2 of Morewood and Roberts (2002).
A. Map showing the surface trace of the fault. Note the dip-slip style at the center and oblique-slip style near both ends. Boxes show the areas of Figures 7. 4 and 7.5.
B. Fault throw as a function of distance along the South Alkyonides fault.

single hanging-wall basin that is also stretching in an along-strike direction (Anders et al., 1990; Morewood and Roberts, 1997, 1999, 2000, 2001; Roberts 1996a, b; Roberts and Ganas, 2000). Both the offsets of Mesozoic rocks and offshore seismic reflection data indicate that the maximum cumulative throw in the center of the fault is 2.5–3.0 km (Myrianthis 1982; Perissoratis et al., 1986; Roberts, 1996b). Estimated maximum cumulative displacement is 2.8–4.7 km. Cumulative throw decreases towards both the western and eastern ends (Fig. 7.3B), as is typical of rift-related normal faults (Schlische et al., 1996). Such displacement gradients are reflected in the landscape, especially in the way they control the directions in which streams flow.

The presence of only three obvious marine terraces near the western tip of the South Alkyonides fault suggests a slow (<0.5 m/ky) uplift of the footwall block (Fig. 7.4). Only the highest Late Pleistocene sea-level highstands are recorded as raised shorelines where uplift is slow. Shorelines are created at lower highstands but they may not rise fast enough to escape destruction. If the next high sea level occurs soon and is above the previous highstand, it erodes and submerges the shoreline created at the earlier time of relatively lower sea level.

^{230}Th/^{236}U analyses of coral and mollusks provide useful age estimates of marine shorelines in Greece. Vita-Finzi, (1993) obtained ^{230}Th/^{236}U dates of 128 ± 3 and 134 ± 3 ka for coral on the lowest marine terrace collected from near the western end of the South Alkyonides fault. Farther east at Alepechori (Fig. 7.3), Collier et al. (1998) collected marine-terrace coral that was ^{230}Th/^{236}U dated as 127 ± 6 ky. Here, the rate of uplift is also a modest ~0.3 m/ky.

Two older linear notches in the Perchora peninsula landscape (Fig. 7.4) have features indicative of marine shorelines. These include coral, mollusk, and gastropod fossils, lithophagid borings into the rocks of the shore platform, algal bioherms, and serpulid worm tubes.

Tilt increases with terrace age. Noting that a nearly constant uplift rate applies to most paleo-shorelines (Lajoie, 1986; Merritts and Bull, 1989; Armijo et al., 1996), Morewood and Roberts (1999) correlated the two higher marine terraces with the 240 and 330 ka times of major global sea-level highstands. They allow for the possibility that the middle terrace could be assigned a time of 200 or 210 ka.

Stream channels have been offset by vertical displacements on the normal faults. This either

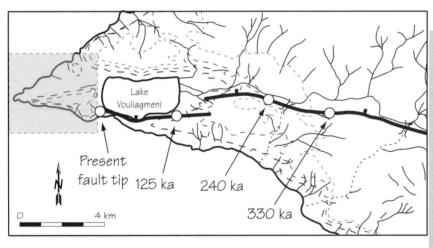

Figure 7.4 Western tip of the South Alkyonides fault, Greece at the present time. White circles show how far the surface trace of the fault had propagated at the times of the present and three Late Pleistocene major global sea-level highstands. Figure 4 of Morewood and Roberts (2002).

⊢ Surface trace of the South Alkyonides fault
)(Wind gap of former channel
▓ Boundary area between two active fault segments

⌐ Stream channels
····· Abandoned stream channels
‑‑‑‑ 330 ka
‑ ‑ · 240 ka Coastlines of sea-level highstands well constrained by marine terraces
‑ ‑ 125 ka

creates knickpoints if the channel downstream from a fault is lowered or escarpments that impede or block streamflow where a reach of the channel is raised. Waterfalls can migrate upstream as knickzones, but blockage terminates flow along a channel. The downstream reach then becomes an abandoned channel. Streamflow is impounded against the rising fault scarp and deposits its sediment load in a small basin of interior drainage. Locations of former stream-channels can be preserved as notches in ridgecrests. These wind gaps contain fluvial gravel.

A marine-terrace that laps onto fluvial sediments deposited in wind gaps on the footwall block post-dates the time of stream-channel abandonment. This is the basis for estimating the lateral propagation rate for the western tip of the South Alkyonides Fault (Fig. 7.4). The estimate is 12–17 m/ky since ~330 ka (Morewood and Roberts, 1999). Uplift rates for a given north–south transect are uniform, but increase progressively towards the east with increasing distance away from the present location of the western tip of the fault (Morewood and Roberts 1999, Figs. 4, 7).

The adjacent segment boundary to the west of the present tip is characterized by dispersed complex normal faulting that accommodates Corinth rifting in the area between the South Alkyonides and Xylokastro faults (Morewood and Roberts, 1997). Note how the style of faulting in Figure 7.3 goes from a single fault trace to many traces. This trend towards dispersion of faulting continues into the segment boundary. The segment boundary style of dispersed tectonic deformation is similar to that mapped by Kirk Vincent for the Elkhorn Creek segment boundary of the Lost River normal fault of Idaho, USA (Vincent, 1995; Bull, 2007, p. 111–113).

Eastward propagation at the other end of the South Alkyonides fault has beheaded stream channels and changed the orientation of drainage networks as the footwall block rose relatively ever higher in the landscape. At 1 Ma (Fig. 7.5A) a large stream, the Megara River, was flowing northwest to the coast where it was depositing a delta. Eastward propagation of surficial faulting by the South Alkyonides fault 1) increased the rate of uplift and tilting of the footwall block, and 2) beheaded tributaries to the Megara River. The trunk stream channel of the Megara River began to flow towards the southeast instead of continuing to flow northwest (Fig. 7.5B). Propagation of the fault tip continues eastward (Fig. 7.5C), beheading stream channels and creating small basins of internal drainage.

The rate of lateral propagation for the eastern tip of the South Alkyonides fault is much less than the 12–17 m/ky of the western tip (Morewood and Roberts, 1999). It cannot exceed 3.7 m/ky and probably is close to 2.8 m/ky. Conglomerates about 5 km to the west have been offset a total of ~170 m and are at least 1.36 My old. This information provides the estimate of maximum propagation rate of 3.7 m/ky. Enclosed internal-drainage basins 1, 2, and 3 of Figure 7.5 are adjacent to the fault and well below their respective wind gaps. Drainage at location 4 flows through the initial stages of a basin, and then around the present fault tip to the Megara Basin. Drainage beheading has not occurred yet, so the eastern tip of the surface trace of the fault presently is between basins 3 and 4.

The approximate 2.8 m/ky lateral propagation rate was used to reconstruct the faulting and drainage history of the area. Active faulting began along the shores of the Gulf of Alkyonides at about 1 Ma (Leeder et al., 1991; Armijo et al., 1996 – as discussed in their section 1.5), an age consistent with the age of the youngest sediments in the now inactive Megara Basin. The gentle tilt of the basin to the northwest directed flow to a fan-delta (Fig. 7.5A) whose bedding confirms this flow direction.

Morewood and Roberts calculate that fault-tip propagation through the north end of the Megara Basin tilted the surface to the southeast and made the Pateras fault inactive. The Megara River flowed southeast instead of northwest. Former south-flowing stream channels became wind gaps in the footwall ridgecrest immediately downstream of where the South Alkyonides fault dammed basins 3 and 4.

Such changes in hills and streams of landscapes that are being deformed by propagation of normal faulting are distinctive and fascinating. We now turn to compressional tectonic settings to assess how uplift produced by folding shifts the locations of streams. Anticlines commonly are associated with propagating blind thrust faults deeper in the crust.

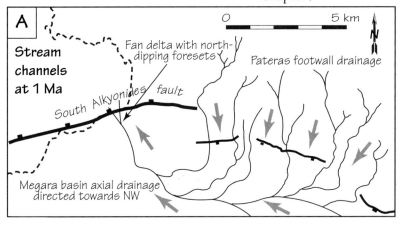

A

Stream channels at 1 Ma

0 ——— 5 km

Fan delta with north-dipping foresets

Pateras footwall drainage

South Alkyonides fault

Megara basin axial drainage directed towards NW

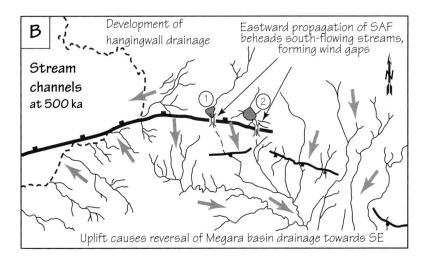

B

Stream channels at 500 ka

Development of hangingwall drainage

Eastward propagation of SAF beheads south-flowing streams, forming wind gaps

① ②

Uplift causes reversal of Megara basin drainage towards SE

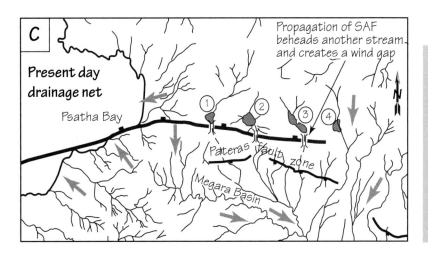

C

Present day drainage net

Psatha Bay

Propagation of SAF beheads another stream and creates a wind gap

① ② ③ ④

Pateras fault zone

Megara Basin

SAF trace

Other normal faults

①

Interior drainage basin with alluvial fill

Stream channels

Abandoned channels

Drainage direction

)(
Wind gap

Poorly constrained coastline

Figure 7.5 Evolution of the eastern end of the South Alkyonides fault (SAF) based on a lateral propagation rate of 2.8 m/ky. Figure 5 of Morewood and Roberts (2002).

A. At 1 Ma.
B. At 500 ka.
C. At the present time.

7.2 Thrust Faulting

Rising folds and drainage diversions associated with their active thrust faults are obvious and accessible. Diverse and careful studies in the Otago fold-and-thrust belt in New Zealand and Wheeler Ridge in California are summarized here.

7.2.1 New Zealand

It might appear unusual to focus on tectonic deformation of basement schist in a study of active tectonics. Actually this discussion is about folding and faulting of the Waipounamu erosion surface (Landis and Youngson, 1996; Landis et al., 2008) – a time-transgressive, nearly planar, wave-cut surface of Cretaceous to Miocene age eroded across schist. Formerly called the Otago Peneplain (Cotton, 1917; Mortimer, 1993; Bishop, 1994), this regional erosion surface provides elegant ways to study Quaternary evolution of drainage nets to better understand fault-propagation folds (Jackson et al., 1996; Markley and Norris, 1999; Bennett et al., 2005, 2006).

Three elements were needed for quantitative appraisal of thrust-fault propagation rates.
1) The Waipounamu erosion surface provides a regional datum that pre-dates folding. Local stripping of a thin mantle of early Tertiary sediments reveals tectonic warping of the regional datum.
2) A means of estimating minimal ages of onset of uplift along the axis of an anticline relies on dating with terrestrial cosmogenic nuclides. Parts of a thin bed of early Tertiary quartzose sandstones and conglomerates deposited on the surface have been silica-cemented to form patches of extremely resistant quartzites (Lindqvist, 1990; Youngson et al., 2005).

"Sarsen stones" (named after similar durable quartzite blocks on the land surface of southern England) from this horizon now are found in widespread patches where erosion of upwarped parts of the landscape have brought them to the surface. Their presence as undispersed groups and as stones capping small tors of schist suggests minimal post-exposure degradation of the surface. Exposure times of sarsen stones were estimated with terrestrial cosmogenic nuclides by Jackson et al. (2002), Youngson et al. (2005), and Bennett et al. (2005, 2006). Cosmogenic ^{10}Be accumulation in quartz

begins when exhumation brings the sarsen stones to within ~2 m of the land surface.
3) Landforms that are sensitive to tectonic deformation. The drainage networks of small streams were ideal in this case, because there appears to be minimal overprinting caused by Quaternary climate-change induced aggradation–degradation events. Strath terraces at two study sites record attainment of stream-channel steady state between pulses of uplift.

The surface has been folded into parallel northeast-trending ridges and valleys (Fig. 7.6) in the 500–1000 m high fold-and-thrust belt of central Otago, New Zealand. A nearby set of GPS measurements indicates a rate of contraction of ~1.5 m/ky (Norris and Nicolls, 2004). These structures accommodate <10% of the oblique convergence between the Pacific and Australian plates; most of the action is closer to the Alpine fault (Norris et al., 1990; Berryman et al., 1992; Bishop, 1994). Folds in compressive tectonic settings form because of movement on thrust faults at depth.

Drainage-net patterns and their evolution on the now exhumed Waipounamu erosion surface were used by Jackson et al. (1996) and Bennett et al. (2005, 2006) to study fault-propagation folding. Headwaters streams that are perpendicular to a ridgecrest divide of anticlines are a consequence of active folding. Belts of short and long watersheds on opposite sides of a divide suggest ongoing asymmetric folding caused by a single thrust fault at depth.

Streams with sufficient unit stream power can maintain their course across a rising fold but their watershed is not centered directly upstream from the water gap. Instead flow from the drainage net progressively shifts to one side. This is because tilting of the land surface lets the stream capture progressively more drainage in the direction of fault propagation. Drainage-net history also shows how simple asymmetric folds develop into box folds when a second thrust fault becomes active, and how topographic ridges and their underlying causative structures coalesce (Jackson et al., 1996). Streams in the broad intermontane valleys that now flow against the regional southwestward slope (Figure 7.6) reflect the consequences of local Quaternary uplift.

Several key landforms were used in these studies. Straths represent situations where a particular

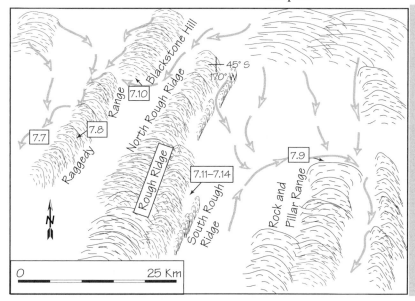

Figure 7.6 Map of part of the fold-and-thrust belt in central Otago, New Zealand. These northeast trending ranges were formed by late Cenozoic contraction that folded a Cretaceous erosion surface beveled in schist basement rocks. Gray flow arrows show streams diverted around or through the rising folds. Numbered boxes show locations of figures discussed in the text.

stream has attained its base level of erosion. Strath terraces resulting from episodes of stream-channel incision may be considered as time lines in a tectonically deforming landscape. This same process in a larger–scale fluvial landscape results in pediments where straths of different fluvial systems coalesce during long time spans of tectonic stability. Regional coalescence of pediments creates a peneplain.

The Waipounamu erosion surface is different, being wave-cut and then raised and folded (Fig. 7.6), deformation that determines where and how streams flow. But how can one ascertain the amount of post-exposure degradation of an erosion-surface datum such as the reference surface shown in Figure 7.8A? Is a broad topographic saddle a genuine tectonic undulation in the erosion surface or is it merely the result of erosion? Only part of the Raggedy Range anticline is shown in Figure 7.7. Bennett et al. (2006) show that the long ridgeline consists of several én echelon anticlinal folds that have coalesced to form a single structure. This is an example of completion of Stage 1 of the Figure 7.1 fault-evolution diagram.

Accordant tops of numerous tors may approximate the schist erosion surface in this landscape of obvious anticlines and synclines (Fig. 7.8B). Tors may also persist due to case-hardening weathering processes or cementation by silica. But tor locations

could just as well be the result of random weathering and erosion processes. Young tors are just as common and are of similar height on the sides of deep valleys incised into the Waipounamu surface (Fig. 7.8C).

The combination of small mountain ranges whose shape mimics anticlines and the presence of tors is reassuring to earth scientists who use undulations in the Waipounamu surface to assess rates of thrust faulting and folding. Most small streams flow directly down the sides of the anticlines as a consequence of Quaternary uplift.

Some streams found their route blocked by a younger growing fold and had insufficient unit stream power to maintain their courses across the rising anticline. These locations of abandoned former streamcourses notched into anticline crests are based on positions of these wind gaps (Fig. 7.8B). Most wind gaps contain fluvial gravels. They are now occupied by tiny streams that could never erode such large valleys or transport bouldery gravel residing in long abandoned streambeds.

Amalgamation of small watersheds by stream-capture processes is a common adjustment. Streams shift their positions to accommodate the folding. Deflected streams join other streams, which increases their ability to keep pace with ongoing folding of the land surface.

Figure 7.7 View northeast towards the Raggedy Range in the Otago fold-and-thrust belt, New Zealand. Consequent streams have incised into the flank of the anticline, whose axis plunges toward the left. Dark specks on the snowy ridgecrest are tors whose accordant tops suggest the position of the Waipounamu erosion surface, which was beveled in the early Tertiary and tectonically deformed in the late Tertiary.

Figure 7.8 Tors (**T**) associated with the Waipounamu erosion surface.
A. View down several kilometers of the sloping ridgecrest of Figure 7.8C. This planar remnant in the St. Marys Range approximates – to within 10 to 20 m – the wave-cut Waipounamu erosion surface in schist and phyllite. Periglacial landforms here include block fields (**B**), stone stripes (**ST**), and a solifluction lobe (**SL**) in the foreground wet area.

Figure 7.8   Tors associated with the Waipounamu erosion surface.
B. View southwest across the tor-strewn landscape of the Raggedy Range, New Zealand. Minimal to modest erosion (E) of the Waipounamu erosion surface began after exhumation from
beneath early Tertiary sediments. The ridgecrest valley at the right side (WG) may be a wind
gap – a stream channel that was abandoned as the ridge was folded into en échelon anticlines.

These larger streams may cut deeply into the weathered schist despite the modest runoff of this semiarid climate. The view shown in Figure 7.9 reveals an informative valley cross section. Straight (uniformly sloping midslopes) valley sides have short concave footslopes that merge with a valley floor that is much wider than the active stream channel. This stream is beveling a strath that becomes wider where streamflow impinges on adjacent hillslopes. This is presently occurring at the right-center of this view.

Figure 7.8   Tors associated with the Waipounamu erosion surface.
C. View south across a valley on the northeast flank of the St. Marys Range. The skylined planar ridgecrest is a tectonically tilted part of the Waipounamu surface (PS). The numerous tors (T) in the foreground are younger, post-dating valley incision, which was a response to uplift of the nearly planar surface relative to the local base level of the Waitaki River.

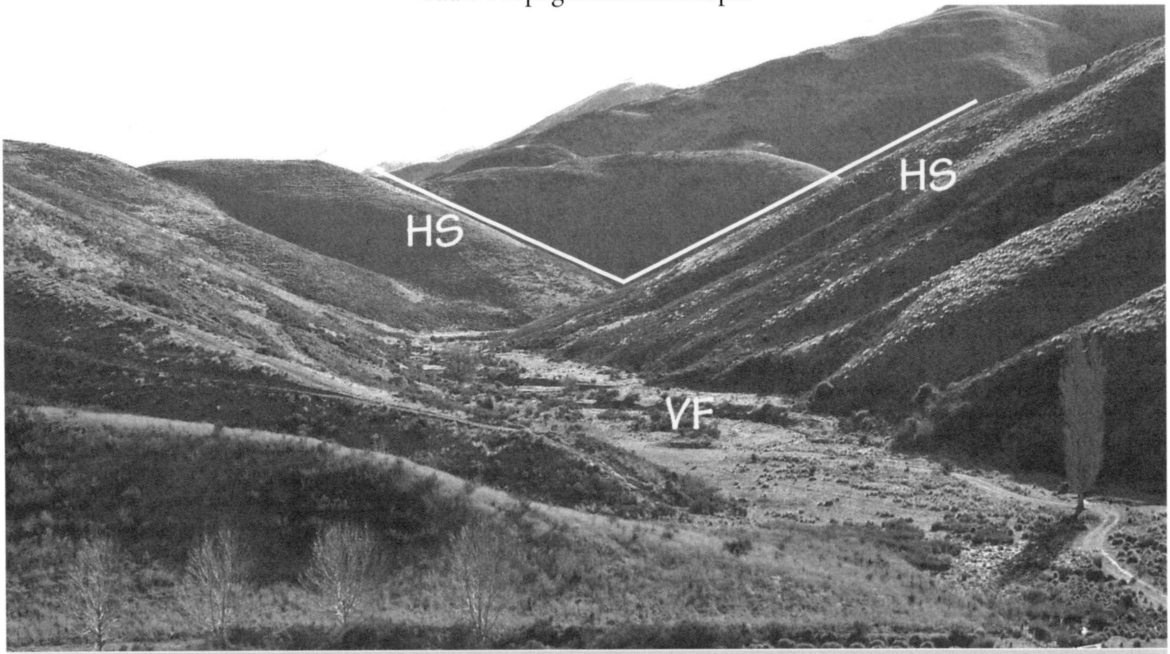

Figure 7.9 Flat-floored valley floor, **VF**, of a stream draining the northern Rock and Pillar Range has no adjacent stream terraces. Simple, straight hillslopes, **HS**, and no inner gorge reveal a slow, constant, rate of rock uplift.

The lack of either an inner gorge, or of strath terraces, is indicative of downcutting that has kept pace with the long-term uniform, slow uplift rate.

The absence of fill-terrace remnants in valleys suggests that Quaternary climatic fluctuations had minimal impact on some fluvial systems. This local lack of stream terraces indicates minimal tectonically induced base-level fall in reaches downstream from the Figure 7.9 reach of the fluvial system.

Not all Otago streams were this complacent. The longitudinal profiles of the larger rivers of the region are characterized by nice flights of stream terraces. Distant parts of their watersheds in the Southern Alps were sensitive to Quaternary climatic changes. Major changes in sediment yield occurred upon change from glacial to interglacial conditions.

Some local streams have terraces. Strath terraces cut by Ida Burn along the southeastern flank of Blackstone Hill (figure 6 photo of Markley and Norris, 1999) may be atypical. Ida Burn flows through a wide valley before dropping through Ida Burn Gorge in the structural saddle between the

Raggedy Range and Blackstone Hill en échelon folds. I suspect that Ida Burn may have beveled straths when it occasionally shifted sideways to impinge on Blackstone Hill. By the time it returned to notch the hillside again thrust-fault induced folding had raised the former strath level well above the active channel.

Strath remnants were tilted toward the valley-floor axis with increase of amplitude of the Blackstone Hill anticline. The highest remnant has now been tilted 6° (Fig. 7.10) and the trend of the plot suggests a uniform rate of uplift. Similar strath terraces occur on the northwest side where the Manuherikia River has impinged on the other side of Blackstone Hill. So both Ida Burn and Manuherikia River appear to be antecedent streams – established before onset of fault-propagation folding. They were able to notch into or cut down through the rising folds.

Drainage-net changes are revealing for Rough Ridge and South Rough Ridge (Fig. 7.11). Both ridges decrease in altitude towards the north

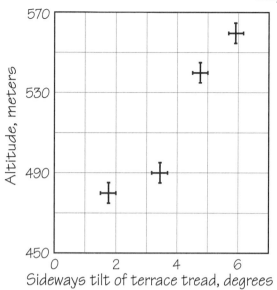

Altitude, meters

Sideways tilt of terrace tread, degrees

Figure 7.10 Relation between altitude and slopes of progressively older straths away from Blackstone Hill. Strath slopes were measured over distances of 30–120 m. Each strath was incised and elevated to become a strath terrace remnant with concurrent tilting from near-horizontal attitude by episodic ruptures on a blind thrust fault beneath the Blackstone Hill anticline. Figure 7 of Markley and Norris (1999).

but Rough Ridge is higher and older. The many wind gaps in South Rough Ridge were once parts of the streamcourses with headwaters in Rough Ridge. Downcutting by these small streams was not able to keep pace with rock uplift rates. Flow from each was deflected along the trough between the two anticlines until collective unit stream power increased sufficiently to allow stream-channel downcutting rates to exceed rates of rock uplift. Streams X, Y, and Z have sustained breaching of the anticline of South Rough Ridge. Each records another step in this drainage-integration process. Watersheds X and Y are highly asymmetric. Watershed Z is less asymmetric, perhaps because its younger stream only has to flow around the tip of the propagating anticline at Oliverburn. The ridgecrest altitudes, wind gaps, and asymmetric drainage basins all support the Jackson et al. (1996) model of northward propagation of South Rough Ridge.

A combination of geomorphic and paleoseismic data can be used to estimate amounts of tectonic displacement per thrust-fault rupture event. The distance between a wind gap and the tip of a propagating anticline can be measured, as can its increase in altitude. The literature regarding the behavior of active thrust faults is a useful basis for assigning a reasonable size of rupture for a thrust fault of a particular length (Scholz, 1982; Beanland and Berryman, 1989; Norris et al., 1994). Table 7.1 suggests that a typical rupture event in the central Otago fold-and-thrust belt increases the length of the fault about 10 to 50 m. These approximations seem reasonable and are similar in scale to the 30 to 50 m estimate for the Pearce normal fault in west-central Nevada of similar length made by Jackson and Leeder (1994).

Interesting work by James Jackson and many colleagues (2002) used [10]Be isotopic analyses of the sarsen stones littering the landscape. Parts of Rough Ridge and South Rough Ridge have these quartzite blocks left on the surface as erosion removed Tertiary cover beds. How long have the exhumed sarsen stones been at the land surface? Is there a systematic trend in sarsen stone exposure ages along the axis of a rising fold? Just the presence of sarsen stones exhumed from the overlying Tertiary section and now perched on the schist surface pedestals (Fig. 7.12) indicates extreme durability of this quartzite.

A study of sarsen stones at the German Hill locality on the lower slopes of the North Rough Ridge anticline (Youngson et al., 2005) confirms the premises of earlier work. Diagenetic cementation of selected parts of quartz sand and gravel beds creates local patches of complete infilling of all pore space by silica. The localized nature of cementation of the parent materials is shown by the common presence of a meter of softer sandstone between the base of a sarsen stone and the underlying weathered schist. The resulting impermeable quartz is exceptionally resistant to chemical solution, mechanical weathering, and abrasion by water and wind. The surrounding materials weather and wash away leaving a sarsen stone perched on a pedestal (Fig 7.12). These

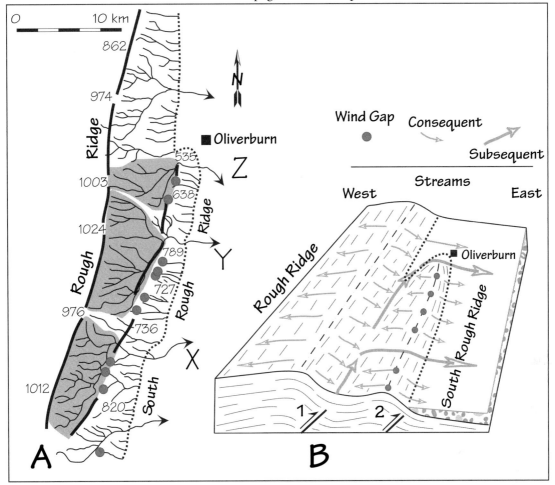

Figure 7.11 Evolution of fault-propagation folds and associated landscapes. From figure 8 of Jackson et al. (1996). Dotted line is the mountain–piedmont junction.
A. Drainage map of the left step in the range front between South Rough Ridge and Rough Ridge. Asymmetric watersheds X, Y and Z are shaded. Gray circles mark the locations of wind gaps where streams previously flowed across the axis of the South Rough Ridge anticline. Ridgecrest spot altitudes are in meters.
B. Structure and drainage schematic cartoon. The asymmetric character of the drainage basins of streams X, Y and Z that cross South Rough Ridge suggests that fault 2 is younger than fault 1 under Rough Ridge (1), and that fault 2 is propagating to the north.

relations suggest that the total erosional lowering of the exposed folded schist surfaces is not much more than heights of schist tors above the hillslopes.

The [10]Be analyses showed that the erosion rates of these quartzite blocks probably are <0.5 mm/ky, amongst the worlds slowest rates of rock erosion. Rough Ridge sarsen stones have minimum [10]Be ages of as much as 1300 ka, which makes them just about the oldest found anywhere – comparable to arid sites in Antarctica (Brook et al., 1995), Africa

Structure	Location	Uplifted height, m	Propagated length, m	Number of earthquakes	Propagation rate, m/event
Raggedy Ridge	Unnamed air gap	100	3,000	100–200	15-30
Rough Ridge	Gimmer Burn	200	7,000	200–400	18-35
Rough Ridge	Omnibus Creek	100	4,500	100–200	23-45
Rough Ridge	Unnamed gorge	40	1,500	40–80	19-38
S. Rough Ridge	Waitoi Creek	160	4,000	160–320	13-25
S. Rough Ridge	Dingo Creek	50	1,500	50–100	15-30

Table 7.1 Gorges of subsequent streams and wind gaps of former consequent streams used by Jackson et al. (1996) to estimate propagation rates in the Otago fold-and-thrust belt, New Zealand. Locations were raised (Uplifted height) while the ridge propagated a distance based on a reasonable number of earthquakes. Propagated length/number of earthquakes = propagation rate, in m per event. From table 1 of Jackson et al. (1996).

(Cockburn et al., 1999); and Australia (Bierman and Caffee, 2002). The extraordinarily low maximum sarsen stone erosion rates are only ~0.0003 m/ky.

The schist tors of central Otago have been dated too. [10]Be measurements on schist indicate erosion rates of 10 mm/ky, about 20 times faster than nearby sarsen stones (Youngson et al., 2005; Bennett et al., 2006).

Sarsen stones can reach 1 Ma exposure ages without being saturated with respect to erosion. An unknown is the amount of time to erode the soil and rock from around a sarsen stone and completely expose it to incoming cosmogenic particles. Sarsen stone NZ97-8, although on a ridgecrest, was partially buried, which resulted in an anomalously young age (Fig. 7.13).

The diagenetically silica-cemented strata range are massive to bedded sandstone and conglomerate. Conglomerate erodes faster and thick, bedded sarsen stones with sandstone and conglomerate layers may be prone to exfoliation that sheds the exposed top layer. The process is obvious where it is ongoing, but is not apparent where loss occurred a long time ago. Exfoliation greatly reduces the [10]Be content, producing an anomalously young age.

A strategy of extracting two samples per stone reduced the factor of variable exposure time by using only the oldest of the two age estimates.

Figure 7.12 Photograph of a sarsen stone in a gully on the footslope of North Rough Ridge at the German Hill site. This stone is resting on a schist pedestal. Other sarsen stones at this site still have 1 m of soft sandstone between the schist and the silica-rich remnant. Similar blocks in the background have toppled. Figure 7C of Youngson et al. (2005).

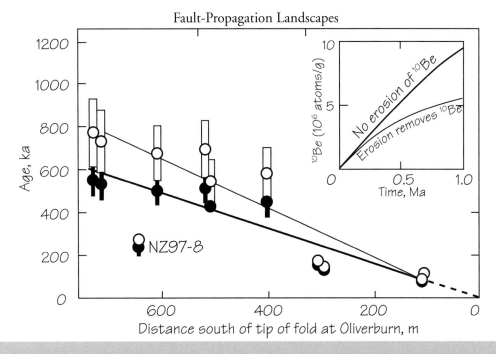

Figure 7.13 Plots of ^{10}Be ages along the nose of South Rough Ridge. From Bennett et al. (2006) after Jackson et al. (2002). Calculations of age estimates use a sea-level, high latitude production rate of 4.53 atoms/g/yr. Black circles are minimum ages assuming no erosion, with error bars reflecting mainly the assigned 10% error in production rate. These data suggest fault propagation rates of 1.3 m/ky. White circles and open error bars assume a maximum erosion rate. These data suggest fault propagation rates of 0.8 m/ky.

Inset graph: Decreasing rates of concentration increase for ^{10}Be with time at summit of Rough Ridge with no erosion (thick line) and allowing for erosion (thin line). Erosion rate was calculated assuming that Rough Ridge samples have reached saturated, steady-state values (input of new ^{10}Be equals loss by radioactive decay of ^{10}Be). Such steady state is not likely.

Even so, these may be minimum ages for sarsen-stone exposure. The time span between initiation of folding of the Waipounamu surface and exposure of the sarsen-stone stratum remains an educated guess.

Sarsen stones were found only on the northern section of the South Rough Ridge, but exposure ages for seven stones are much younger as one goes from south to north over a 600 m distance (Fig. 7.13). These data let Jackson et al. (2002) estimate fold-propagation rates that can be compared with the results of the drainage-network analysis. Assuming a steadily propagating ridge, anticline extension rates during the past 0.5 My would be 1.3 ± 0.4 m/ky with no erosion or 0.8 ± 0.4 m/ky allowing for erosion (Fig. 7.13). Scatter about this regression might also

be the result of fold growth that occurred in two episodes, one at about 0.6 to 0.4 Ma ago and another at 0.15 to 0.1 Ma.

The oldest sample in Figure 7.13 is 60 m above the valley floor. Assuming that the stream in this valley floor has been at the base level of erosion, the estimated uplift rate would be about 0.1 to 0.15 m/ky. This slow rate agrees with the results of paleoseismology studies by Beanland and Barrow-Hurlbert (1988) and Litchfield and Norris (2000). A mean propagation rate that is 10 times faster than the uplift rate concurs with other studies (Jackson and Leeder, 1994; Jackson et al., 1996; Keller et al., 1998; Benedetti et al., 2000), and with theoretical estimates by Jackson et al. (2002, p. 191).

The initial studies (Jackson et al., 1996, 2002) suggested that additional dating of elements of this landscape would improve our understanding of rates of tectonic and geomorphic processes. Isotopic analyses of ^{26}Al as well as ^{10}Be contents has reduced dating uncertainties (Bennett et al., 2006).

Assessment of the rates of Waipounamu surface degradation and how it varies on active folds needs crosschecks. One different approach would be to describe and date the soils chronosequence for the loessial blanketed landforms in central Otago. Soils chronosequences have a low precision and accuracy of dating (Bull, 2007, fig. 6.1), but can be can be used to efficiently identify the rapidly and slowly changing parts of a landscape (Amit et al., 1995).

Stream terraces are excellent sites for describing soil profiles for a soils chronosequence (Bull, 1991). Alluvium provides a uniform parent material. Watershed lithologies are represented by well drained sand and gravel. Gently sloping stream-terrace treads typically have minimal rates of erosion.

Many New Zealand terrace treads are so stable that loessial dust accumulates, especially during Quaternary times of full-glacial conditions (Pillans, 1994). Interglacial times are when soil profiles form in the loess, which continues to accumulate on both terraces and hills, but at a slower rate (Raeside, 1964; Young, 1964; Tonkin et al., 1974; Eden and Hammond, 2003). The Otago chronology would be a loessial soils-stratigraphic chronology as much as a fluvial surfaces soils chronosequence. See the excellent recent studies by Roering et al. (2002, 2004), Hughes (2008), and Hughes et al. (2009).

The 14 ^{10}Be age estimates used in the Figure 7.14 analysis clearly depict much different ages for the three stream terraces. T1 may be so old that much of the loess and soil may have been eroded from it, but the best preserved sections will still be quite useful. Soil characteristics of this dated "type section" can be correlated with soils on other flights of stream terraces, piedmont surfaces, and even flat ridgecrests to estimate approximate times of folding and creation of tectonic landforms.

The Oliverburn locality appears to undergo a pulse of uplift at about roughly 160 ky intervals. Each tectonically induced downcutting event creates a stream terrace. This geomorphic record

further supports the conclusions of Jackson et al. (1996, 2002) that these thrust-fault-cored anticlines are of Quaternary age but fault propagation events of modest size only occur infrequently. Uplift rates appear to be only 0.25 m/ky, or less.

7.2.2 California

Studies at Wheeler Ridge illustrate the advantages of using the soils-chronosequence approach to tectonic geomorphology in a California fold-and-thrust belt. It also provides an opportunity to consider how climate changes may have influenced wind and water gap processes in a rapidly propagating fold. As in New Zealand, wind and water gaps in the plunging Wheeler Ridge anticline record Late Quaternary extension of the fold tip and its associated thrust fault, as do the hillslope and drainage-net morphologies.

Tectonic deformation of Wheeler Ridge is rapid and young. Stream gravels deposited at only ~60 ka in the adjacent San Emigdio study area now dip as much as 50° (Keller et al., 2000).

The climate is different than in the Otago district of New Zealand. Wheeler Ridge has a strongly seasonal arid, moderately seasonal thermic climate (Appendix A). So the amount of moisture available for plant growth and streamflow is much less than in the weakly seasonal semiarid, and moderately seasonal mesic climate of Otago, New Zealand.

Rates and types of soil-forming processes, and the behavior of fluvial systems, varied greatly between full-glacial and interglacial climates in both New Zealand and California (Bull, 1991, Chapters 5, 2, 4). Climate-change induced sediment-yield increases were large enough to temporarily overwhelm the transport capacity of streams.

The result of Late Quaternary climate change was an aggradation event, even in the fluvial systems of steep rapidly rising mountain ranges. Keller et al. (1998) summarize this concept nicely: "Investigations of the tectonic geomorphology of Wheeler Ridge support the hypothesis that climatic perturbations are primarily responsible for producing geomorphic surfaces such as alluvial fan segments and river terraces – tectonic perturbations deform the surfaces." Keller used soils to document Late

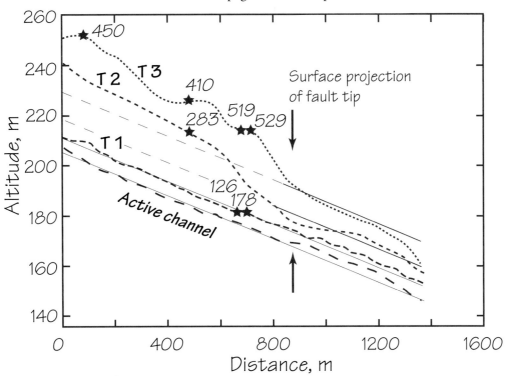

Figure 7.14 Longitudinal profiles of the strath terrace treads and Oliverburn active channel where the valley swings around the nose of the South Rough Ridge plunging anticline. The black stars are where sarsen stone sample locations are projected onto the profiles. Only the oldest [10]Be age, in ka, for each pair of [10]Be samples is shown here. Thin dashed lines are projected average slopes of terrace surfaces to show the progressively increasing offset of the present and former valley floors. From figure 7 of Bennett et al. (2007).

Quaternary tectonic and climatic controls on fluvial systems in the Wheeler Ridge–San Emigdio area.

Episodic deposition of fill terraces and al-luvial-fan segments was caused by major changes in Late Quaternary climate. By major I mean that cli-mate changed so much that annual stream power for a given reach of a stream was insufficient to con-vey all the bedload supplied by its watershed. This requires perturbations with time spans much longer than mere destruction of plant cover by fire. The impact of such climate-controlled perturbations was to accelerate alluvial-fan deposition, or to switch a stream channel's mode of operation from degrada-tion to aggradation to produce a fill terrace. Such processes typically require >1 ky. Careful work and

correlations by Ed Keller and his team placed the Wheeler Ridge soils chronosequence in a regional context that includes central and southern California (Davis et al., 1983; Dinklage, 1991; Keller et al., 1999; Laduzinsky, 1989; Seaver, 1986; Zepeda, 1993; Zepeda et al., 1986).

Aggradation events in the watersheds of the nearby Sierra Nevada coincided with times of Pleistocene glaciation. But unglaciated Wheeler Ridge had pulses of maximum hillslope sediment yield during times of rising sea level (Table 7.2), much like other arid and semiarid parts of southern California (Bull, 1991, Chapters, 2 4). Having a soils chronosequence helped answer questions about how hillslope length and steepness affected evolution of

Alluvial geomorphic surface and soil-profile thickness, m		Moist color (Munsell, 1990)	Soil structure Clay films on the pebble-matrix surfaces	Carbonate stage (Machette, 1985)	Approximate age, ka	Lower Colorado River correlative surfaces
Q1	0.5–0.8	10 YR 3/3	Massive None	Weak I	<10	Q3b
Q2	>2.4	10 YR 4/4	Moderate coarse angular blocky Many, thin	I to III	14	Q3a
Q3	>2.7	10 YR 4/4	Weak medium angular blocky Many, thick	II to III	60	Q2c
Q4	>3.1	7.5 YR 4/6	Massive to subangular blocky Many, thin	III	125	Q2b
Q5 soil profile has been largely stripped by erosion				IV	185	Q2a

Table 7.2 Soils chronosequence for alluvial surfaces on the Wheeler Ridge anticline, California. From table 2 of Keller et al. (1998). Correlative Lower Colorado River surfaces are described by Bull (1991, Chapter 2).

new drainage networks of ephemeral streams during the past 125 ky. Stream channels are initiated on alluvial surfaces being raised by propagation of the Wheeler Ridge anticline. Threshold values for slope and contributing runoff area need to be defined. Climate-change induced aggradation events may also cause water gaps to be abandoned, leaving them as wind gaps (much like the stream-channel avulsion described in Section 2.4).

Magnitudes and rates of Wheeler Ridge tectonism are related to movements on the nearby San Andreas fault (Fig. 7.15A inset). The 350 m high and 10 km long Wheeler Ridge is a plunging Late Pleistocene asymmetric anticline 22 km north of the San Andreas fault. Part of the obvious asymmetry shown in Figure 7.15D is related to a pre-tectonic-deformation piedmont that sloped northward at about 0.02. The north side is steeper than if a horizontal surface had been folded.

Thick Pleistocene alluvial-fan gravels of the Tulare Formation were generated by uplift of the San Emigdio Mountains. Continued compression generated by right-lateral displacement of the left step (the Big Bend) of the San Andreas fault resulted in parallel partitioning of strain (Fig. 1.10) and created new thrust-fault mountain ranges that encroached into the southern San Joaquin Valley.

Recent creation of the Wheeler Ridge thrust fault shunted alluvial-fan deposition still further north. These thrust faults are active. The magnitude Mw 7.5 Kern County earthquake in 1952 (Fig. 1.16, Bull, 2007) occurred 15 ± 6 km below Wheeler Ridge (Gutenberg, 1955) and raised the surface about 1 m (Stein and Thatcher, 1981). The magnitude Mw 5.1 earthquake of April 16, 2005 occurred beneath the incipient Los Lobos folds (Keller et al., 2000) that are starting to deform the surface of the alluvial fan of San Emigdio Creek.

Wheeler Ridge consists of West, Middle, and East Segments that have distinctive structural and geomorphic characteristics. The topography of Wheeler Ridge clearly documents its formation (Fig. 7.15). Anticline width, height, and dissection by streams undergo marked decreases to the east. Instead of gradual change, the landscape characteristics change abruptly at the large wind gap and again at the large water gap (the two transverse valleys in Fig. 7.15B).

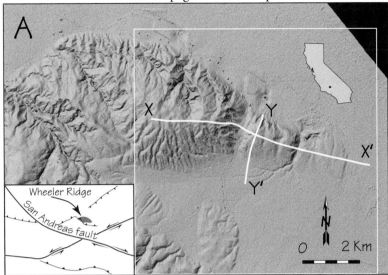

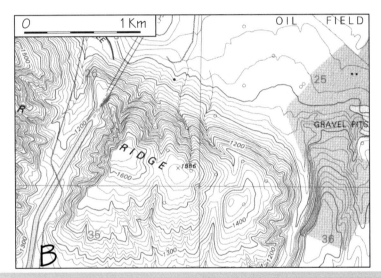

Figure 7.15 Topography of Wheeler Ridge, a rapidly propagating anticline encroaching into the southern end of the San Joaquin Valley, California. Inset map from Dibblee (1973) shows relation of Wheeler Ridge to thrust and strike-slip faults of the immediate region.

A.   Shaded radar image of Wheeler Ridge. Progressive diversion of streamflow around the nose of the eastward propagating anticline is recorded by wind gaps and water gaps. White box is area of Figure 7.17 geologic map. X–X' and Y–Y' are cross-sections shown in Figure 7.15 C, D. Image is NASA TOPSAR data, courtesy of Scott Miller.

B. Topographic map with a 20 foot contour interval of the middle segment. The California Aqueduct passes through the major wind gap. Small wind gaps are 200 m east and 100 m west of hill 1586 and south of "gravel pits" label. The stream issuing from the deep water gap is depositing an alluvial fan in NW ¼, section 25.

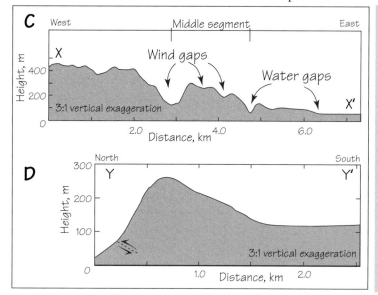

Figure 7.15 Topography of Wheeler Ridge, a rapidly propagating anticline at the southern end of the San Joaquin Valley, California. C. Topographic profile along line X–X' of Figure 7.15A shows characteristics of the West Segment, the Middle Segment between the largest wind and water gaps, and the East Segment.
D. Topographic profile along line Y–Y' of Figure 7.15A shows asymmetric anticlinal folding of the Middle Segment since 125 ka. Break-in-slope on north side coincides with emergence of an active thrust fault.

Both gaps are the locations of vertical tear faults oriented perpendicular to thrust faults beneath the ridge crest (Mueller and Suppe, 1997; Mueller and Talling, 1997, Figure 2). The linear scarp 1–2 km southwest of X' in Figure 7.15A marks the location of another tear fault bordering the East Segment. The locations of the tear faults coincide with the boundaries between different ages of alluvial surfaces deposited during Late Pleistocene aggradation events. Of course small wind gaps are present too (Fig. 7.15B), clearly recording a range in persistence of stream channels at different locations.

Unusually dense subsurface information from oil exploration (averaging 10 wells/km^2) provided Medwedeff (1992) many resistivity and self-potential well logs to analyze thrust-fault geometry (Fig. 7.16). Topographic expression of the tectonic deformation is obvious only where the thrust-fault ramps are stacked. The Middle Segment of the plunging anticline formed above a thrust wedge. His interpretation fits a model of upward increase in structural relief and suggests the presence of a second south-verging thrust ramp. About 320 m of structural relief since ~100 ka (125 ka is my preferred estimate) suggested an uplift rate of 3.2 m/ky (2.5 m/ky if age is 125 ka). Using soils chronosequence age estimates of Zepeda (1993), Medwedeff calculated rates of fault-slip as 4.3 m/ky, and of deposition as 1.8 m/

ky. The fold tip is propagating eastward at about 25 m/ky, five times the fault displacement rate. These rapid rates are typical of active collisional tectonic settings (Davis et. al, 1983).

Of course degradation of the rising anticline is highly variable (Shelton, 1966). Medwedeff divided the volume of incised stream channels by the surface area of eastern Wheeler Ridge to estimate a mean erosion rate of only about 0.37 m/ky.

The south-dipping thrust that offsets strata in the uppermost part of the fold crops out on the northern flank of the anticline (Fig. 7.15D), coinciding with the sharp break in slope approximately at the 1,100 foot contour in Figure 7.15B. Fault geometry for the East Segment is simpler. A straight north-trending escarpment is indicative of the tear-fault location.

Stream terraces and surfaces of alluvial fans were used as time lines passing through this tectonically active fold-and-thrust belt. Ed Keller and his students dated soil profiles in alluvium on hillslopes of the Wheeler Ridge and adjacent San Emigdio study areas. Strength and type of soil-profile formation were described at 23 sites and then used to correlate discontinuous remnants of tectonically deformed alluvium, and to estimate ages of distinct pulses of alluviation (Keller et al., 2000). Little remains of the Q4 deposits on Wheeler Ridge (Fig.

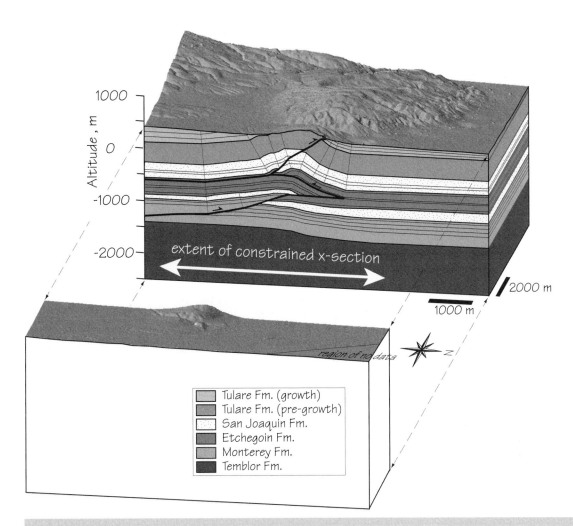

Figure 7.16 Block diagram structure with no vertical exaggeration of Wheeler Ridge. After Medwedeff (1992, figure 1.18) with added topography from NASA TOPSAR data, as compiled by Scott Miller.

7.17) but Q4 is the most extensive aggradation-event surface in their adjacent San Emigdio study area to the west.

Soil-age estimates were constrained by limited radiocarbon and uranium-series analyses made mainly on pedogenic pebble coatings of calcium carbonate (Table 7.2). Some estimates might be minimum ages because of the time needed to accrete the carbonate sample. Age estimates are approximate because Q5 soils are partially stripped, Q4 might date to either 105 ka or 125 ka, and the age of Q3 was estimated by interpolation assuming a constant rate of fold propagation (Fig. 7.18B).

This chronosequence correlates well with times of aggradation events in arid watersheds elsewhere in southern California. The Wheeler Ridge

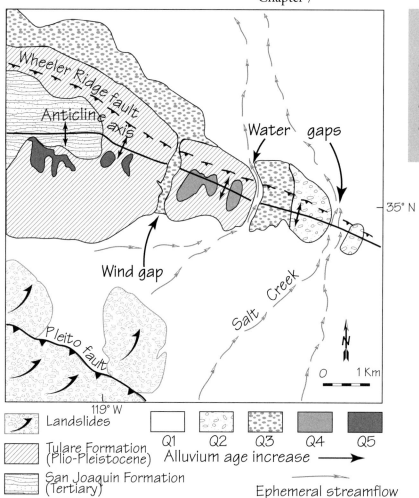

Figure 7.17 Generalized geologic map of Wheeler Ridge from figure 4 of Keller et al. (1998). Table 7.2 describes the soils chronosequence.

Landslides

Tulare Formation (Plio-Pleistocene)

San Joaquin Formation (Tertiary)

Q1 Q2 Q3 Q4 Q5
Alluvium age increase ——→

Ephemeral streamflow

soils chronology is part of a regional picture of Late Quaternary aggradation events.

Progressive tectonic tilting of the south side of the anticline converted northward sloping piedmont gravels to south-facing hillslopes (Fig. 7.18A) capped with remnants of alluvium (Fig. 7.17). These data have been corrected for the pre-uplift northward slope of the piedmont using a mean value of 0.021 estimated from the present piedmont. The correction is not small; it accounts for one-third of the tilt for the Q3 surface. The Figure 7.18A tilt rate appears quite uniform for the past 125 ky, but appears much less before 125 ky even if the minimum age of 185 ka is used for hillslopes with remnants of

Q5 alluvium. This is an artifact of the mechanics of folding. Uniform shortening that results in folding will reach a geometric threshold where uplift rates must decrease (Rockwell et al., 1988).

A rough approximation of the Wheeler Ridge thrust-fault propagation rate is estimated in Figure 7.18B. Q2 and Q4 are dated alluvium with known locations along the anticline crest. The Y axis error bars describe the distribution of each of the four mapped surfaces. The X axis error bar for Q4 is the 83 to 125 ka estimated age range based on uranium-series dating. The Q3 assigned age of about 60 ka is simply a reasonable interpolation between the Q2 and Q4 control points. The fold propagation rate (and the fault slip rate) seems to have been constant during the Late Pleistocene. The average rate of lateral propagation since Q4 time is approximately 30 m/ky, which is about 10–12 times the rate of rock uplift.

Were deflections of streams – creation of wind gaps – only the result of anticline propagation, or were climatic perturbations important too? Streamflow from the San Emigdio Mountains crosses the Pleito fault (Fig. 7.17) into a piggyback basin of piedmont deposition that became narrower and was tilted eastward with continued growth of Wheeler Ridge. Such tectonism caused streams that once flowed NNE to be diverted eastward before crossing the tip of the Wheeler Ridge anticline (Burbank et. al, 1996b).

Eastward diversions may coincide with times of Late Pleistocene aggradation events. Flow in entrenched stream channels, such as those funneling flow to the present large water gap (Fig. 7.15A, B, C), will continue until an aggradation event of sufficient magnitude completely backfills the channel thus allowing the stream in the piggyback basin to head off in a new direction.

Two scenarios are presented here. First is the case where an aggradation event lacks the intensity and/or duration to completely backfill the active channel in a water-gap reach. Second is the case where an aggradation event determines the time of abandonment of a water gap. A third possibility is that the stream channel continues to degrade during a regional aggradation event, but that the rate of incision is slowed as a water-gap reach is moved closer to the threshold of critical power during a period of increased bedload conveyance (or decreased unit stream power). Switching to an aggradational mode of operation within a water-gap reach is a function of tectonic base-level fall at the fold axis, base-level rise caused by alluvial-fan deposition downstream from the anticline, and how close the stream is to the threshold of critical power in the fold axis reach.

Viewed in terms of base-level processes, that promote aggradation of an incised active channel in a water-gap reach:

1) *ca*, strength of aggradation caused by the watershed sediment-yield increase and/or decrease in annual unit stream power in the water-gap reach. Conceptually, this would be the magnitude of

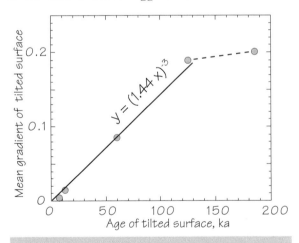

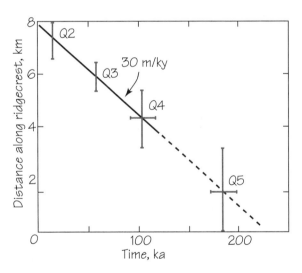

Figure 7.18 Rates of tectonic deformation of the Wheeler Ridge anticline during the Late Quaternary.
A. Increase of gradient of the south flank of the eastward plunging anticline. See Figure 7.18 for age error bars.

Figure 7.18 Rates of Late Quaternary tectonic deformation of Wheeler Ridge anticline. B. Propagation of Wheeler Ridge. Use of age control for Q4 and Q5 assumes constant vertical rates of propagation of the fold. From figure 8 of Keller et al. (1998).

departure below a threshold of critical power ratio of 1.0.

2) *pa*, is the stream-channel rise in aggrading alluvial-fan reach downstream from the water-gap reach. Exponent **n** is <1.0 because of the exponential decrease in this depositional base-level rise with increasing distance upstream from the fan apex. This assumes that the fan-apex reach is more strongly on the aggradational side of the threshold of critical power (is aggrading more rapidly) than is the upstream valley floor. *u* is the rate of rock uplift of the Wheeler Ridge anticline axis during the interval of an aggradation event. Uplift tends to cause continued incision by the stream in the water-gap reach, but only if this tectonic perturbation is strong enough to overcome climate-change perturbations that promote aggradation of the reach. The case for aggradation at the anticline axis is defined as

$$\frac{\Delta u}{\Delta t} < \left(\frac{\Delta ca}{\Delta t} + \left[\frac{\Delta pa}{\Delta t} \right]^{n} \right) \qquad (7.1)$$

Salt Creek passes through a water gap east of the lowest segment of the plunging anticline (Figs. 7.17, 7.19). The anticline here is a simple fault-bend fold and the thrust fault dips 36° to the south flattening with depth. Folding of the Q2 surface is obvious. An inset fill terrace was dated by Keller et al. (1998) and was considered as being the result of a weak Holocene aggradation event. Note that even this low terrace surface mimics the anticlinal warping of the older Q2 surface. The Q1 stream terrace is folded but the channel of Salt Creek is not. This water-gap reach briefly met the requirements of equation 7.1, and then returned to a degradational mode of operation.

The water gap did not become a wind gap, perhaps because:

1) A thrust-fault event increased uplift, *u*, sufficiently to quickly reverse the mode of operation in a reach that was already fairly close to the threshold of critical power.

2) The aggradation event lacked the duration, as well as intensity, to backfill the channel up to a spillover point height.

3) A new spillover direction might require major deposition because the trend of the anticline here, relative to the piedmont slope, does not favor stream-channel avulsion. The channel would shift only if the Q2 surface was overtopped.

Fluvial excavation of the large wind gap (Fig. 7.15B) was pre-Q3 and excavation of the large water gap was post-Q3. The mass removed in the creation of the large wind gap, combined with the weight of adjacent deposition, may have been sufficient to influence uplift processes though the self-enhancing feedback mechanism (Fig. 6.14) described by Simpson (2004). The Q3 climate-change perturbation was both strong and persistent relative to the local tectonically induced tendency for stream-

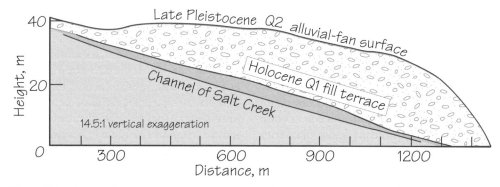

Figure 7.19 Topographic profiles of the active channel and incised alluvial geomorphic surfaces of Salt Creek, which crosses the eastern tip of the Wheeler Ridge anticline. See Figure 7.17 for distribution of the alluvial surfaces. From figure 6 of Keller et al. (1998).

channel downcutting, thus setting the stage for permanent shift of streamflow to the east.

The large wind gap is floored with alluvium of the extensive Q3 aggradation event (Fig. 7.17). Here is a case where aggradation played an important role in diverting streamflow to a new location. A reasonable model is that aggradation promoted by climate change plus base-level rise resulting from Q3 alluvial-fan deposition (equation 7.1) shunted streamflow eastward where additional Q3 alluvium was deposited east of the present large water gap.

Termination of the climate-change induced Q3 aggradation event coincided with initial entrenchment of the large water gap. The rate of aggradation in the piggyback basin upslope from Wheeler Ridge may also control drainage patterns and shifts (Van der Beek et al., 2002). Streams leading to the new water gap were steeper than the former north-trending stream to the wind gap. The resulting gradient-induced increase of stream power was sufficient to induce and maintain persistent stream-channel downcutting. Flows through the large water gap have remained far to the degradational side of the threshold of critical power The Q2 and Q1 aggradation-event perturbations brought the water gap reach closer to the threshold of critical power (1.0) but not across the threshold.

Salt Creek (Fig. 7.17) may have been more sensitive to diversion during Q2 time because of a low spillover point. It presumably flowed around the tip of the propagating anticline and through the now large water gap after the Q3 aggradation event. A shift to its present course would involve minimal change in direction, so it may have been more susceptible to diversion caused by aggradation during the Pleistocene–Holocene climatic change. Temporary backfilling of any reach upstream from the large water gap would create a channel-fan environment (Bull, 1997), like those along the present channel, thus facilitating shift eastward to the termination of the East Segment.

The potential for stream capture is greater between aggradation events. New channels being created by steepening near the nose of the fold may eventually capture drainage passing through a water gap (Edward Keller, written communication, May, 2008).

The deep and persistent water gap between the Middle and East Segments is partly a function of a lithologic control. The occasional nature of streamflow events relative to a rapid uplift rate, should have elevated the water-gap reach sufficiently to divert the active channel around the nose of the propagating anticline. But the loose to weakly cemented sands and gravels of the Tulare and San Joaquin formations are easily eroded – a lithologic control that more than offsets the low annual unit stream power.

Drainage density of stream-channel networks reflects the processes governing landscape dissection (Schumm, 1997). Wheeler Ridge may be unsurpassed for understanding how drainage networks evolve with the passage of time. Drainage networks along the ridge become younger and progressively less tilted towards the anticline's eastern tip. The small range in altitude and small area indicate negligible spatial variations in climate. Gravelly and weakly cemented alluvium, although of different ages, provides a uniform lithologic substrate with sparse plant cover to prevent runoff from eroding channels. A surface gradient of 5° to 10° is needed to initiate headwaters rills (Keller et al., 1998, and previous workers cited by Talling and Sowter, 1999).

Drainage density involves subjective interpretations as to what constitutes a stream channel, whether one uses digital elevation models, topographic maps, or aerial photos. Channel density interpretations by Talling and Sowter based on aerial-photo interpretations (1:16,000 scale) may have more imponderables than their valley-density analysis based on topographic maps (1:24,000 scale).

Three different ages of stream channels are shown in Figure 7.20A. Deeply incised valleys in the upper-right part of the oblique aerial view record the long-term initial incision of stream channels into the rising anticline. Such primary channels define the shapes of triangular facets into which a younger set of channels is well established. The ephemeral rills leading to the headwaters of the triangular facet channels may be temporary. They probably formed during the past 150 years in response to climatic perturbations (Bull, 1964) or livestock grazing practices. Such relatively minor channels would be included in a drainage-net analysis using aerial photos, but not if

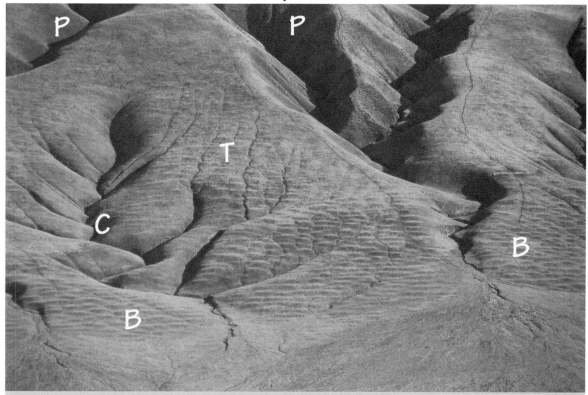

Figure 7.20 Drainage density variations on the Wheeler Ridge anticline.
A. Temporary and permanent stream channels on the north side of Wheeler Ridge. **P,** Primary
channel between triangular facets; **C,** consequent drainage net eroded into triangular facet; **T,**
temporary rills and channels related to the current episode of arroyo cutting. Small hillslope
benches, **B,** are 3–5 m high and 7–11 m wide, and are soil-creep features on this wetter north-facing slope (Bielecki and Mueller, 2002). Photograph courtesy of Karl Mueller, University of Colorado.

one were using topographic maps. The prominent topographic benches on these hillslopes are thought to be the result of mass movements of an overthickened A-horizon. This process is intermittent, occurring during exceptionally wet winters. It might be accentuated by strong earthquakes if they occurred during the rainy season (Bielecki and Mueller, 2002).

The north-facing slopes of Wheeler Ridge are more complex; their slope decreases below the thrust fault (Fig. 7.20C). Their cooler, wetter microclimate results in a spatially variable mixture of mass movement and surface flow geomorphic processes. The presence of local bedrock badlands on the more rugged north-facing slopes also complicates use of north-side data. We would prefer only soil-mantled

hillslopes whose rates of change are dictated largely by the magnitude and frequency of rainfall-runoff processes. It may be difficult to decipher tectonic controls on the origins of stream channels on either the north or south sides of Wheeler Ridge if parts of the drainage net (such as **T** in Figure 7.20) reflect nothing more than the result of short-term climatic variations (Bull, 1964; Bull and Kirkby, 1997), or human impacts of construction and grazing.

Talling and Sowter recognize possible influences of these many variables and present a nice comparison of drainage nets in the north- and south-side settings. I minimize potential complications in this summary by using only valley density to illustrate the drainage-net evolution on the south-facing flank of

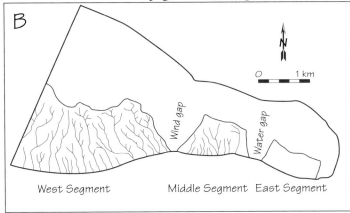

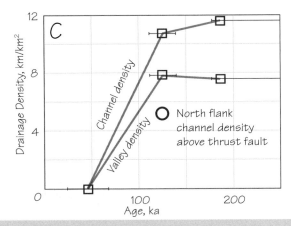

Figure 7.20 Drainage density variations on the Wheeler Ridge anticline.
B. Map of small ephemeral streams based on crenulations on 20 foot contour interval, 1:24,000 scale topographic maps.
Age of the West segment is >125 ka, Middle segment 60 to 125 ka, and East segment <60 ka. From figure 6 of Talling and Sowter (1999).
C. Graphs of changes in stream channel density and valley density with passage of time. Circle shows much lower channel density on the north flank. From figure 8 of Talling and Sowter (1999).

the anticline (Fig. 7.18B). Drainage density of valleys is markedly different for the West, Middle, and East Segments. The hillslopes of the young East Segment have yet to develop stream channels. Slopes are gentle but tectonic deformation has been enough to reverse the slope direction from north to south. The channels of the Middle Segment are short, but density of drainage appears similar to that of the West Segment. Both channel density and valley density increase rapidly from 0 in the East Segment to 11 and 8 km/km² in the Middle Segment. Then as noted

by many other studies in the literature (see Talling and Sowter's figure 3), the trend flattens (Fig. 7.18C). Once the slopes have been tectonically steepened to a threshold value, drainage density quickly increases to an optimal value for a given setting. Note that the steep, shady north flank of the Middle Segment has only half the channel drainage density as the adjacent south flank with less plant cover.

Fault displacement rates at Wheeler Ridge are compared with six other thrust-fault and normal-fault study sites in Table 7.3.

Fault, Locality	Fault age, ky	Type of fault	Length, km	Vertical displacement rate, m/ky	Lateral displacement rate, m/ky	Vertical rate/ lateral rate ratio
South Alkyonides, Gulf of Corinth, Greece		Normal	38	2.6–4.7	West tip 12–17	0.13–0.30
	800 ±200	Normal			East tip 2.8–3.7	
Pearce fault, Nevada	5,000 ±2,000	Normal	30	1.0	1*	0.05
Raggedy Ridge, New Zealand	2,000 ±1,000	Thrust	20	>1.0	3–6**	0.03–0.07
Rough Ridge, New Zealand	2,000 ±1,000	Thrust	20	>1.0	2.6–9**	0.02–0.08
Wheeler Ridge, California	200 ±50	Thrust	10	2.5–3	30	0.17
Montello, Italy	400 ±100	Thrust	15	1.8–2.0	10–20	0/05–0.1

* Assuming an earthquake recurrence interval of 10 ky
** Assuming an earthquake recurrence interval of 5 ky

Table 7.3 Comparisons of vertical displacement rates and lateral propagation rates for normal and thrust faults. Prehistoric events are not dated for the Nevada and New Zealand faults; instead estimated earthquake recurrence interval is used. Patterned after table 1 of Morewood and Roberts (2002). Locations and source articles are cited in the text, except for the Italian study of Benedetti et al. (2000).

7.3 Transtensional Faulting

Active strike-slip faults play a key role in the tectonics of plate boundaries because of offsets that dwarf lengths of those by normal and thrust faults. Geomorphic appraisal of propagation rates of the tips of strike-slip faults is restricted to a few sites that are available to observe. The San Andreas fault of California is ruled out because it ends in the Mendocino triple junction and the rhombochasms of the Gulf of California. The Alpine fault zone of New Zealand terminates in two subduction zones.

So we go to an area where strike-slip faulting has commenced in the more recent geologic past – the Walker Lane–Eastern California shear zone. Here, the Death Valley–Fish Lake Valley fault zone extends northward ~300 km from the Garlock fault to its termination on the eastern flank of the lofty White Mountains of California and Nevada. Activity on the impressive Owens Valley right-lateral fault (Sections 1.2.2, 8.2) is surpassed by the Death Valley–Fish Lake Valley fault zone, which may have been picking up slip transferred from these nearby fault zones (Reheis and Dixon, 1996). Geodetic surveys suggest that this fault zone has 3 to 8 mm/yr of the 9.3 ± 0.2 mm/yr relative motion between the Pacific and North American plates (Humphreys and Weldon, 1994; Dixon et al., 1995, 2000, 2003; McClusky et al., 2001; Wernicke et al., 2004).

An estimate of the magnitude of tip-propagation events of this impressive fault would be routine if there had been a historical surface rupture. Lacking that, geologists have turned to geomorphic

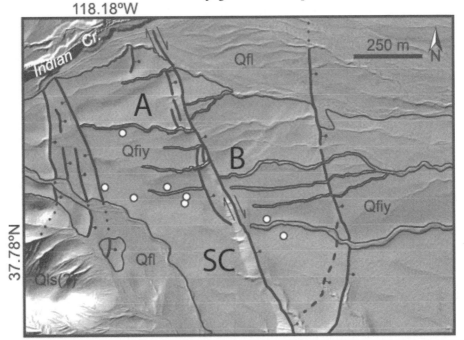

Figure 7.21 Faulted alluvial fan of Indian Creek, Nevada, on the Fish Lake Valley side of the White Mountains (Fig. 1.2). Image of this northern end of the Death Valley–Fish Lake dextral fault system was created with LiDAR data. Normal faulting prevails here except for the center fault, which has mainly dextral slip, despite the high scarp, SC. A and B mark a dendritic stream channel that has been offset ~178 m. The white circles are the locations where surficial boulders were sampled for [10]Be dating. Qfiy is Late Pleistocene alluvium, and Qfl is latest Pleistocene and Holocene alluvium. Figure 3a of Frankel et al. (2007b). Geology is after Reheis et al. (1993).

ways of studying the characteristics of the northern tip of the Death Valley–Fish Lake Valley fault zone. Mapping recently faulted landforms and dating ruptured alluvial surfaces with a soil-profile chronosequence by Marith Reheis and Tom Sawyer (1997) provided an estimated slip rate of 1–3 m/ky.

This work included the terminal area of the Death Valley–Fish Lake Valley fault zone at Indian Creek. Recent dating of the ruptured fan surface using terrestrial cosmogenic nuclides and mapping the fine details of offset stream channels using LiDAR (airborne laser swath mapping) images confirmed and refined the initial estimates (Frankel, 2007; Frankel et al., 2007a, b, 2008). The main points of their work are discussed here.

Obvious surface ruptures only extend about 5 km further north from Indian Creek and normal faulting is the dominant termination style. All but one of the prominent scarps on the alluvial fan of Indian Creek are the result of normal faulting. The strike-slip component of slip is partitioned, occurring only on the center fault zone of Figure 7.21. One characteristic of transtensional fault propagation here is for normal faulting to arrive first, perhaps changing rock mass strength characteristics. In a more regional sense, the westward encroachment of the extensional faulting may have preceded, and set the stage for, development of the new dextral plate-boundary fault zone that we call the Walker Lane–Eastern California shear zone.

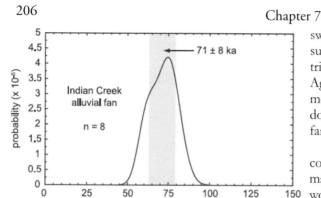

Figure 7.22 Probability density plot of
[10]Be terrestrial cosmogenic nuclide ages
from surficial boulders on the Indian Creek
fan. The white circles on Figure 7.21 are the
sample locations. The vertical grey bar is the
1 σ standard deviation about the mean age
for the Qfiy surface. Figure 4b of Frankel et
al., 2007b.

Normal faulting of fan surfaces on the east side of the White Mountains created local base-level falls that caused streams to incise their channels. The age of this depositional surface and the factors controlling subsequent development of dendritic patterns of stream channels is a key part of this story about slip rates and fault-tip propagation processes.

This mountain front had characteristics of class 5 tectonic activity (inactive in terms of base-level fall) before the recent arrival of faulting. Absence of early and middle Pleistocene fanhead trenches here indicates that the faulted surface is young. It is the result of an episode of aggradation related to a large sediment-yield increase in the White Mountains source watershed of Indian Creek. The tip of the transtensional fault zone propagated onto a thin alluvial fan whose surface was created during a single aggradation event. This single age surface is ideal for those using cosmogenic [10]Be to date gneissic boulders on the fan surface.

The style of deposition for a tectonic alluvial fan would be different. Space for deposition would be created by repeated base-level falls. Deposition would raise the surface along one radial line and then

switch to a different radial location. A time span of substantial length might pass before the braided distributary channels returned to the initial radial line. Ages of different parts of the fan would vary still more if climate-change episodes of stream-channel downcutting delayed renewal of deposition on the fanhead.

The best sampling strategy, for terrestrial cosmogenic nuclide dating of either tectonic or climatic fans, is along a single radial line. Ideally this would be the path of a single flow event from the apex to the toe of the fan. All but one of the boulders dated by Frankel et al. (2007b) are along the same radial line (Fig. 7.21). A modeled age estimate for these eight boulders (Fig. 7.22) indeed has minimal spread, being 71 ± 8 ka. Clearly, this surface is quite young and its age indicates that it was deposited in the Pleistocene during a time of full-glacial climate. We would like to answer questions such as:
1) When did the propagating fault tip arrive?
2) Did normal faulting precede initiation of dextral faulting?
3) If so, what happened in the interval between arrival of normal faulting and initiation of dextral faulting?
Details of the fan topography (Fig. 7.23) and large dextral offsets (A to B on Figures 7.21 and 7.23) answer these questions.

An aggrading fan has braided distributary channels. Streams split into ever smaller channels in the downslope direction as bedload is deposited. The present fan surface has dendritic erosional channels. Tributaries feed into ever larger channels in the downslope direction in order to be able to transport the bedload that erosion supplies from the upstream reaches of the stream. This change in style of fan-surface processes coincided with 1) the end of the climate-change induced aggradation process, or 2) a change in tectonic setting.

Incipient dendritic channels of streams on the fan surface were deepened as a result of normal faulting. This occurred before right-lateral faulting began. Strike-slip faulting then offset headwaters reaches from the downstream reaches – it beheaded stream channels from their source areas. Such removal of headwaters streamflow, such as at location B of Figures 7.21 and 7.23, makes the stream

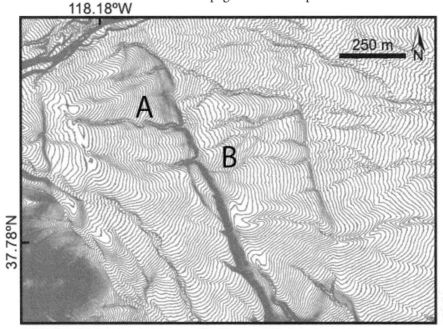

Figure 7.23 A 1-m contour interval map of the area of the Figure 7.21 map showing the details of erosion of the surface of the Indian Creek fan, and the scarps created by normal and oblique-dextral faulting. Figure 3e of Frankel et al. (2007b).

channel a relict landform. Little change occurs after a channel loses its source of streamflow.

Normal faulting began shortly after deposition of the climatic fan surface, and may have been the reason for cessation of fan deposition. Rocks fail most readily under tension, which would seem to explain why normal faulting occurred before dextral faulting upon arrival of the transtensional fault tip. Large tectonic base-level falls would have increased the gradient of Indian Creek sufficiently to cross over to the erosional side of the threshold of critical power. The depositional mode – fan aggradation – would switch to erosion as Indian Creek (upper left corner of Figure 7.21) cut a channel into the fan surface.

Each downfan-facing zone of normal faulting would be a base-level fall that would facilitate convergence of fan-surface runoff into channels. Maximum channel gradients and incision would occur just upstream from these fault scarps (Figs. 7.21, 7.23). Such local base-level falls eventually would

convert the planar fan surface into gentle hillslopes draining to channels. An example is shown by the contour lines near location A in the northwest part of Figure 7.23.

The boulders sampled for [10]Be dating are on the ridgecrests between these small streams. Being on those parts of the fan surface least affected by subsequent erosion tends to minimize the dispersion of ages in Figure 7.22.

Estimation of the rates of horizontal slip depends on what is offset and how it is dated. Reheis and Sawyer (1997) noted 83–165 m of right-lateral offset of a debris-flow channel. The Qfiy Indian Creek fan was assigned a Late Pleistocene age based on soil-profile characteristics. The soil has a 5–10 cm thick silty vesicular A horizon, an argillic Bt horizon with moderate clay films, and stage II to III carbonate development in the Bca horizon (Reheis and Sawyer, 1997). The Qfiy surface age was thought to be created between 50 ky and 130 ky. This provided an estimated fault-slip rate of 1.1–3.3 m/ky.

The fan surface age of Frankel et al. (2007b) falls within the Reheis and Sawyer estimate, being 71 ± 8 ka (Fig. 7.22). Using the LiDAR images they noted consistently similar amounts of right-lateral offset of four stream channels. Such features that extend across an active strike-slip fault are called *piercing points*. Their estimate of Late Pleistocene displacement at Indian Creek is 178 ± 20 m, which yields a slip rate of 2.5 ± 0.4 m/ky.

This slip-rate estimate is a minima because fan-surface creation pre-dates the initiation of faulting. It would be nice to know more about the timing of the four events:

1) Cessation of fan deposition. This climate-change aggradation event occurred during full-glacial stade 6-7 (78 to 68 ka) of Bischoff and Cummings (2001). The fan surface records termination of aggradation. The range of fan-surface ages is assumed to be short for the small fan of Indian Creek that was sampled. [10]Be dating of other parts of the fan might reveal that 71 ka is the universal age estimate, or the probability density plot might have closely spaced peaks.

2) Creation of dendritic stream channels. New channels would develop slowly on a gravelly alluvial fan because high infiltration rates would minimize runoff. Soil-profile development would increase runoff/infiltration ratios. Perhaps formation of dendritic channels would be delayed until a runoff threshold value had been reached.

3) Onset of normal faulting would result in rapid stream-channel incisement. Such base-level fall is efficient and quick, so much so that one might assume that obvious, deep channels would not form on the fan surface until then. Ideally, the onset of normal faulting would terminate fan deposition and initiate rapid stream-channel downcutting. In this case the 71 ka timing would apply to both onset of normal faulting and channel incision.

4) Onset of right-lateral offset of the stream channels. This did not occur until after the channels were created. Channel incision would cease where a stream lost its headwaters source of streamflow. Not much has happened to the channel below the letter B on Figure 7.23 after it was disconnected from its headwaters, which is the channel below the letter A.

The other key study site of Frankel et al. (2007b) is at Furnace Creek, which is 29 km southeast of Indian Creek. Similar dating and mapping yielded a slip-rate estimate of ~3.1 m/ky, which is less than ~4.5 m/ky estimate for the Death Valley section of the fault (Frankel et al., 2007a), 130 km from Indian Creek.

Noting that the 2.5 m/ky slip rate is a minimum, we can use a faster rate to deduce the time span between cessation of fan deposition and onset of dextral faulting at Indian Creek. Let us assume that the arrival of the Death Valley–Fish Lake Valley fault zone at Indian Creek was abrupt with no diminishment of slip rate as it propagated towards the northwest. By using the Furnace Creek value for slip rate, 3.1 m/ky, the onset of strike-slip faulting at Indian Creek would have been at ~57 ka (2.5/3.1 × 71 ka = 57 ka). This inferred 14 ky time span after cessation of fan deposition would be long enough for incipient soil profiles to form to enhance runoff, and for normal faulting to initiate the base-level fall process essential for incising the channels of streams originating on the Qfiy fan surface.

How did arrival of the Fish Lake fault affect landscapes on the east side of the White Mountains? Normal and transtensional faulting arrived so recently at Indian Creek that the larger features of the landscape have been little changed by a tectonic base-level fall of >50 m. Instead of being straight (active-fault-controlled sinuosity of 1.0) the mountain front just northwest of Indian Creek has sinuosity values of >4. Instead of a V-shaped canyon upstream from the mountain front, Indian Creek has been at the base level of erosion long enough for the stream to bevel a wide valley floor (Fig. 7.24). Note the broad, concave footslopes. Ratios of valley-floor width to mean valley height are 3 to 6. Such aspects of the landscape would rate a tectonic activity classification of 5, were it not for the recent arrival of the Death Valley–Fish Lake Valley fault zone.

Normal and transtensional faulting are active at McAfee Creek, 19 km to the southeast, but the landscape is much different (Fig. 7.25). A narrow canyon has been incised upstream from a rapidly accumulating tectonic (meaning thick) alluvial fan. Concave footslopes are absent, except where the remnants of the former valley floor, FVF, have been preserved. Presumably the alluvial-fan reach crosses several normal faults, and the main strand of

Figure 7.24 View up the broad valley of Indian Creek from the fan apex. Qfiy and Qfi are the mapped units of Figure 7.21. The white line parallels a markedly concave footslope that together with the width of the valley floor is indicative of >1 My of tectonic quiescence.

Figure 7.25 View of McAfee Creek and alluvial fan from near the town of Dyer, Nevada. Gentle concave footslope below outcrop is a remnant of the former valley floor (FVF). Present alluvial fan is inset below old fan levels (OF), now dark with rock varnish. Entrenched fan and the upstream trunk stream record a large Pleistocene base-level fall associated with the Fish Lake Valley fault zone.

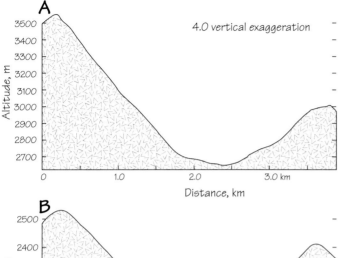

Figure 7.26 Topographic profiles across the valleys on the east flank of the White Mountains of California and Nevada.
A. Cross-valley shape at 7 km upstream from the mouth of Indian Creek.
B. Cross-valley shape at 4 km upstream from the mouth of McAfee Creek

the transensional fault. This appears to have caused sufficient lowering of the area of active deposition that Late Pleistocene surfaces, OF, are isolated off to the sides. Their darker hue is the result of accumulation of rock varnish on the surficial materials. This front has a class 2 tectonic activity rating, or class 1 if we consider the unusual reasons (not a simple range-bounding fault zone) for fanhead entrenchment.

A pair of representative cross-valley topographic profiles for Indian and McAfee Creeks illustrate the significant contrast between these two tectonic settings. The creation of an inner gorge in McAfee Creek suggests that the Late Pleistocene tectonic perturbation has migrated up to the midslopes on the valley sides.

How long does it take to make such changes (Fig. 7.26A, B) in landscape? Fault-tip propagation rate of the Death Valley–Fish Lake Valley fault zone to the northwest is not likely to exceed the rate of slip along the new trailing dextral fault.

Lateral propagation rates for normal and thrust faults range from 3 to 30 m/ky (Table 7.3). Using a propagation rate of 30 m/ky, the Death Valley–Fish Lake Valley fault zone would propagate the 19 km from McAfee Creek to Indian Creek in only 630 ky. This is not enough time to accomplish the Figure 7.26B scale of erosion in the mountains, considering the present arid to subhumid range of climate with increasing altitude. At the low end of the Table 7.3 range, fault-tip propagation at 3 m/ky would require 6,300 ky to reach Indian Creek. This seems too long because inception of dextral faulting in the Walker Lane–Eastern California shear zone probably began after that time.

A fault-tip propagation rate in the range of 6–9 m/ky would equate to 1.9–3.2 My. The arrival of the Death Valley–Fish Lake Valley fault zone at McAfee Creek in this interval seems reasonable and would post-date the Sierra Nevada batholithic delamination event at 3.5 Ma. If so, strike-slip fault propagation rates would not be greatly different than the mean values estimated for normal and thrust faults (Table 7.3).

Creation of the 300 km long Death Valley–Fish Lake Valley fault zone in the late Cenozoic would involve more than fault-tip propagation at the rates noted above. The Figure 7.1 model can be used here. Time −1 would be the rapid creation of the long fault zone as many separate faults coalesced to create a long master fault. Then Times 0 to 4 style of tectonic controls would apply, using dextral offset instead of throw.

7.4 Summary

Geomorphic investigations have added much to our understanding about the magnitudes and rates of propagation of faults, and of Quaternary slip rates. This diverse discussion focused on Nevada, Greece, New Zealand, and California. The costal landform of marine terraces helped assess fault propagation rates and constrain ages of wind gaps in the Gulf of Alkyonides of Greece. Morewood and Roberts (2002) compared fault-displacement rates for these sites, and the Montello, Italy site of Benedetti et al. (2000).

Fault lengths equal or exceed the local seismogenic depths. Vertical displacement rates vary by tenfold with the faults in the Nevada and Otago study areas being much less active. Three sites have fault tips that propagate > 10 m/ky. The ratio of vertical to lateral propagation appears to be a function of fault age. This supports the Figure 7.1 model for two stages of fault formation and propagation.

Tectonic deformation directs streamflow. Changes in streamflow position record formation of new faults and continuing propagation of the tips of existing active faults. Initial rapid lengthening of short normal and thrust faults creates a single fault zone with a decreasing propagation-rate/uplift-rate ratio. A series of 30–50 m increases in fault length results in a cumulative displacement sufficiently large to be obvious in a drainage-net analysis.

Small streams may lack sufficient annual unit stream power to degrade their channels in a reach that is being raised by faulting or folding. Such local faulting or folding creates a barrier that forces

the stream laterally to a new location on a piedmont, or allows capture by a larger master stream in mountains. Former stream channels are more likely to be preserved in ridgecrests where erosion is slowest. Such fluvial notches in ridgecrests, wind gaps, usually contain fluvial gravel.

Larger streams can maintain their courses across rising folds. Their watersheds are not centered directly upstream from the water gap – where the active channel passes through a rising ridge. This is because shifting and steepening hillslopes help the stream capture progressively more drainage in the direction of fault propagation.

A Cretaceous erosion surface that pre-dates folding was used to appraise thrust-fault propagation rates in new Zealand. Modest size uplift events occur infrequently, uplift rates are only ~0.25 m/ky. Onset of anticline uplift was dated using exposure timespans of durable quartzite blocks exhumed from thin early Tertiary strata capping the Waipounamu erosion surface. ^{10}Be accumulation in quartz begins when exhumation brings these "sarsen stones" within ~2 m of the land surface. These quartzite blocks erode at <0.5 mm/ky, which is amongst the worlds slowest rates.

In contrast to Otago, thrust-faulted Wheeler Ridge in California is rising fast. The Late Pleistocene fold propagation rate is ~ 30 m/ky and Late Quaternary gravels dip 50°. Landscape characteristics, and thrust-fault patterns change abruptly at the large wind gap and again at the large water gap farther east. Anticline width, height, and dissection by streams decrease to the east. Northward flowing stream channels entrench into the axis of a rising fold. Eastward diversions of streamflow seem to coincide with times of Late Pleistocene aggradation events. Large aggradation events completely fill entrenched stream channels, thus allowing diversion of flow to the east and around the tip of the propagating anticline.

Propagation rates of strike-slip faults are best studied in relatively young and expanding tectonic settings (Stein, et al. 1988). Northwest propagation of the transtensional Death Valley–Fish Lake Valley fault zone in the Walker Lane–Eastern California shear zone is preceded by propagating normal faults for perhaps 10–15 ky. Streamflow rapidly incises the fan surface dendritic stream channels where active normal faulting creates large base-level falls. The process creates narrow stream channels that are piercing-point landforms to estimate —rates of slip by subsequent strike-slip faulting. Having a transtensional component provides sufficient base-level fall to cause the associated fluvial systems draining the east side of the White Mountains to change. This allows assessment of landscape change as related to the rate of fault-tip propagation. Approximate transtensional propagation rates of 6 to 9 m/ky may be about the same as average propagation rates for normal and thrust faults.

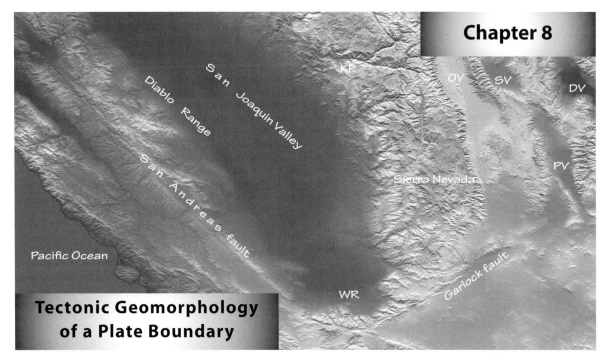

Chapter 8

Tectonic Geomorphology of a Plate Boundary

Geomorphologists like to study Holocene tectonic activity and how it affects geomorphic processes. Tectonically active and inactive landscapes are part of a plate-tectonics framework with longer time spans and large spaces. We need to know how plate-movement rates and perturbations influence geomorphic studies.

Shifting crustal plates create majestic mountain ranges such as the Andes and Himalayas. Large-scale tectonic processes within plates influence lesser landforms such as mountain-front escarpments, drainage basins, lakes, and stream channels.

In this chapter we examine the tectonic geomorphology of the North America–Pacific plate boundary in California and western Nevada. Major changes occurred during the Neogene in the two primary features of this broad boundary, the San Andreas fault system and Sierra Nevada microplate. Inflow of hot asthenosphere changed mountain-building forces for both. Upwelling of asthenosphere under the microplate was caused by loss of dense rocks of the Sierra Nevada batholithic root. Migration of the Mendocino triple junction along the northwest coast of California created an ever-longer San Andreas fault system and a trailing pulse of asthenospheric upwelling. High rates of rock uplift were not sustained in either study area so landscape changed as uplift rates decreased.

This digital elevation model shows diverse landscapes of the Pacific–North America plate boundary in central California. Death Valley (DV), Panamint Valley (PV), and Saline Valley (SV) are in the Basin and Range Province at the upper right. Owens Valley (OV) borders the eastern front of the Sierra Nevada. KR is the Kings River. The broad Sierra Nevada dips under the San Joaquin Valley as a plutonic microplate. The encroaching Diablo Range borders the west side of the San Joaquin Valley. The straight trace of the right-lateral San Andreas fault is in the Coast Ranges northeast of the Pacific Ocean. Wheeler Ridge (WR), Section 7.2.2, is at a big bend in the fault, near the junction with the left-lateral Garlock fault.

Tectonically Active Landscapes, 1ˢᵗ edition. By W. B. Bull. Published 2009 by Blackwell Publishing, ISBN 978-1-4051-9012-1

Tectonic setting discussions in Chapter 1 described these two plate-tectonics perturbations for the San Andreas transform boundary. Here we consider how delamination of the Sierra Nevada crustal root and migration of the Mendocino triple junction resulted in distinctive mountains, and how this uplift influenced landforms as small as valley floors.

8.1 Walker Lane–Eastern California Shear Zone

About 20 to 25% of the dextral shear between the Pacific and North American plates (Frankel et al., 2008) split off from the San Andreas fault at 3 to 5 Ma. Diffuse, and appearing spatially intermittent, this is called the "Eastern California shear zone" in the south and the "Walker Lane shear zone" in the north. Two prominent active dextral faults in the south are separated by 100 km – the 310 km long Death Valley–Fish Lake and the 110 km long Owens Valley fault zones (Reheis and Dixon, 1996). Two active dextral fault zones in the north are separated by 60 km – Honey Lake and Mohawk Valley (Wills and Borchardt, 1993). Field studies, global positioning system velocities (Oldow, 2003), and seismic analyses reveal widespread dextral faulting in the 820 km long Walker Lane–Eastern California shear zone.

The relative importance of the San Andreas as a plate boundary fault is obvious because of its 315 km cumulative displacement since 17 Ma (Graham et al., 1989). Magnitude Mw >7 San Andreas fault earthquakes occurred in A.D. 1812, 1838, 1857, and 1906.

In marked contrast the Walker Lane–Eastern California shear zone initially appeared so spatially intermittent that it has taken a century to recognize its importance. Nevertheless, historic earthquakes attest to its tectonic significance. It too has had four Mw >7 dextral earthquakes since A.D. 1800, including the magnitude Mw 7.6 Owens Valley earthquake of 1872, the Mw 7.2 Cedar Mountain event of 1932 in western Nevada, the Mw 7.3 Landers, 1992 and Mw 7.1 Hector Mine, 1999 earthquakes; both in the central Mojave Desert.

Geodetic measurements indicate right-lateral shear between the Pacific and North American plates of about 10 to 14 m/ky (Bennett et al., 1999; Dixon et al., 2000; Argus and Gordon, 2001; Frankel

et al., 2008). The Sierra Nevada microplate has been rotated counterclockwise in response to deformation of adjacent tectonic provinces.

Extension of the Basin and Range Province began at 35 Ma in the north and had propagated to the south by 20 Ma (Dilles and Gans, 1995). Extensional faulting encroached into the Sierra Nevada microplate. This contributed to uplift and eastward tilting of the lofty White Mountains block (Guth, 1997) beginning at 12 Ma (Stockli et al., 2003), which is the same time as encroachment near Reno created the Carson Range (Henry and Perkins, 2001). This early episode of synchronous extension occurred along nearly 300 km of the Sierra Nevada–Basin and Range boundary. The Panamint Range may have been created at this time. Recent range-bounding normal faulting after ~3.5 Ma created the eastern escarpment of the present Sierra Nevada and greatly increased the relief of the White–Inyo, Panamint, and other mountain ranges. Neither synchronous episode of tectonic rejuvenation was related to migration of the Mendocino triple junction (Unruh, 1991).

Diverse recent studies provide a clearer tectonics picture for the Coast Ranges and the Sierra Nevada, but tectonic deformation of the western Basin and Range Province is not as clear. Local changes from oblique-slip extension to nearly pure strike-slip faulting at Honey Lake and Owens Valley (Monastero et al., 2005) underscore changes in the importance of dextral faulting. Rates of strike-slip displacement on individual faults of 5 to 9 m/ky (Klinger, 1999; Frankel et al., 2007a) are not obvious unless mountainous relief is created.

Unlike the obvious San Andreas fault, the much younger Walker Lane–Eastern California shear zone seems diffuse and disconnected. The change from regional extension to regional transtension east of the Sierra Nevada microplate suggests early stages of a new plate-boundary shear zone. This interesting possibility is based on seismic (Wallace, 1984), geologic (Wesnousky, 1986, Dokka and Travis, 1990 a, b), and geodetic studies (Sauber et al. 1986, 1994; Savage et al., 1993). Wallace speculated that future earthquakes would fill gaps in historical seismicity. The subsequent north-trending dextral surface ruptures of the Landers earthquake of 1992 and

Figure 8.1 Pedimented landscapes of the Walker Lane–Eastern California shear zone. A. Displacement on slices of the Johnson Valley strike-slip fault zone have changed a tectonically stable pediment surface to steps and ramps in the central Mojave Desert. B. Prominent pediment bench of the Diamond Mountains east of the northern end of the Sierra Nevada rises above the dextral Honey Lake fault zone, which is concealed in the foreground basin fill. Photo taken from just below the tilted pediment at the crest of the Fort Sage Mountains.

(Fig. 8.1). Such pedimented landscapes still are typical of the tectonically inactive landscape to the east of the Salton Trough of southeastern California (Section 2.2.1). Uplifted pediments occur in the Eagle Mountains of the eastern Transverse Ranges (Bull, 2007, fig. 4.20). Pedimented landscapes of the central Mojave Desert have been recently ripped apart by parallel right-lateral faults. This has resulted in mountain ranges with a stair-stepped appearance, with each tread or ramp being part of the dismembered pediment surface (Fig. 8.1A).

The northern and eastern flanks of the San Bernardino Mountains (highest in the Transverse Ranges) also have similar, larger scale, pediment benches. The Alabama Hills, next to the dextral Owens Valley fault zone, are a remnant of a pedimented landscape (Fig. 1.4). A combination of dextral and normal faulting has left a tilted pediment at the crest of the Fort Sage Mountains east of the northern end of the Sierra Nevada. The view across the valley that contains the Honey Lake dextral fault zone (Fig. 8.1B) shows another remnant of the same pediment surface, which is now part of the Diamond Mountains fault block. This was a regional change from tectonically dormant, pedimented to active landscapes.

Of course the mere presence of a raised pediment does not tell us whether normal or transtensional faulting terminated pediment formation and elevated fragments of these diagnostic piedmont landforms. The Section 7.3 discussion illustrates

Hector Mine Lake earthquake of 1999 corroborate his 1984 prediction. Some workers include these surface-rupture events in a model of a new plate-boundary fault zone (Nur et al., 1993; Sauber et al., 1994; Du and Aydin, 1996).

Several key questions need to be answered in regard to the possible continuity of a Walker Lane–Eastern California shear zone. Where did the shear zone begin and what tectonic event created conditions favorable for its propagation? What evidence supports increase of dextral shearing rates at 3 to 5 million years ago? How does one explain areas of apparent tectonic inactivity along some mountain fronts? Has increased tectonic activity continued to the present along the entire shear zone?

A fairly recent change in landscape style characterizes the Walker Lane–Eastern California shear zone. Tectonically inactive pedimented landscapes were typical of much of the future location of the Walker Lane–Eastern California shear zone how fault-zone propagation changes landscape characteristics from class 5 (inactive) to class 1 or 2 (active). Normal faults arrived first followed by transtensional faulting. The overall geomorphic characteristics and ages of tectonically rejuvenated landscapes are remarkably similar throughout the Walker Lane–Eastern California shear zone.

Extension on low-angle normal faults in the Death Valley region ended as oblique-dextral faulting (Fig. 8.2) became more important. Complex basement topography far below smooth playa surfaces contains steep-sided sub-basins (Burchfiel and Stewart, 1966; Klinger and Piety, 1996; Blakely et al., 1999; Klinger, 1999) suggestive of pull-apart basins along active right-stepping, dextral faults (inset map of Figure 8.2).

Castle (1999) takes this concept forward a major step by suggesting that the southern Death Valley fault zone has a position and orientation indicative of being the northern end of a line of plate-boundary rhombochasms extending from the Gulf of California and the Salton Trough. If we accept Castle's hypothesis, and include the Walker Lane, a nascent broad shear zone extends from the Gulf of California and passes along the east side of the Sierra Nevada microplate. Prominent strike-slip faults (Fig. 8.2) may be much more than an embryonic shear zone. They are highly active at present and probably were important in Neogene horizontal and vertical displacements of the Death Valley region.

Timing of changes in tectonic style in the Gulf of California fit well with subsequent evolution of the postulated Walker Lane–Eastern California shear zone. Splitting off of Baja California from the North American plate apparently was a rapid transition from merely local extension and minor dextral motion prior to 6.3 Ma, to status as the principal dextral plate boundary by 4.7 Ma (Oskin et al., 2001). The timing of this major shift in the plate boundary does not conflict with the 3 to 5 Ma initiation dates proposed for dextral faulting at many sites to the north, and may have been responsible for creation of a northward propagating dextral shear zone.

Endorsement of a continuous Walker Lane–Eastern California shear zone has its skeptics. It passes through areas of prolonged active tectonism such as Death Valley. These active faults locally are concealed beneath aggrading valley floors flanked by spectacular mountain fronts – a situation of parallel partitioning of active faulting (Fig. 1.10).

Some areas have east–west sinistral faulting instead of north–south faulting. One locality is in the eastern Transverse Ranges bordering the Salton Trough, and another is in the northern Mojave Desert just south of the left-lateral Garlock fault shown at the bottom of Figure 8.2.

Do the active fault zones of the central Mojave Desert continue to the Garlock fault? Jenning's (1994) map of active faults of California indicates that they do. However, the geomorphic evidence presented by Bull (2007, Chapter 4) describes mountain fronts indicative of minimal Quaternary uplift. Minor Holocene surface ruptures are present locally. More than 100 km of mountain fronts do not have geomorphic characteristics indicative of active faulting. Is there a lack of continuity in the proposed regional shear zone, or is this a "seismic gap" that will be filled by dextral faulting in the near future?

Analysis of synthetic aperture radar interferometry (Peltzer et al., 2001) yields some surprises that may resolve this quandary. Their map of surface strain averages eight years of satellite radar data. It suggests the presence of a vertical fault below a depth of 5 km with a right-lateral slip rate of 7 ± 3 mm/yr. This observed dextral shear zone extends through the major sinistral (left-lateral) Garlock fault and continues north to the southern end of the Owens Valley dextral fault zone. It passes through the Coso geothermal field (Fig. 8.3) where Monastero et al. (2005) describe a blind left-stepping dextral fault that connects with the Owens Valley fault. Global-positioning data, 1993 to 2000, suggest ~6.5 m/ky rate of dextral offset across the Coso Range volcanic complex (McCluskey et al., 2001).

Peltzer et al. (2001) ascribe the lack of historical surface ruptures on either the Garlock fault or on the northwest-trending faults of the northern Mojave Desert to complicated strain between interacting right-lateral and left-lateral fault systems. They conclude that their observed anonymously fast slip rate and lack of surface manifestations might be revealing the birth of a new fault zone. As such, this might be considered as a seismic gap between

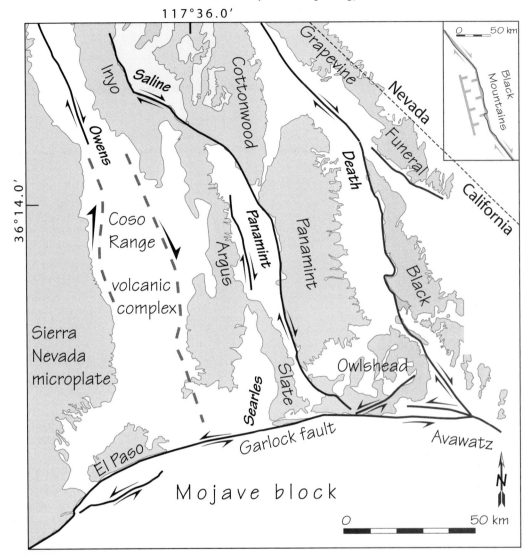

Figure 8.2   Important Neogene strike-slip faulting in the southwest corner of the Basin and Range Province. Mountains are gray areas and intervening valleys are named with bold italic lettering. Garlock and associated sinestral faults provide necessary block rotation to impart north–south continuity of Walker Lane–Eastern California shear zone (Dickinson, 1996). The Death Valley fault system has a releasing right step that results in a rapidly subsiding pull-apart basin (upper-right inset map). Owens Valley fault is here included as the broad shear couple proposed by Monastero et al. (2005) who consider the resulting Coso Range volcanic complex to be a nascent metamorphic core complex. The sharp releasing bend in Saline Valley is responsible for the impressive mountain front shown in figure 2.20A of Bull, 2007. Panamint Valley may have active strike-slip faults on both sides of the valley and the class 1 mountain front has many similarities with the Black Mountains on the east side of Death Valley. Figure 4.2 of Bull (2007).

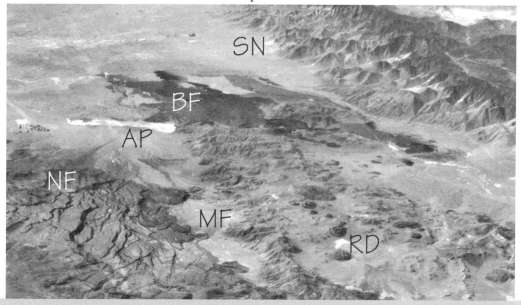

Figure 8.3 High-altitude oblique photo of the Coso Range volcanic complex, Eastern California. Airport playa (AP) is 7.4 km long. Basalt flows (BF). Rhyolite domes and flows (RD). Normal-fault collapses (NF). Diagonal mountain front (MF) records active wrench-fault tectonics. This broad dextral shear couple (Monastero et al., 2005) is by the Sierra Nevada (SN). From U.S. Air Force photo, 374R supplied courtesy of Malcolm Clark, U. S. Geological Survey.

the 1872 Owens Valley and the 1992 Landers surface ruptures. In any case, their work supports continuity of the Walker Lane–Eastern California shear zone across the northern Mojave Desert. Oskin and Iriondo (2004) urge caution in using initial tectonic geodesy results that contradict slow slip rates on the Blackwater fault that decrease towards the north.

Extending the dextral Walker Lane–Eastern California shear zone through sinistral zones of faulting, such as the Garlock fault and the eastern Transverse Ranges, is not a problem if we use a model of conjugate fault blocks in this broad plate-boundary transition. Dickinson (1996) and Dickinson and Wernicke (1997) use a "transrotation" model to describe where crustal fault blocks rotate about vertical axes to contribute to the translation of terrains across the North America–Pacific plate boundary. Their model resolves a post-16-Ma San Andreas fault-slip discrepancy where ~320 km of slip on the fault is only half the ~740 km of transform displacement inferred from global plate models. "Oblique extension," the lateral component

of Basin and Range Province rifting, and areas of sinestral faulting reconcile this apparent discrepancy.

The vector of oblique basin-range post-Miocene extension was only 30°–40° more westerly than the strike of the San Andreas transform boundary, so contributed to relative Pacific–North America plate movements (Atwater, 1970; Wernicke et al., 1988; Harbert, 1991). Models that include transtensional rotations of crustal blocks in the Death Valley region about vertical tectonic axes conclude that the Sierra Nevada shifted about 270 km N 75° W relative to the Colorado Plateau and motion parallel to the N 40° W strike of the San Andreas fault was a surprising 221 km (Holm et al., 1993; Snow and Prave, 1994; Snow and Wernicke, 1994).

Transrotational shear in the eastern Transverse Ranges is estimated to be about 58 km (Dickinson, 1996), so sinestral faulting in this southern edge of the Mojave block contributed to basin-range transtension. The component of transrotational shear thus generated is transmitted through the Mojave block (Dokka and Travis, 1990a,b;

Dickinson, 1996).

Recent earthquakes underscore the importance of transrotational tectonics. The Walker Lane–Eastern California shear zone must cross the sinistral Pinto Mountain fault on the north side of the eastern Transverse Ranges in order to connect with the San Andreas fault in the Salton Trough. The April 1992 magnitude Mw 6.1 Joshua Tree dextral earthquake was the precursor to the magnitude Mw 7.3 June 1992 Landers rupture. It occurred on a fault

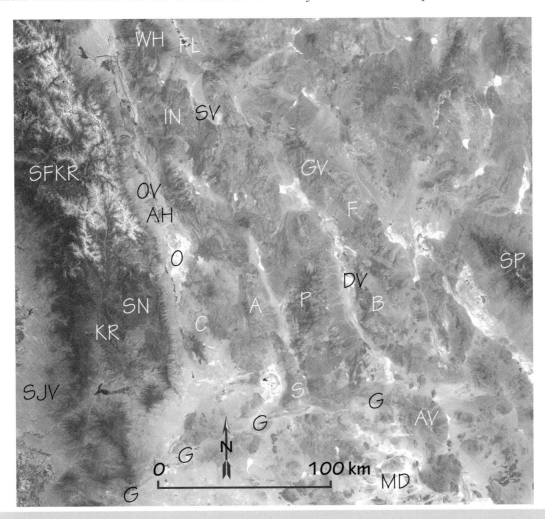

Figure 8.4 Location map showing the locations of mountain ranges and valleys of the Death Valley region of southeastern California and the figures used in this book. Width of view is 325 km. Base image was provided courtesy of Tom Farr, Jet Propulsion Laboratory of the California Institute of Technology and NASA. A, Argus Range; AH, Alabama Hills; AV, Avawatz Mountains; B, Black Mountains; C, Coso volcanics and geothermal area; DV, Death Valley; F, Funeral Mountains; FL, Fish Lake Valley; G, Garlock fault; GV, Grapevine Mountains; IN, Inyo Mountains; KR, Kern River; MD, Mojave Desert; O, Owens Dry Lake; OV, Owens Valley; P, Panamint Range; S, Slate Range; SFKR, South Fork Kings River; SJV, San Joaquin Valley, SN, Sierra Nevada; SP, Spring Mountains; SV, Saline Valley, and WH, White Mountains.

in the eastern Transverse Ranges and played a key role in propagating strain from the San Andreas fault zone to the Walker Lane–Eastern California shear zone. Another transrotational domain of left-lateral faulted mountain fronts in the northeastern Mojave Desert and the sinestral Garlock fault may play a similar role in linking the adjacent sections of the Walker Lane–Eastern California shear zone. Sinestral faults that define the north and south sides of the Mojave block contribute to, instead of thwarting, the regional dextral overprint imposed by an ever more important Walker Lane–Eastern California shear zone.

A dynamic Walker Lane–Eastern California shear zone (Unruh et al., 1996) fits well with my geomorphic analyses of landscape evolution for the Death Valley–Sierra Nevada and Mojave Block regions. The geomorphic consequences of initiation of wrench-fault tectonics into tectonically inactive pedimented terrains, or normal faulted mountains, is the central theme of the following examples.

8.1.1 Panamint Range

The plutonic rocks of the Inyo Mountains and Panamint Range (Fig. 8.4) are part of the Sierra Nevada batholith (Ross, 1962), but these lofty mountain ranges are now topographically part of the southwestern Basin and Range Province. The larger-scale landscape features discussed here provide clues about the changing plate-tectonic scene since ~5 Ma.

Late Cenozoic tectonic activity in the Panamint Range should be considered in the context of regional plate tectonics as influenced by delamination of the Sierra Nevada crustal root (Jones et al., 2004). Similarities shared by the Sierra Nevada and Panamint Range include:
1) Basement plutonic rocks are of the same age and origin,
2) Both mountain ranges have lost their crustal roots and their lofty summits are supported by upwelling hot asthenosphere,
3) Ancestral terrain with nominal relief is now perched 1 km above the adjacent valley floor as a result of major late Cenozoic normal faulting,
4) Dextral faulting continues to be active in the adjacent valley.

Oblique-normal faulting continues to be highly active on the western front of the Panamint Range. The next section concludes that the Owens Valley part of the eastern escarpment of the Sierra Nevada has been much less active since 2 Ma.

Sierra Nevada crustal delamination was accompanied by a brief pulse of ultrapotassic volcanism and initiation of 1 km of uplift along the eastern escarpment. Times of volcanism are clustered at 3.5 ± 0.25 Ma (Fig. 1.5). Panamint and Argus Range crustal-root detachment may have occurred synchronously with the Sierra Nevada.

Alternatively, pulses of crustal delamination may have moved westward as did the encroachment and acceleration of Basin and Range Province style of extensional faulting. In this model extensional faulting would play a major role by triggering delamination events, each step in the delamination process resulting in a burst of volcanism. Detailed studies by Manley et al. (2000) support the possibility of volcanism related to delamination before 4 Ma.

Geomorphic characteristics of two tilted mountain range blocks are compared in this chapter. The Panamint Range block is tilted east (Hunt and Mabey, 1966; Hooke, 1972), opposite to the tilt direction for the Sierra Nevada microplate. The Panamint Range block extends beneath the playa of Death Valley. Shorelines of Quaternary pluvial Lake Manly are higher on the west side of Death Valley and tilting influences the mineral zoning of the accumulating valley floor playa salt deposits.

Geomorphic characteristics of alluvial fans on the east and west sides of the Panamint Range clearly identify different classes of mountain front tectonic activity. Fans actively aggrade as soon as streams leave the precipitous west-side front. The east-flank piedmont has incised alluvial fans and an irregular mountain front (Fig. 8.2). Minor faulting has ruptured fans older than 120 ka. Local faulting of Late Quaternary alluvial-fan deposits near the toes of the fans is a response to recent wrench-fault tectonics of a pull-apart basin hidden beneath the playa. Lack of zones of tectonic base-level fall within the Panamint Range block implies minimal active faulting. Tilting at $0.0036 \pm 0.0007°$/ky (Roger Hooke, email of July 7, 2008) has increased the piedmont slope.

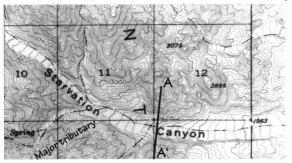

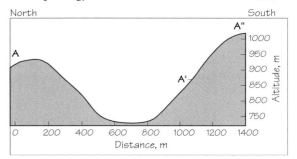

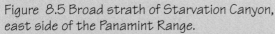

Figure 8.5 Broad strath of Starvation Canyon, east side of the Panamint Range.
A. Contour map, interval 40 feet (12.2 m). Valley floor widens downstream from major tributary junction, being 170 to 200 m wide upstream and 230 to 280 m wide downstream from the junction. Terrace south of spot altitude 1963 feet (598 m) is ~39 m above present strath and is a remnant of an old fanhead embayment. A–A` is part of the Figure 8.5B topographic profile. Section line grid for scale; 1 mile = 1.61 km.

Figure 8.5 Broad strath of Starvation Canyon, east side of the Panamint Range.
B. North–south cross-valley profile along A'–A of Figure 8.5A (plus extension to A") reflects a long time span without significant tectonically induced downcutting caused by a downstream base-level fall.

The net effect is that long reaches of east-side streams have remained at the base level of erosion during much of the Quaternary. Attainment of persistent type 1 dynamic equilibrium conditions for >1 My probably were necessary to create long reaches of prominent straths (Fig. 8.5), considering the low total annual stream power available to widen the valley floors during either glacial or present interglacial climates. Let us examine the characteristics of an east-side stream channel, and later on a west-side stream channel.

Broad straths beneath the stream channel of Starvation Canyon, and nearby canyons, are noteworthy. Their smoothness contrasts with adjacent rugged hillslopes (Fig. 8.5A). They are remarkable because they extend so far upstream from the mountain front (Fig. 8.5C). Straths define reaches of equilibrium conditions; here they begin at a basin-position coordinate of only 0.16. The arithmetic plot of the streambed longitudinal profile reveals two types of equilibrium conditions. Reach A–B is the typical headwaters disequilibrium reach. Reach B–C has alternating strath sub-reaches (type 1 dynamic equilibrium) and gorge reaches (assumed to

be at type 2 dynamic equilibrium), which combine to form a smooth concave longitudinal profile. Straths prevail throughout reaches C–D–E whose arithmetic longitudinal profiles plot as nearly straight lines (Fig. 8.5C). Such lack of profile concavity suggests that channel-forming flood flows do not become progressively larger downstream at the same rate as for humid-region rivers. These strath reaches should plot as straight line(s) in a "gradient-index" analysis.

A semi-logarithmic plot of reach B–E should plot as a straight line whose slope is the inclusive gradient-index (section 2.6.1 of Bull, 2007). Regression provides nice linear results with virtually no scatter, but only when considered as two reaches (Fig. 8.5D). The change from an inclusive gradient-index of 801 to 1,100 occurs at the junction of the major tributary shown in Figure 8.5A. This adjustment is not the result of abrupt increase in size of flood discharges because the inclusive gradient-index increases downstream from this tributary junction. Furthermore, gradient-index analyses generally are not sensitive to where tributaries enter the trunk channel (Hack 1973a, 1982, fig. 2.25 of Bull, 2007). The double-reach plot gives the impression that this tributary is responsible for increasing hydraulic roughness by introducing large boulders into the trunk stream channel, or perhaps it is the source of an increase in bedload transport rate. Equilibrium

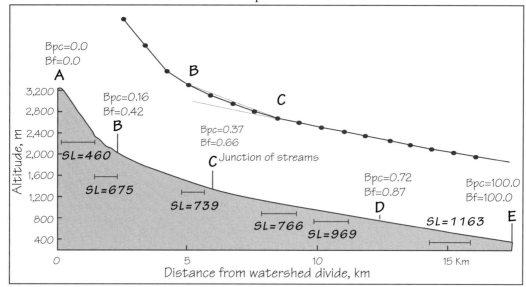

Figure 8.5 Broad strath of Starvation Canyon, east side of the Panamint Range.
C. Longitudinal profile. Bpc and Bf are basin-position and basin-fall coordinates for trunk stream
channel of Starvation Canyon. A–B is a disequilibrium reach. B–C is an alternating type 2 and type 1
dynamic equilibrium reach. C–D and C–E are type 1 dynamic equilibrium reaches. D is the major tributary junction of Figure 8.5A. Inclusive stream-gradient (SL) indices are shown for bracketed reaches.

conditions are maintained but with a different relation to logarithmic distance from the watershed
divide.

The west flank of the Panamint Range is a
dramatically different. A rugged mountainous escarpment has a conspicuous break in slope (Fig. 8.6)
at approximately 900–1000 m above the range base.
The ancestral landscape consisted of hills and broad

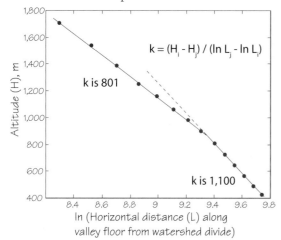

valley floors. Uplift of this relict landscape records
an increase in relief during tectonic extension of the
southwestern Basin and Range Province.

Consider the implications of the three types
of terrain (Fig. 8.6). A postulated older contractional
stress field is represented by the gentle uplands. This
landscape represents a starting point for subsequent
changes in the landscape produced first by extensional and then by strike-slip faulting. This landscape, as
well as its plutonic rocks, may share a common history with the Sierra Nevada. If so, the contractional uplift dates to the times of Mesozoic plutonism.
The gently rolling terrain is the paleorelief remaining
after lengthy denudation of the Sierra Nevada microplate in the early and middle Cenozoic.

Figure 8.5 Broad strath of Starvation Canyon, east side of the Panamint Range.
D. Semi-logarithmic plot of longitudinal profile
reveals of two reaches with different gradient
indices. Change of longitudinal profile characteristics coincides with entry of flow from
major tributary stream shown in Figure 8.5A.

Figure 8.6 Dissected low-relief upland on the west flank of the Panamint Range. A. The lower left half of the view is a highly dissected escarpment rising 1,300 m above the active range-bounding fault. The upper right half of the view is a gently rolling terrain that rises gradually to a hilly range crest and is being dissected by ephemeral streams as a result of the tectonic base-level fall at the range front. Happy Canyon watershed is at HC and Telescope Peak (3,368 m) is at TP. The horizontal distance between TP and HC is 11 km.

Formation of a separate Panamint Range block may have begun at the same time as the nearby White Mountains at 12 Ma (Stockli et al., 2003). These authors note "Right-lateral faulting on the western side of the White Mountains occurred at ~3 Ma and is distinctly younger than the faulting in the Fish Lake Valley area, indicating a westward migration of transcurrent deformation through time." The 1,000 m high escarpment on the west flank of the Panamint Range formed as a sequence of low-angle fault zones promoted extension (Cichanski, 2000). Normal faulting along both the eastern and western fault zones was followed by a greater emphasis on tilting when range-bounding fault displacements virtually ceased on the Death Valley side of the Panamint Range. Uplift may have increased, beginning at 3 to 5 Ma, when regional wrench faulting began. Low-angle normal faulting was terminated recently in both the Black Mountains and the Panamint Range, and steep range-bounding faults became the norm.

The V-shaped canyons that characterize the Panamint Range west-side fluvial systems (Fig. 2.1A)

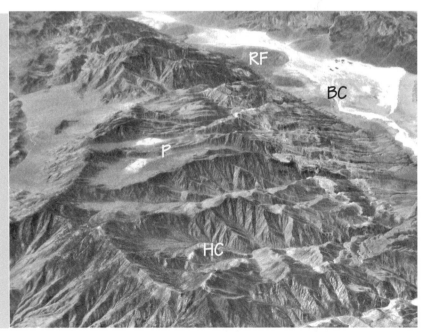

Figure 8.6 Dissected low relief upland on the west flank of Panamint Range. B. High-altitude oblique view towards the southwest. Happy Canyon watershed is at HC. Middle Park and South Park playa lake basins at P have yet to be dissected. Bighorn Canyon alluvial fan is at BC, and Redlands Canyon fan is at RF. From U.S. Air Force photo, 374R supplied courtesy of Malcolm Clark, U. S. Geological Survey. The horizontal distance between RF and HC is 19 km.

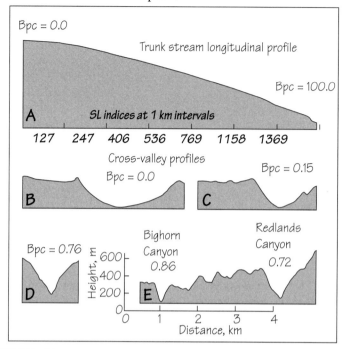

Figure 8.7 Topographic profiles of the Redlands Canyon drainage basin. All sections have same scale and vertical exaggeration of 1: 6.68. Bpc is Basin-position coordinate.
A. Longitudinal profile of trunk stream channel from headwaters divide to the range-bounding fault.
B. Cross-valley profile at the headwaters divide. Bpc is 0.0.
C. Cross-valley profile at Bpc of 0.15.
D. Cross-valley profile at Bpc of 0.76.
E. Cross-valley profile from Bighorn Canyon (Bpc of 0.86) across Redlands Canyon (Bpc of 0.72).

are responses to large, rapid base-level fall generated by active mountain-front fault zones. The convex longitudinal profiles of these streams (Figs. 8.7, 8.8) indicate extremely long response times for these cumulative tectonic perturbations to migrate to headwaters reaches that have diminutive unit stream power. Even in the arid Panamint Range, erosional power is concentrated along valley floors where streams gather progressively more force as they flow down valley. Erosion rates are infinitely slower for hillslopes than for valley floors as generally weak geomorphic processes gradually remove huge volumes of rock. Watershed divides should have the longest landform reaction time to a mountain-front tectonic perturbation.

The shape of a drainage basin changes very slowly as watershed divides migrate in response to differential erosion rates of opposing hillslopes (Horton, 1945). Capture of a headwaters drainage net by the stream of an adjacent watershed represents a pulsatory impulse of expansion by the dominant drainage basin as it becomes an ever more circular shape.

The contrast between the longitudinal profiles of trunk streams on the east and west sides of the Panamint Range (Figs. 8.5, 8.7) suggests that the planimetric shapes of the drainage basins also should be different. Indeed they are (Fig. 2.3). The six narrow west-side drainage basins are crowded into the same distance along the main divide as the four east-side basins. The degree of watershed divide circularity varies greatly from nearly circular for Hanuapah Canyon to excessively long and narrow for Pleasant Canyon. Basin shapes can be described

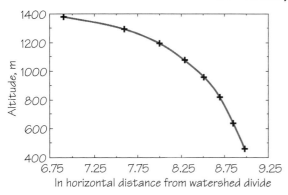

Figure 8.8 Semi-logarithmic plot of longitudinal profile of Redlands Canyon. Convexity underscores lack of attainment of equilibrium conditions for this stream channel.

by an elongation ratio, Re of equation 2.1 (Cannon, 1976).

Mean values of elongation ratio are 0.68 for the east-side and 0.53 for the west-side drainage basins. This larger Re value, like the extent of straths of east-side stream channels, indicates that >1 My has passed since base-level fall was significant at the east-side mountain front. These markedly different drainage-basin shapes are an introductory glimpse into the tectonics of long-term landscape evolution. The next step would be to use the "Hack equation" (1957) to evaluate relative increases of drainage-basin area with increasing length for the two classes of drainage basins.

What is the rate of uplift along the west-side range front? The 1,000 m escarpment height is a minimum tectonic displacement, to which we add an assumed 1,000 m of concurrent basin deposition for an estimate of throw of approximately 2000 m. A 3.5 Ma basalt capping the rims of the northern Panamint Valley pre-dates the initiation of low-angle faulting (Burchfiel et al., 1987). So if uplift began at approximately 3 Ma, the west front of the Panamint Range would have risen at 0.7 m/ky (or as low as 0.35 m/ky if rates of deposition of basin fill have been minimal). A mean uplift rate of 0.5 ± 0.2 m/ky is all that is needed here to create high relief and spectacular tectonic landforms. This moderate rate of uplift is the same as de Polo and Anderson (2000)

describe for the more active parts of the Walker Lane part of the shear zone. They note that the faster normal fault slip rates in the Walker Lane are in the 0.5 to 2.0 m/ky range at seven localities.

8.2 Sierra Nevada Microplate

8.2.1 Present Topography

The imposing Sierra Nevada presents a different tectonic scene that contrasts with the much younger Southern Alps (Section 5.3.1) and soil-mantled hillslopes of the eastern margin of the California Coast Ranges (Sections 2.3.2.2, 3.2, 3.7).

This introduction illustrates a surprising diversity of Sierra Nevada landscapes. The east-side escarpment has impressive scenery (Fig. 1.4), and will be the focus of this discussion of the effects of the post-4-Ma uplift pulse. The escarpment and westward tilted microplate are obvious in Figure 8.9A.

Recent tilting of the Sierra Nevada microplate raised tectonically inactive pedimented terrain and created a 2 km high escarpment (Fig. 8.9B, C). What was the duration and rate of the post-4-Ma uplift of the Sierra Nevada crest? Has uplift been uniform in time and space? Is fastest uplift occurring now, or was rock uplift only a strong, but brief tectonic pulse that has come and gone? Geomorphic responses to this plate-tectonics-scale perturbation provide interesting clues about rates, duration, and causes of rock uplift.

The topography of the western mountain front is subdued. Westward tilt of the Sierra Nevada is uniform sloping north of the San Joaquin River. Old soils on the alluvial geomorphic surfaces in the adjacent San Joaquin Valley date back to the Pliocene (Arkley, 1962; Marchand, 1977; Marchand and Allward, 1981; Harden, 1987).

The Sierran foothills mountain front has characteristics of a tectonically inactive terrain (class 5). It records mountain-front erosional retreat under conditions of Cenozoic tectonic inactivity (Fig. 8.9D). A low relief and highly embayed mountain–piedmont junction contrasts greatly with the lofty tectonically active east-side landscape. Lack of landscape signatures indicative of tectonic uplift suggests that the foothills margin is the "hinge line" between a rising Sierra Nevada and subsiding Central Valley.

The western mountain front south of the Kings River is different. The overall initial slope of the range is gentler than to the north, and includes the prominent highlands of the Great Western Divide. Then steeper terrain drops down to the San Joaquin Valley. Miocene marine rocks and younger strata are being faulted and folded and young alluvial fans are being deposited next to the rising mountain front (Fig. 8.9E). The present tectonically active nature of this Sierra Nevada mountain front was underscored by a Mw magnitude 5.0 earthquake on 29 September 2004 in the mountains 23 km NNE of Bakersfield.

An exceptionally straight western front at the extreme north end of the Sierra Nevada, in the northeastern Sacramento Valley (west of FR on Fig. 8.9A), also appears to be tectonically active. Such active range-front faults and folds may have been initiated by tectonic events at ~4 Ma.

Rising Coast Ranges have encroached onto the west side of the microplate. This caused the rate of deposition in the adjacent San Joaquin Valley to double as a result of the large increase (perhaps a doubling) in sediment yield (Section 3.3). Accelerated deposition on the west side forced the

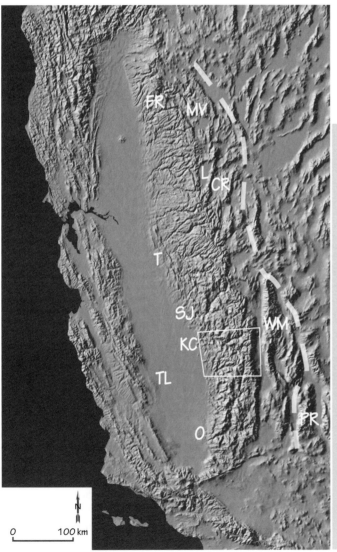

Figure 8.9 Topography of the Sierra Nevada microplate

A. Digital image of central California and west-central Nevada showing the westward tilted Sierra Nevada microplate. Note the irregular, rugged east-side escarpment, relatively subdued western mountain front, the low-relief adjacent San Joaquin Valley. The Coast Ranges border the Pacific Ocean. Main discussion area — Kings Canyon–Owens Valley — is outlined; it includes Kaweah and upper Kern watersheds and the highest part of the range crest Mt. Whitney, altitude 4418m. CR, Carson Range; FR, Feather River; KC, Kings Canyon; L, Lake Tahoe; MV, Mohawk Valley; O, Oildale mountain front; PR, Panamint Range; SJ, San Joaquin River; T, Tuolumne River mountain front; TL, Tulare Lake bed; WM, White Mountains. Area west of the dashed line indicates the area of accelerated extensional faulting since 5 Ma (Jones et al., 2004).

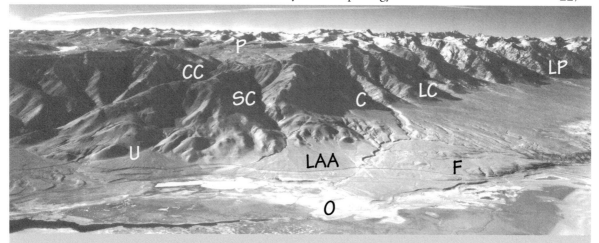

Figure 8:9 Topography of the Sierra Nevada microplate.
B. Oblique aerial view of eastern escarpment of the Sierra Nevada looking west across the bed of Owens dry lake (O). Range-crest flat Chagoopa Plateau (P) at the head of Cottonwood Creek (CC) to Slide Canyon (SC) was a pediment before uplift along the range-bounding fault. Minimal Pleistocene uplift along the range-bounding fault has resulted in deep fanhead trenches, Carroll Creek (C), Lubkin Creek (LC), and Lone Pine Creek (LP). Splays (F) of the highly active Owens Valley fault zone bound the rapidly subsiding lake basin. The prominent multiple-rupture event fault scarp on the Pleistocene alluvial fan (U) probably was active during the Holocene. Continuous thin line is the Los Angeles Aqueduct (LAA). This 1955 Photograph GS–OAI–5–15 provided courtesy of U. S. Geological Survey.

contact between Sierra Nevada and Coast Range basin fill far to the east (Fig. 1.6), thus accelerating rates of deposition on the east side of the San Joaquin Valley. Pleistocene fluctuations in Sierra Nevada sediment yield are recorded by full-glacial aggradation events on the piedmont along the east side of the San Joaquin Valley.

Degradation during the Cenozoic lowered range-crest altitudes to 600 m in the north and 1500 m in the south. Then the microplate tilted.

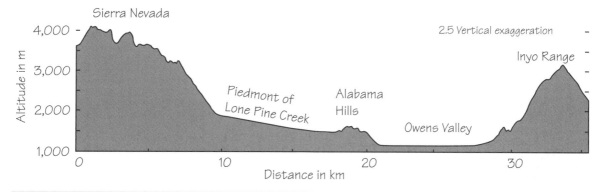

Figure 8:9 Topography of the Sierra Nevada microplate.
C. Topographic profile from the Sierra Nevada crest to the crest of the Inyo Range showing the relation of the Lone Pine Creek piedmont and Owens Valley floor to the adjacent erosional landscapes.

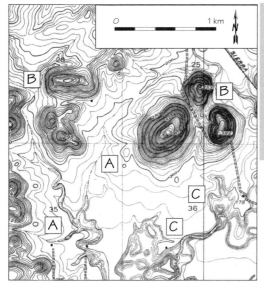

Figure 8.9 Topography of the Sierra Nevada microplate D. Topographic map of the western mountain front showing a tributary of Dry Creek; location is 10 km north of the Tuolumne River. Pedimented terrain (A) with inselbergs (B) is indicative of a tectonic activity class 5. Recent incision by Dry Creek is exhuming beveled bedrock knobs (C). Paulsen Quadrangle, 5 foot (1.52 m) contour interval from 1913 plane-table surveys.

the range-bounding fault system, and the other 600 m may record displacements along the Owens Valley fault zone and related faults bordering the lake basin. A recent treatise (Phillips, 2008) about the Owens River region drainage rearrangements focuses on post-4-Ma responses to uplift of the Sierra Nevada.

Uplift of the Mt. Whitney section of the Sierra Nevada began at 3.5 ± 0.25 Ma, which was the time of a brief pulse of potassium-rich basaltic vulcanism (Figs. 1.5, 8.10) in a 200 km diameter circular area (Manley et al., 2000).

The Figure 8.10 model is one of several possible alternatives and was chosen on the basis of current geomorphic, geophysical, and structural information. It centers about a crustal-release threshold–a tectonic perturbation on a plate-tectonics scale (Section 1.2.2). Encroachment by Basin and Range Province faulting into the Sierra Nevada microplate may have played a role by inducing delamination of the crustal root of the microplate, perhaps

8.2.2 Geomorphic Responses to an Uplift Event

Microplate tilt created the eastern escarpment and raised the Sierra Nevada range crest by 1.5 to 2.5 km (Huber, 1981; Unruh, 1991; Wakabayashi and Sawyer, 2001). The magnitude of uplift that followed prolonged landscape degradation can be seen in the Chagoopa Plateau pediment surface at the top of the escarpment (Fig. 8.9B), which is now 1,500 m above the mountain front at Carroll Creek, 1,900 m above the mountain front at Slide Canyon, and 2,100 above the dry bed of Owens Lake. The 1,500 m estimate is the vertical displacement along

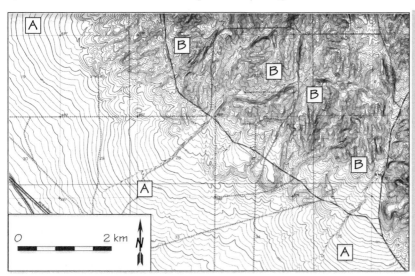

Figure 8.9 Topography of the Sierra Nevada microplate
E. Topographic map of the western mountain front 10 km north of the city of Bakersfield. Active deposition of alluvial fans (A) and valleys with low Vf ratios, defined in equation 3.14 (B) are indicative of a tectonic activity class 2, or perhaps class 1. Oildale Quadrangle, 5 foot (1.52 m) contour interval from 1929 surveys.

including both the Panamint Range and the Sierra Nevada. Generation of new normal faults that disrupt the massive nature of the batholith might also reduce lithospheric strength. Subsequent tilting of the microplate was enhanced by accumulation of > 1 km of basin fill, chiefly on the west side of the San Joaquin Valley. Delamination caused a brief, distinctive pulse of vulcanism followed by rock uplift.

This seems to have been an abrupt pulse of uplift based on geomorphic evidence presented later in this section. I presume that initiation of uplift matched an abrupt detachment of the crustal root. Rock uplift is still continuing slowly in an isostatic sense, but not at the previous fast pace when inflow of hot asthenosphere to replace the lost crustal root was the dominant control on uplift rates.

Quaternary isostatic uplift may have been sufficient to maintain or slowly raise the range crest (Small and Anderson, 1995; Pinter and Brandon, 1997). Sierra Nevada ridgecrests are being degraded so slowly (Stock et al., 2004, 2005) that even minor isostatic uplift can gradually increase range-crest altitudes. Quaternary isostatic uplift was maintained or accelerated because of renewed downcutting of valley floors by glaciers, which created the spectacular scenery of today's Sierra Nevada.

Creation of a lofty Sierra Nevada resulted in several climatic changes. Timing of changes in orographic controls of a progressively higher range crest mimic the postulated rock-uplift curve of Figure 8.10 – rapid after delamination, then slowing after the conclusion of the initial pulse of asthenosphere influx. Quaternary orographic variations essentially paralleled the difference between ridgecrest and valley-floor degradation rates – the isostatic uplift generator. Rain-shadow effects have been approximately constant during the past 2 My.

The present climate varies greatly with altitude, and between the rain-shadow and up-wind sides of the mountain range. Occasional large streamflow events occur in large watersheds under the present semiarid to humid, thermic to frigid, strongly seasonal climate (Appendix A). Compared to the New Zealand Southern Alps, these massive plutonic rocks are more resistant to erosion and much less annual unit stream power is available for erosion and transport of detritus. Sierra Nevada Quaternary sediment

yields are probably two orders of magnitude less than those of the Southern Alps.

Winograd et al. (1988, 1992) believe that uplift rates increased sufficiently since 4 Ma to create a rain shadow east of the range. But there must have been a considerable previous rain-shadow because the southern Sierra Nevada was lofty before rejuvenation (Wakabayashi and Sawyer, 2001, Poage and Chamberlain, 2002; Stock et al., 2004, 2005; Figure 8.18 in this chapter).

Pliocene uplift of the Sierra Nevada set the stage for enhanced Pleistocene glaciation. Glaciers complicate geomorphic analyses (Starkel, 2003; Knuepfer, 2004). Glaciers are capable of doing more erosional work than rivers (Montgomery, 2002). They do not behave like rivers. The fluvial concept of the base level of erosion does not apply to longitudinal profiles of valley floors because glaciers scoop out depressions where basal ice flows uphill. They trim valley sides, especially footslopes, setting the stage for increased landslide activity after melting removes the lateral support provided by thick ice (Section 6.1). The U-shaped valleys remain the most landslide-prone part of the Sierra Nevada landscape even after a 12 to 15 ky absence of glaciers (Bull, 2007, section 6.2.2.3). Glaciers remove the signatures of prior fluvial landscapes, erasing clues about processes that tectonic geomorphologists would prefer to study.

The spectacular canyon of the South Fork of the Kings River is a representative large stream draining the western flank of the Sierra Nevada. Forest now blankets the wide valley floor (Figure 8.11A) and the thick soils that once supported dense vegetation have been scraped off the hillsides above the break-in-slope that defines the U-shaped portion of the valley cross section. Periods of heavy rainfall or rapid snowmelt presumably are much more likely to generate large runoff and floods than on a pre-ice-age topography with deep soils and continuous forests. Large amounts of glacial outwash detritus delivered to Kings River shifted the mode of fluvial system operation across the threshold of critical power in the reach downstream from glacier termini. This also induced rapid aggradation on the Kings River alluvial fan in the adjacent San Joaquin Valley (Atwater et al., 1986; Weissman et al., 2002).

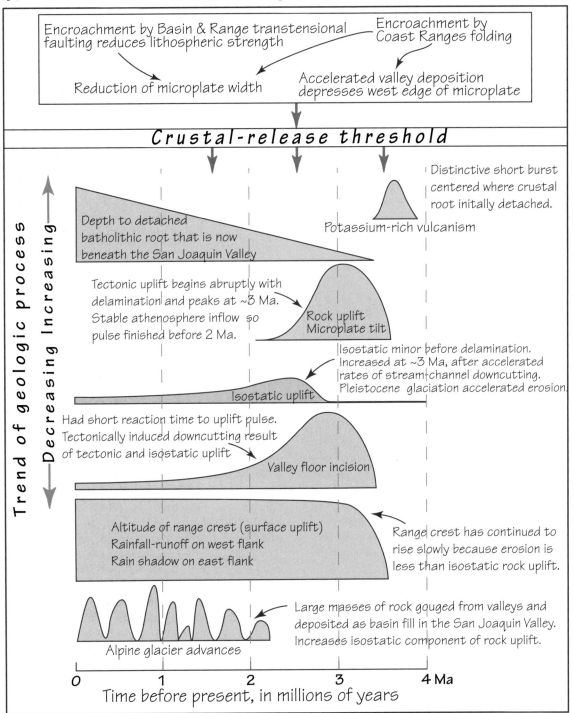

Figure 8.10 Model of late Cenozoic tectonic and geomorphic events, southern Sierra Nevada.

Breaks-in-slope

Figure 8.11 Glacial alterations to the landscape of the South Fork of the Kings River, Sierra Nevada, California.
A. Sides of U-shaped glacial valley rise above nearly flat valley forested floor opposite Roaring River tributary. Note prominent breaks in slope marking the edge of the inner gorge.

Such alternation of glacial and fluvial conditions during the Quaternary is an extreme result of climatic changes. Prevailing geomorphic processes are replaced by radically different processes. Normal river gradients become ever steeper upstream, but the upstream half of the 36 km profile of the South Fork of the Kings River (Fig. 8.11B, Section A–A') has only one-third of the gradient of the downstream half. This anomaly is the result of deep scouring by valley glaciers, which created the U-shaped valley cross sections shown in Sections B–B' and C–C'. Tributary streams commonly are left high above the trunk valley as "hanging valleys" that decouple (Bull, 2007, Section 2.5.2) upstream from downstream reaches. But the powerful Roaring River tributary has responded to the glacial-erosion base-level fall by cutting a steep-sided V-shaped canyon (section D–D'). The maximum extent of Pleistocene valley glaciation was 1-2 km downstream from Section E–E' and may have been sufficient to create a small waterfall on Grizzly Creek. Alternatively, the waterfall was the result of lateral cutting by a river raised by glacier outwash aggradation. Aggradation may also explain why the low-gradient reach extends a short distance downstream from E–E'. The V-shaped valley cross section implies dominance by fluvial processes.

Incision by western flank rivers created inner gorges with steep hillsides that plunge down to narrow valley floors. Post-4-Ma fluvial downcutting created a 1/2 km deep V-shaped canyon at E–E' and 1 km deep at F–F'. These comparisons between glaciated and unglaciated reaches underscore how tectonic signatures in the pre-glacial fluvial landscape were eradicated by Pleistocene glaciation.

These landscape elements were created in response to range-crest surface uplift, which tilted entire river drainage basins strongly to the west. The lack of broad straths in many reaches of these rivers (Fig. 8.12) suggests several possibilities. These include 1) stream-channel downcutting that continues to be the dominant process, 2) modest amounts of annual unit stream power require still longer time spans to create wide valley floors, or 3) strong lithologic control is responsible for long response times to create landforms indicative of tectonically stability.

Ridgecrest breaks-in-slope (Figs. 8.11A, B, 8.12) that are now 500 to 700 m above the valley floors, constrain estimates of maximum channel downcutting since 4 Ma. They are maxima because breaks-in-slope also migrated upslope while rivers continued to lower the adjacent valley floors.

Lava flowed down some river valleys and subsequent tilting records the amounts of bedrock uplift caused by displacements on the east-side range-bounding fault zones. Post-lava-flow stream-channel incision is progressively less downstream. Tilted lava flows in the valleys of the Tuolumne and San Joaquin Rivers provide rates of post-Miocene tilt, as do tilted Cenozoic sedimentary strata in the adjacent San Joaquin Valley (Huber, 1990; Unruh, 1991).

The San Joaquin River extended well into the present Basin and Range Province prior to westward migration of extensional faulting and shortly before volcanism at 3.2 Ma that filled its ancestral valley at the present range crest (Huber, 1981). This constrains the time for establishment of the east-side escarpment. Escarpment height is the sum of surface uplift of the range crest and subsidence of Owens Valley. Tilt analyses using lava flows involve upstream extrapolation of inferred river gradients (Wakabayashi and Sawyer, 2000, 2001), but even so are a better way to estimate uplift magnitude than east-side escarpment height.

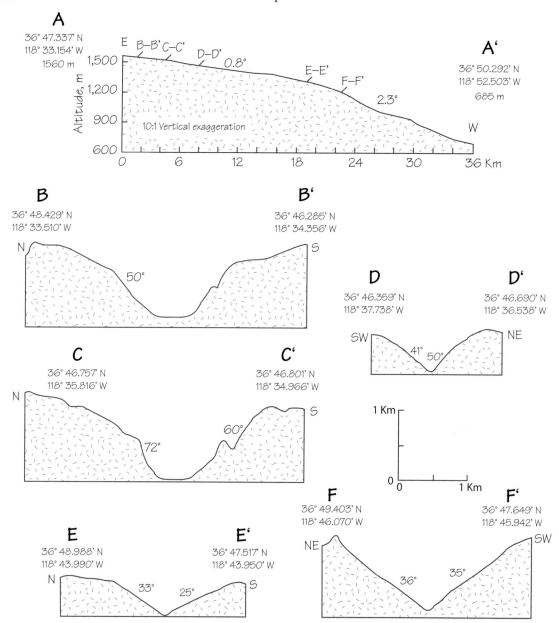

Figure 8.11 Glacial alterations to the landscape of the South Fork of the Kings River, Sierra Nevada. B. Topographic profiles. A-A' is along the active channel of the South Fork between Bubbs Creek and the Middle Fork of the Kings River. Valley floors at B–B' and C–C' have been greatly deepened and changed to a U-shape cross section by valley glaciers. D–D' shows response of the Roaring River tributary to glacial erosion base-level fall in the trunk valley. V-shaped canyons E–E' and F–F' reveal prolonged fluvial downcutting by the South Fork River. No vertical exaggeration is used here, except for A–A' which is 10:1.

Breaks in slope to inner gorge

A

Strath

B

Strath

Gorge

Figure 8.12 Inner gorge of the South Fork of the Kings River.
A. Silhouettes of the breaks in slope are ~1000 m above the active channel. Strath terrace is at the Horseshoe Bend reach of the river.
B. Rugged inner gorge channel is 300 m below Horseshoe Bend strath surface. Lower Convict Flat strath terrace is hidden but is at the level of the highway at the right center of this view.

Part of the relief of Owens Valley occurred as extensional processes promoted collapse of formerly higher terrain in much same way as Basin and Range Province valleys in Arizona fell away from the Colorado Plateau (Lucchitta, 1979; Mayer, 1979). Uplift played a minor role in creating relief in Arizona as compared to subsidence of valleys resulting from extensional faulting.

Geophysical surveys reveal that the basement rocks beneath the east side of Owens Valley now are below sea level (Gillespie, 1991; Hollett et al., 1991). Much of the Quaternary tectonic deformation has been transtensional graben formation east of the Alabama Hills (Figs. 1.4, 8.9C), which is 14 km from the range-bounding fault of the Sierra Nevada. This belt of active transtensional faulting broadens greatly to the south (Monastero et al., 2005) creating the Owens Lake basin (Fig. 8.9B) and the intensely faulted complex of the Coso volcanic field. The Owens Valley fault ruptured in a Mw 7.6 earthquake in 1872 (Beanland and Clark, 1994), but one should be cautious in assuming that the Sierra Nevada range-bounding fault is just as active.

Tectonic geomorphologists can estimate rates of base-level fall fairly easily. It is harder to separate out the components of mountain uplift and basin subsidence.

The Sierra Nevada piedmont discussed here has unusual features, perhaps because it is changing in a geomorphic and tectonic sense. This piedmont lacks old, inset fan surfaces, such as the east side of the Panamint Range (Fig. 2.2A). It lacks continuing deposition of thick, Late Quaternary, alluvial fans where streams leave the mountains, such as the west side of the Panamint Range (Figs. 2.2B, 4.21A).

Why does the Sierra Nevada piedmont seem anomalous when compared to nearby mountain ranges? This 4 km high range is the major barrier to airmass circulation – winter cyclonic storms or summer monsoon thunderstorms. Pleistocene glaciation did more work here, changing hillslopes and providing detritus and dams for outburst-flood events whose deposits mantle the piedmont. More rain and snow here, as compared to arid-realm piedmonts, means more annual stream power to shift rocks, entrench channels, and greatly shorten return times for streamflow to return to a given part of the piedmont. These highly active streams created a smooth piedmont, but is it a depositional or an erosional surface?

Scrutiny of mountain-front landforms suggests two different conclusions – active and fairly inactive. Short, discontinuous fault scarps and rapid Holocene fan deposition indicate recent tectonic base-level fall at the mountain front, but only locally.

The most active site has high, continuous, fault scarps cutting across thick, young alluvial fans (left-center part of the Figure 8.9B view). It deserves a local tectonic activity class rating of 2.

Elsewhere, both south and especially north of Owens Lake, mountain-front landscapes with fanhead trenches and small, discontinuous fault scarps suggest much less displacement on the Sierra Nevada frontal fault – class 3. This is an order of magnitude less active than where extensional faulting is encroaching west of Owens Lake. For example, the front between Carroll and Lone Pine Creeks might appear inactive, except for a 300 m long normal fault scarp cutting young debris-flow deposits at the apex of the alluvial fan of the North Fork of Lubkin Creek (LC on Fig. 8.9B).

Let us consider three models.
1) Virtually inactive between 2 Ma and 0.1 Ma.
2) Same as the more active parts of the Basin and Range Province since 0.1 Ma
3) Post-4-Ma listric style of faulting explains the rather unusual geomorphic characteristics of this piedmont (an ongoing study by Fred Phillips). A reasonable case can be made for each of these models.

Virtually Inactive Model The mountain-front landscape has characteristics indicative of slow rates of base-level fall where it passes lofty Mt. Whitney. It lacks features indicative of a class 1 tectonic activity (Fig. 2.2A normal fault; and figure 3.5C, Bull, 2007 thrust fault) such as steep, >40°, triangular facets on spur ridges that end abruptly at the mountain–piedmont junction.

The front between Bairs and Shepherd Creeks is typical. Exposures of a white siliceous fault zone dip 35° to 40° according to Fred Phillips. The fault trace is marked by spring-fed vegetation, except where buried by colluvial wedges. About 1 km of colluviated mountain front has a uniform gradient

of only 16° (Fig. 8.13). Pedimentation south of Shepherd Creek has beveled a surface across the Sierra Nevada frontal fault. Such landforms in a normal-faulted terrain would clearly be indicative of prolonged tectonic quiescence in a geomorphic base-level context.

Other landforms suggest occasional range-front tectonic base-level fall events. The mountain–piedmont junction is fairly straight, especially between George and Symmes Creeks (Fig. 1.4), which includes the Figure 8.13 location. Valley floors downstream from terminal glacial moraines are moderately narrow, and the hillslopes are straight (Fig. 8.14).

The piedmont between Lubkin and Symmes Creek is smooth with stream channels incised below extensive bouldery surfaces created by flood events. Fanheads are small and steep, created at times of maximum sediment yield during the Pleistocene. Debris-flow levees are common. Fanhead deposition was not sustained as indicated by the present 30–60 m deep fanhead trenches.

Remnants of pediments are common along this side of the Sierra Nevada. The best exposures of a pediment surface are in the Alabama Hills where thin alluvial cover has been partially stripped to expose beveled bedrock. Continuing tectonic subsidence along the Owens Valley fault zone enhances this nearby exhumation of very thin alluvial cover. Inselbergs rise above truncated plutonic rocks halfway up the piedmont (AH on Fig. 8.15), clearly indicative of a widespread erosional surface beneath young piedmont alluvium. Similar features occur in the Independence Hills about 17 km north of the Alabama Hills. The intervening piedmont without pediment exposures suggests burial caused by local faulting related to the active Owens Valley fault.

The geologic map of Stone et al. (2000) has several features that suggest minimal Quaternary

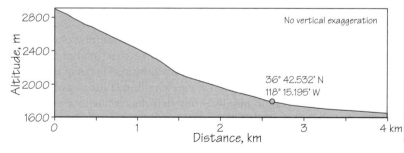

Figure 8.13 Topographic profile down a ridgecrest, across the Sierra Nevada range-bounding fault (location noted), and ending on young piedmont alluvium.

Figure 8.14 View upstream of Sierra Nevada range front at the mouth of Lone Pine Canyon. Unfaulted grussy Pleistocene alluvial-fan deposits in the foreground (FAN) and ~30 m deep fanhead trench (FT) at right side. Abundant 4 to 8 m boulders on fill terrace (STR) tread date to 1.2 to 2 ka (Bierman et al., 1995). V-shaped canyon upstream from range-bounding fault zone is 1 km downstream from Pleistocene glacial moraines. 4, 420 m high Mt. Whitney (WHITNEY) is at center skyline. Photo by Adolf Knopf (1918, Plate 16A), courtesy of the U. S. Geological Survey.

displacements on the Sierra Nevada frontal fault (Fig. 8.15) where it passes the highest peak in the range, 4,418 m high Mt. Whitney. Faulted alluvium is only locally present along this mountain front. Deposition of thick alluvial fans is not occurring next to the mountains, but much of the piedmont surface has young channels and huge boulders suggestive of large flood flows capable of overtopping 10–30 m deep fanhead trenches. Deposition during the latest Quaternary is thin, commonly being only 5 to 15 m thick. Exposures of grussy Pleistocene alluvium occur as dendritic patterns (Fig. 8.15) where 6 to 25 m deep stream channels have incised through a few meters of young bouldery gravel. This setting is much different than nearby tectonic activity class 1 mountain fronts (Figs. 2.2A, 4.21A) or the class 2 front west of Owens Lake (Fig. 8.9B).

Rapid accumulation of class 1 thick alluvial fans is directly related to rates of base-level fall on range-bounding fault zones. The rate of mountain-front base-level fall ($\Delta u/\Delta t$) has to equal or exceed the sum of stream-channel downcutting ($cd/\Delta t$) upstream from the range-bounding fault, and aggradation on the adjacent piedmont ($pa/\Delta t$).

$$\Delta u/\Delta t \geq \Delta cd/\Delta t + \Delta pa/\Delta t \qquad (8.1)$$

A tectonically active (class 1) mountain front would have thick, rapidly accumulating, alluvial-fan deposits next to the range-bounding fault despite continuing downcutting by the stream upstream from the fault. Such alluvial fans are more strongly linked to the rate of slip on steeply dipping range-bounding

faults than to climatic controls that also influence output of water and sediment from the source drainage basin (Densmore et. al. 2007).

How active have the Sierra Nevada piedmont streams been and where has deposition and downcutting occurred? The streams presently flush detritus through incised piedmont channels to the Owens River floodplain. Deposition on fanheads occurred during the transition between full-glacial and Holocene climates – a time of paraglacial geomorphic processes (Ballantyne and Benn, 1996; Ballantyne, 2002). Lakes were impounded behind flimsy terminal ice masses and moraines in trunk or lateral valleys, only to burst downstream as glacial outburst floods (O'Connor and Costa, 1993; O'Connor et al., 2001; Cenderelli and Wohl, 2003).

The last multiple-advance glaciation in the Sierra Nevada occurred from ~25 to 15 ka (Phillips et al., 1996). Five tightly clustered age estimates of boulders on the north lateral moraine in the Lone Pine valley upstream from the mountain front (Fig. 8.15) have a mean age of 17.8 ka. This Tioga

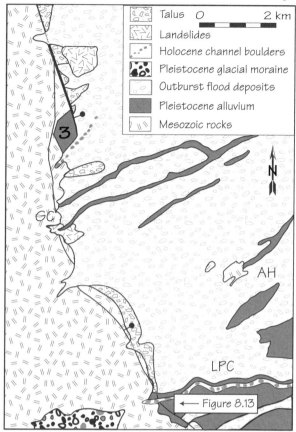

Talus	
Landslides	
Holocene channel boulders	
Pleistocene glacial moraine	
Outburst flood deposits	
Pleistocene alluvium	
Mesozoic rocks	

O   2 km

N

3

GC

AH

LPC

← Figure 8.13

Figure 8.15–Geologic map, Lone Pine Quadrangle; legend is modified from Stone et al.
(2000) map. This map has several features
indicative of a low rate of displacement on
the Sierra Nevada frontal fault during the
Pleistocene. AH are Alabama Hills inselbergs that are part of a pediment mantled
with alluvium. LPC is Lone Pine Creek: GC
is George Creek. 3 marks the location of
detailed figure 3 map of Le et al. (2007).

the piedmont, quickly backfilled fanhead trenches,
and spread out on the fans (Blair, 2001, 2002).

A hallmark study by Benn et al. (2006) describes channel dimensions and processes. Boulder
sizes were used to estimate entrainment velocities
as being 5–9 m/sec. [10]Be, a terrestrial cosmogenic
nuclide, was used to date 40 surficial boulders from
the blankets of outburst flood deposits on the adjacent alluvial fans of Lone Pine and Tuttle Creeks.
Intervals of outburst-flood events occurred at 9–13,
16–18, 23–32, and 32–44 ka. Holocene channel incision preserves sheets of outburst flood deposits.

These floods created U-shaped channels
on the fanheads that are 50–100 m wide and 5–10
m deep, flanked with bouldery levees indicative of
debris flows. These debris-flow channels change to
large braided stream channels in midfan reaches.

This style is characteristic of debris-flow
events. The initial surge of water efficiently entrains
most of the available channel-floor deposits and adjacent hillslope sediment slumps as it moves down a
valley. A water flood scoops up so much sediment
that it becomes a debris flow. Continuing outflow of
water from upstream source areas changes the flow
characteristics back to water flooding as sources of
sediment influx become depleted. This second stage
flows through newly created debris-flow channels
and reworks the just-deposited debris-flow terminal
sheets.

The fans mapped by Stone et al. (2000) and
studied by Benn et al. (2006) are climatic, not tectonic fans. Cessation or marked reduction of base-level
fall along the range-bounding fault would be conducive for stream-channel entrenchment and erosional
beveling of previously deposited tectonic alluvial
fans.

Aggradation events induced by climate
change, be they single outburst floods or ~10 ky of
accelerated erosion of a hillslope sediment reservoir,
would temporarily reverse the tendency for fanhead
entrenchment. Surficial characteristics would resemble those of a tectonic alluvial fan, but thicknesses
of these climatic-fan deposits would be only a veneer
of braided stream and debris-flow deposits.

The main process, over time spans of 100–
1,000 ky seems to have been redistribution of materials laid down during previous brief episodes of

glaciation maximum advance age is the same as elsewhere (James et al., 2002).

Waning of the Tioga glaciation exposed detritus that glaciers had scraped off mountains and
moraines containing abundant granitic boulders.
Such sources of abundant sediment supplied glacier
outburst floods with huge quantities of bouldery detritus. Floods that exceeded 1,000 m³/sec flowed to

deposition. In contrast, deposition on tectonic fans is so rapid that only incipient soil-profiles can form. Thick tectonic alluvial fans may have been deposited in the Late Pliocene. These fans were truncated by pedimentation processes, and deposition of a few meters of bouldery deposits records climatically induced episodes of deposition and reworking.

The change to Holocene climates, with the return of summer thunderstorms, caused rapid runoff of water from bare, glaciated terrain that exceeded the amount of stream power needed to convey reduced amounts of bedload. The fluvial style was changed for all the streams, and colluvium in the hillslope sediment reservoir is being stripped. Streams have incised their piedmont reaches, with the occasional sediment-laden flow event creating bouldery inset fill terraces (Figs. 8.14, 8.15).

Thin blankets of recent alluvium along this part of the Sierra Nevada front indicate minimal mountain-front tectonic base-level change for these fluvial systems. The thicker deposits of prominent alluvial fans of the Late Pliocene and Early Pleistocene record times of much faster base-level fall on the range-bounding fault. These were times when the landscape had characteristics of class 1 tectonic activity.

Such tectonically induced deposition has been minimal since the Early Pleistocene. Base-level fall seems to have decreased greatly along the range-bounding fault during the past million years. This landscape may have lost the characteristics of a class 1 tectonic activity class mountain front by 2 Ma. The mountain front, despite its impressive height, now has piedmont characteristics, and a degraded mountain front, indicative of formerly rapid uplift followed by lengthy quiescence or slow uplift.

The first school of thought considers the long-term tectonic–landscape relations (>500 ky). Quaternary interrelations between base-level processes are best described as

$$\Delta u/\Delta t << \Delta cd/\Delta t > \Delta pd/\Delta t \qquad (8.2)$$

where $pd/\Delta t$ is long-term piedmont degradation.

Channel downcutting, $\Delta cd/\Delta t$, currently exceeds piedmont degradation rates, hence equation 8.2 describes the present prevalence of incised channels. The $<<$ notation merely indicates that these two piedmont degradation processes are much more important base-level controls on streams flowing through the mountain front than base-level change – vertical displacement on the Sierra Nevada frontal fault. This tectonically inactive description of overall geomorphic processes since 2 Ma describes net erosion that lowered a piedmont capped with thin alluvium, and degraded the mountain front.

A tectonic activity class rating of 3 for this section of the Sierra Nevada front seems reasonable (Bull, 2007, fig. 4.21). This indicates a rate of base-level fall of approximately 0.1 to 0.3 m/ky, but does not provide information about the proportions due to uplift of the Sierra Nevada and due to subsidence of Owens Valley. Even if it was 100% rock uplift, the Pleistocene rock-uplift rate would be much less than the rate of uplift after the delamination event, which is discussed later.

Short-term stream-channel downcutting and deposition of climatic alluvial fans are superimposed on long-term pedimentation of this piedmont. This Quaternary perspective of landscape evolution overlooked the importance of short, discontinuous fault scarps at the range front. We need to add them to the mix.

<u>Moderately Active Model</u>   The second school of thought is that the range-bounding fault has been active, at least since the latest Pleistocene. The Le et al. (2007) team described a Holocene stream terrace at George Creek just 6 km north of Lone Pine Creek that has a fault scarp with ***apparent throw*** (see the glossary) of ~7 m. A similar short (300–500 m long) fault scarp is at the apex of the fan of the North Fork of Lubkin Creek, 9 km southwest of Lone Pine Creek. At 16 km to the north of Lone Pine Creek a degraded multiple-fault scarp ruptures the fanhead of Symmes Creek (Fig. 8.16A). A recent event faulted a glacial moraine at Independence Creek at 21 km north of Lone Pine Creek.

Range-front faulting of the Late Pleistocene Qf2a surface (Table 8.1) at Symmes Creek (Fig. 8.16A) is spread out over 500 km but totals 41 m of apparent throw. Throw is 16% more than the apparent throw if corrections are made using a 9° fan slope and a fault-plane dipping in the same direction

Fan surface	Number of boulders sampled	Range of ^{10}Be ages, ka	Mean age, ka, 1 σ	Apparent net throw, m
Qf1	3	106–140	123.7 ± 16.6	41
Qf2a	none			41
Qf2b	3	53–66	60.9 ± 6.6	24
Qf3a	5	18–37	25.8 ± 7.5	10
Qf3b	none			6
Qf3c	5	3–6	4.4 ± 1.1	7

Table 8.1 Ages and maximum values of apparent throw of faulted piedmont surfaces downslope from the frontal fault on the east side of the Sierra Nevada near Mt. Whitney. From tables 1 and 3, Le et al. (2007).

Figure 8.16 Landforms indicative of Late Quaternary tectonic activity of the Sierra Nevada frontal fault.
A. Multiple-rupture-event fault scarp on the Qf2a fanhead of Symmes Creek. The lines approximate slope and offset of the fan surface.

as the fan at 50° – throw for Qf2a would be ~47 m. A +50% correction increase in throw estimate would be needed for the steepest (24°) dated landform of Le et al. (2007), and for all piedmont surfaces where the fault dip is 35° (Fred Phillips, email communication, 2008). Estimation of apparent and true throw of faulted alluvial fans is discussed on p. 108–112 in Bull (2007). The large recent 40–60 m displacement on the Sierra Nevada frontal fault was a sufficiently large base-level fall to enhance deposition of the Qf3b Holocene alluvial fan (Fig. 8.16B).

Ages of four of the alluvial surfaces were estimated by Le et al. (2007) using terrestrial cosmogenic nuclides. The age estimates for the Qf1 and Qf2b surfaces, 124 and 61 ka, match the 125 and 60 ka age estimates for climate-change alluviation events in the nearby Mojave Desert of California (Bull, 2007, table 1.2). Major Mojave Desert piedmont aggradation also began at about 12 ka, which fits the characteristics (Fig. 8.16B) and constraints on the age of the Qf3b surface of Table 8.1. Return of hot summers that cause monsoonal inflow of moist air were the cause of accelerated Holocene watershed

erosion at these times. This Sierra Nevada piedmont also had aggradation events at 26 ka (Qf3a) and perhaps at the time of deposition of the Qf2a alluvium. Sediment yields probably were increased at these times by glacial erosion. Deposition on this Sierra Nevada piedmont appears to be as discrete events that are controlled mainly by climate-change events.

The Le et al. (2007) preferred vertical slip rate on the Sierra Nevada frontal fault since the Late Pleistocene is 0.2–0.3 m/ky (corrected would be 0.3–0.45 m/ky), a range that is typical of active normal faults in the Basin and Range Province (U.S. Geological Survey, 2004; Wesnousky et al., 2005). Vertical slip rates during the Late Pleistocene may have been slow but appear to have increased fivefold (Le et al. (2007, table 4).

The "two schools of thought" presented here – slow, minimal uplift of the Sierra Nevada during the past 2 My, and local acceleration of normal faulting since 125 ka – are not mutually exclusive. Encroachment by normal and/or transtensional faulting seems to have begun the process of reactivation of parts of the frontal fault zone. Reactivation of the range-bounding fault zone is an acceleration of Owens Valley subsidence, not uplift of the Sierra Nevada. This process began west of Owens Lake and now is underway farther north.

Figure 8.16 Landforms indicative of Late Quaternary tectonic activity of the Sierra Nevada frontal fault.
B. Bouldery Holocene surface of the piedmont of Symmes Creek has a gentle cone-shape indicative of deposition as an alluvial fan. The road in the foreground is 15 km from the Alabama Hills, AH. SN is the Sierra Nevada. AC is the present active stream channel of Symmes Creek.

<u>Listric Model</u>   Fred Phillips has a keen interest in the Owens Valley section of the Sierra Nevada. His recent fieldwork focused on the range-bounding fault, which was nicely exposed as a result of extensive fires. The range-front fault dips only 35° to 40°, which is important from a geomorphic standpoint. Components of horizontal displacement are larger than vertical components (45° would be 1/1). Another key observation was finding an inactive listric fault in Bairs Creek 0.5 km upstream from the mountain front that decreases in dip from 80° to 70° to 55°. This suggests that the range-bounding fault may also be listric, steadily decreasing in dip towards Owens Valley.

Low-angle faulting has become his post-4-Ma working model. The Sierra Nevada crustal root delamination event caused a westward jump of Basin-and-Range extension to the Sierra Nevada front (Jones et al., 2004). The Fred Phillips model presumes that a range-bounding listric fault was created at ~3.5–4 Ma.

Delamination resulted in a postulated listric fault extending from the range crest to beneath old pedimented (tectonically inactive) terrain on the west side of Owens Valley. Bedrock that now is east of the fault formerly was at the range crest at 4 Ma. Extension rates may have been uniform, but the vertical component of faulting would decrease. The vertical/horizontal ratio of displacement would decrease eightfold as the position of the mountain front progressively shifted down the curving fault plane from an initial surface dip of 80° to the present 35°. Recent vertical-component displacement across the fault would be ~0.5–0.7 m/ky, close to the Le et al. (2007) estimate for the past 100 ky.

Acceptance of possible listric faulting puts a much different slant on how one would view Quaternary tectonic geomorphology. Early-stage dominance of vertical faulting favored rapid deposition of thick alluvial fans adjacent to the mountain front. Dominantly horizontal faulting now favors thin, extensive alluvial fans (Fig. 8.17) because low dips on an active range-bounding fault do not provide sufficient space for thick fans (Fig. 1.7, Bull, 2007). Active low-angle faulting would result in a smooth piedmont with only a thin surficial layer of bouldery alluvium. This style of tectonic control explains the virtual lack of older alluvium (such as shown in Fig. 2.2B) on this Sierra Nevada piedmont. The Bairs Creek "hump" (location 3 on Figure 8.15) is anomalous, being tectonic in origin and presumably due to flexural bending over the fault zone.

The resulting low potential fanhead aggradation rate, using a listric style of faulting, allows Late Quaternary climatic perturbations to become more effective in determining where piedmont deposition and channel downcutting are occurring. The piedmont reaches of these fluvial systems would remain close to the threshold of critical power as compared to when a dominant vertical style of displacement put them strongly on the depositional side of

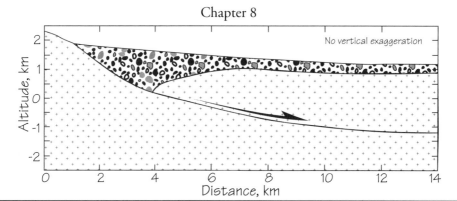

Figure 8.17 Listric faulting controls on space available for accumulation of piedmont alluvium. The depression between 2 and 4 km of this schematic generalization is quickly filled when the hanging-wall block shifts down the fault plane. Rotation on a curving fault plane limits the amount of deposition that can occur between 7 and 14 km. It is a piedmont surface where detritus is flushed through to the trunk stream, the Owens River. Figure courtesy of Fred Phillips.

the threshold. Stream power/resisting power ratios formerly were much less than 1.0. Present ratios are > 1.0 on the entrenched upper part of the piedmont and close to 1.0 on the lower part of the piedmont (a surface of transportation instead of either erosion or deposition).

Post-4-Ma renewal of piedmont deposition terminated a long period of Cenozoic tectonic inactivity during which I suggest the pediment surface of the Alabama Hills–Independence Hills was beveled. Rock uplift of ~2 km has raised parts of this Cenozoic pediment to the crest of the Sierra Nevada; the Chagoopa Plateau (Fig. 8.9B) is an example.

The three different viewpoints presented here illustrate how diverse geologic approaches can be applied, and cross-checked, to better understand this rather unusual geomorphic and tectonic setting. All three seem reasonable and correct. The geomorphic analysis of mountain and piedmont landscapes makes the best case for minimal base-level fall at the mountain front during the past 1–2 My. Recent work a few kilometers to the west (discussed next) confirms a decreasing rate of post-4 Ma uplift of the Sierra Nevada. Dating the young fault scarps is best for assessing the present level of seismic hazard, and for documenting a post-100-ky change in magnitude and perhaps style of faulting. The listric-fault structural approach best explains why this piedmont is so different than other nearby piedmonts, such as

those in Death Valley. It also explains why climatic perturbations are a more important control than tectonic base-level changes on the style and location of fan-aggradation events. The Fred Phillips approach should be considered in studies of other piedmonts in the actively extending parts of the Basin and Range Province.

One should be cautious in assuming that the mountain front discussed here is representative of other segments of the complex eastern front of the Sierra Nevada. Let us also allow for the possibility that a delamination-induced pulse of uplift may have started in the south and then migrated northward. River downcutting rates in the northern Sierra Nevada may have been increasing, a trend opposite that of the south-central part of the range that is west of Owens Valley. Accordingly, my remarks about post-4-Ma landscape evolution concern mainly the Kings River country and adjacent Owens Valley. Next, let us see how the rivers draining the west side of Mt. Whitney responded to post-delamination uplift.

A brilliant way to assess rates of stream-channel downcutting was devised by Greg Stock who determined times when streams stopped flowing into marble caves (Figs. 8.18, 8.19). Cave formation was initiated at active-channel level of the five Sierra Nevada rivers studied by Stock et al. (2004). Times of deposition of granitic stream gravels swept into the caves were estimated using burial dating (Granger and Musikar, 2001), which is based on changes in ^{26}Al and ^{10}Be terrestrial cosmogenic nuclides. Prolonged exposure of detritus to cosmic-ray bombardment while weathering on hillslopes and during transport down drainage networks increases the amounts of these terrestrial cosmogenic nuclides to about the same rate as they are being lost by radioactive decay – termed *secular equilibrium*. Fortunately, ^{26}Al is produced six times faster than ^{10}Be. Kings River gravels should, and do, have a ^{26}Al/^{10}Be ratio of about 6. The ^{26}Al/^{10}Be ratio decreased exponentially with time after the gravel was swept deep into caves (Fig. 8.18), where it was shielded from further cosmic-ray bombardment. Then, only radioactive decay would occur. ^{26}Al decays twice as fast as ^{10}Be. ^{26}Al/^{10}Be ratios would be about 5.2 after 300 ky, 4.9 by 400 ky, and 3.6 after 1 My of burial.

The incision history for trunk rivers and large tributaries is based on terrestrial cosmogenic nuclide isotope analyses of stream gravels in caves and seems reasonable (Fig. 8.19A). Larger tributary streams had sufficient unit stream power to plot on the same downcutting curve as trunk rivers such as the South Fork and Middle Fork of the Kings River. Rock-uplift rates may have been so slow that tributaries had sufficient time to downcut and attain type 2 dynamic equilibrium while the larger rivers locally attained type 1 dynamic equilibrium.

Stock et al. (2004) note that the overall mean rate of post-Miocene stream-channel downcutting of ~0.15 m/ky is about the same as the long-term rate estimated by dating Miocene valley floor lava flows in the central and northern Sierra Nevada at similar basin-position coordinates. The lower dashed gray line in Figure 8.19A is for cave data adjacent to river gorges and describes an exponential decrease in stream-channel downcutting. Channel incision rates, although a fairly slow ~0.6 m/ky at 2.8 Ma, decreased considerably. Downcutting rates were only about 0.18 m/ky between 2.5 and 2.0 Ma, and then decreased to only about 0.04 m/ky during the past 1 My. This suggests that rock uplift of range crest had ceased by 1 Ma. Minimal scatter about the exponential line suggests that the 240 km of the Sierra Nevada between the Kaweah and Stanislaus Rivers responded similarly to the post-4-Ma uplift pulse.

Figure 8.18 An artistic early morning view of the inner gorge cut in Jurassic marble, which has several caves with river gravels. Greg Stock scaled the cliff at the right side of this view to sample ~2.7 Ma river gravels that are now 395 m above the Kings River.

Maximum rates of rock uplift between 3 Ma and 2 Ma were not as dramatic as the Kings–Kaweah data of Figure 8.19A might suggest. Maximum rock uplift rates occurred before 2.7 Ma. Rock uplift was greatest at the range crest, decreasing to zero at the hingeline of the tilting microplate (perhaps at the foothill belt–Great Valley transition). Range-crest uplift would be only 500 m even if estimated rock uplift is increased to 0.5 m/ky between 3 and 2 Ma. So most of the uplift occurred before 3 Ma. The maximum possible rate would be for a scenario where 2,000 m of rock uplift occurred between 3.5 and 3.0 Ma; in which case the range crest would rise at an average of 4 m/ky. The value for rock uplift would be only slightly more if we take erosional lowering of the range crest into account. An uplift rate of 4 to 5 m/ky is comparable to the Southern Alps of New Zealand, Taiwan, and much of the Himalayas.

Tectonic-geomorphology analyses of the Owens Valley mountain–piedmont junction and the cave study in the adjacent Kings–Kaweah Canyons point to the same conclusion. The pre-3-Ma crustal delamination event caused a vigorous pulse of rock uplift, instead of prolonged uplift that has continued to the present time. This model applies to the southern and central Sierra Nevada, but the northern Sierra Nevada may be different.

Rates of inner-gorge formation in the upper reaches of the Feather River valley downstream from Lake Almanor (Fig. 8.19B) at the north end of the Sierra Nevada were estimated using dated lava flows described by Wakabayashi and Sawyer (2000, 2001). A ~1.2 Ma basalt flow is at a break-in-slope suggestive of the top of a 400 m inner deep gorge. The magnitude of post-3-Ma valley-floor incision is only slightly more than that for the southern Sierra Nevada. But the rates of stream-channel downcutting are increasing, in contrast to the opposite trend for the southern Sierra Nevada. Incision of valleys appears to be in the initial stages, especially for the smaller streams. Examples with drainage-basin lengths of ~50 km are Butte Creek and Big Chico Creek (Fig. 8.19B).

The tectonic setting is different than for the cave studies in the southern Sierra Nevada. This Feather River reach is near the crest of the Sierra Nevada, which is being actively encroached upon by Basin and Range and Walker Lane styles of faulting from the east. Minor post-basalt normal faulting cuts across the Feather River.

Opposite trends of river incision in the northern and southern Sierra Nevada imply different tectonic histories. Perhaps 2 My was needed for the delamination-induced pulse of uplift to shift 400 to 500 km northward. Perhaps recent range-crest uplift in the north is the result of a second delamination event, or it may be the result of another tectonic process, such as encroachment into the Sierra Nevada microplate by faulting.

We should consider other tectonic factors besides delamination. The Feather River area

is only 57 km southeast of Mt. Lassen, which is at the southern end of the Cascade volcanic chain and represents the edge of the slab window forming as the Mendocino triple junction continues its northward migration. The northern edge of the expanding slab window would not have reached the Feather River until after 2.5 Ma, well after the postulated delamination of the southern Sierra Nevada batholith. A delay in the uplift of the northern Sierra Nevada might be a situation where cold subducted lithosphere is replaced with lower density hot asthenosphere (W.R. Dickinson, email communication, 2007). Subduction-related volcanism is replaced by upwelling of hot asthenosphere into the slab window (Sections 1.2.4, 8.3), which as along the Mendocino coast, may accelerate rates of uplift and stream-channel downcutting.

Dickinson (1997) relates pulses of coastal California Tertiary volcanism to episodes of mantle upwelling generated by several rise–trench encounters during plate subduction. Such slab-window vulcanism becomes younger towards the north, being only ~2 Ma at Clear Lake and Sierra Buttes southwest of the Feather River (Dickinson, 1997, fig. 8). He concludes that Pacific plate shear became more organized in the late Tertiary after demise of several subducting microplates allowed integration of the San Andreas transform into a coherent plate boundary. This would set the stage for initiation of the Walker Lane–Eastern California shear zone, which has continued to propagate northward.

An important corollary to these tectonic alternatives is the timing of Walker Lane strike-slip faulting. Being a strand of the San Andreas transform system, it can only extend northward as the Mendocino triple junction migrates north. Strike-slip faulting along the Honey Lake and Mohawk Valley fault zones is impressive (Adams et al., 2001) but almost surely is younger than the Walker Lane–Eastern California shear zone further south. Some faulting in the area of the North Fork of the Feather River began after ~600 ka (Wakabayashi and Sawyer, 2000, 2001; Adams et al., 2001). Faulting began as the Walker Lane encroached into the microplate.

Post-4-Ma stream power was increased in two ways throughout the microplate. Gradients were increased, especially in upstream reaches. Second,

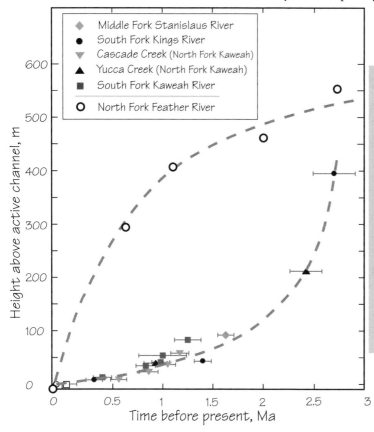

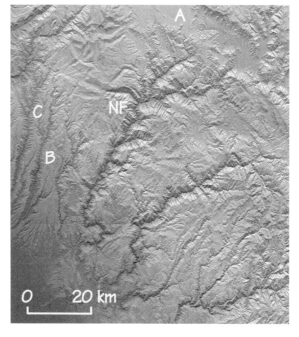

Figure 8.19 Sierra Nevada river incision history since 3 Ma.
A. Gray dashed lines depict trends in tectonically induced downcutting in the southern and northern Sierra Nevada.
Ages of burial of fluvial gravels in caves are from the middle to southern part of the range are from Stock et al. (2004); 1 σ error bars.
Feather River basalt-flow data in the northernmost part of the range are from Wakabayashi and Sawyer (2000).

surface uplift increased orographic precipitation and the resulting increase of storm runoff further increased annual unit stream power. Modeling (Stock et al., 2004) suggests that stream-channel downcutting began in distal downstream reaches and migrated upstream as knickpoints. If so, the threshold of critical power was crossed first in reaches of maximal unit stream power fairly close to the western mountain front.

Continued slow stream-channel downcutting may be the result of another way to raise the

Figure 8.19 Sierra Nevada river incision history since 3 Ma.
B. Recently incised low-relief terrain in the northern Sierra Nevada. NF, North Fork of the Feather River; B, Butte Creek; C, Big Chico Creek; A, Lake Almanor.

Ancestral highlands

Inner Gorge

Figure 8.20 View across the valley of the South Fork of the Kings River from Deer Cove. Flat, high ridgecrest on the opposite side is part of the gently rolling upland topography of the ancestral Sierra Nevada. 400 m deep inner gorge is largely hidden below the break in slope. Most of the relief shown here was present at 4 Ma.

range crest. Isostatic models for promoting uplift, such as that used by Small and Anderson (1995), exploit the mass removed by glaciers and rivers during the Pleistocene as a way to cause isostatic rebound. Net changes in surface altitudes would vary spatially under conditions of regional isostatic rebound. The range crest would increase in attitude, further enhancing precipitation gradients and potential for glaciation (Fig. 8.10). Valley floors would become lower in altitude because tectonically induced downcutting would exceed isostatic uplift.

Inner-gorge width increases were much slower than stream-channel downcutting. River reaction time and response times tend to be short because that is where power is concentrated in fluvial systems. Most hillslope processes disperse erosional forces and breaks-in-slope retreat at exponentially progressively slower rates. Retreat is not uniform, being greater where incision by streams that are tributary to the trunk channel have steepened adjacent hillsides. Steepened slopes can converge from several directions on a ridgecrest break-in-slope. Inner-gorge width increases do not match the brief-pulse uplift model because of exceedingly long response times needed to change the shapes of hillslopes.

The ancestral uplands inherited from pre-4-Ma times have not changed much at the lower altitudes. Elevation of the western flank of the

microplate resulted in glacial erosion at the higher altitudes, but deeply weathered granitic rocks at the lower altitudes attest to little change in denudation rates. The ridgecrests are still fairly flat and the rolling uplands are little incised by streams. Either the breaks-in-slope at the top of the inner gorge essentially decouple the high surfaces from inner-gorge incision, or response times exceed 4 Ma. A casual observer on the road to the South Fork of the Kings River, or even from the high viewpoint of Figure 8.20, could easily conclude that the entire landscape assemblage reflects long-term and continuing rock uplift. But it is just the inner gorge landscape that marks the brief pulse of uplift starting at ~3.5 Ma.

Indeed, it is the sensitivity of the footslope and valley-floor landforms to base-level fall that catches the eye of the tectonic geomorphologist. The higher parts of the landscape attest to substantial relief of the Sierra Nevada before the 3.5 Ma plate-tectonic perturbation. It is almost as if the broad interfluve parts of these fluvial systems failed to sense tectonic perturbations caused by the crustal-root delamination.

Indeed reaches of these streams are disconnected from one another (Fig. 8.21). A prolonged lack of continuity of these smaller fluvial systems records a sustained inability of the largest peak discharges to overcome the substantial rock mass

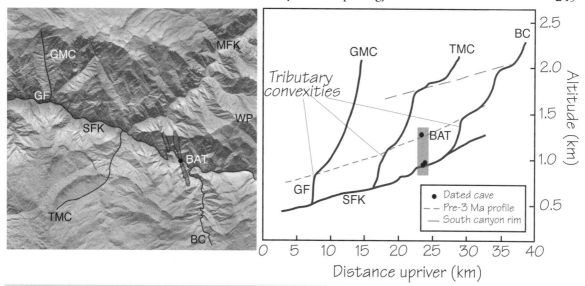

Figure 8.21 Longitudinal profiles of streams tributary to the South Fork of the Kings River suggest a prolonged period of stable base level followed by rapid incision beginning at ~3.5 Ma. Figure 10 of Stock et al. (2005).
A. The canyon downstream of the confluence of the Middle and South Forks of the Kings River is the deepest in the Sierra Nevada dropping 2,391 m from Spanish Mountain to the river. BAT, Bat Cave; BC, Boulder Creek; GF, Garlic Falls; GMC, Garlic Meadow Creek; MFK, Middle Fork Kings River; SFK, South Fork Kings River; SM, Spanish Mountain; TMC, Tenmile Creek. Marble belt across the South Fork is shown in dark gray.
B. The steep terminal reach of each tributary was created by post-3.5-Ma stream-channel downcutting of the trunk river, but these knickpoints have not migrated far upvalley; being at heights similar to that of 2.7 Ma Bat Cave, which is 395 m above the river. Still higher knick-points are at the canyon rim.

strength of quartzites and metavolcanic rocks beneath these valley floors. The result is hanging valleys with waterfalls in some cases. Tectonically induced downcutting by the trunk river records tilting of the Sierra Nevada microplate but has yet to be transmitted far up the tributary valleys, far less to the hillslopes and ridgecrests of the broad interfluves.

Stock et al. (2004) measured granitic ridge-crest erosion rates using concentrations of ^{26}Al and ^{10}Be and concluded that weathering-limited denudation rates are very slow. They average only 0.012 m/ky over a 75 ky time span. ^{26}Al and ^{10}Be concentrations yield an average erosion rate of 0.003 m/ky since ~145 ka for Beetle Rock, a granitic dome on an interfluve between two canyons of the Kaweah River (Stock et al., 2005). Stream-channel downcutting

during the late Cenozoic was one to two orders of magnitude faster than erosional lowering of nearby ridgecrests. Many features of this landscape support a model of non-steady-state landscape evolution for the Sierra Nevada (Section 5.3.2).

8.3 Mendocino Triple Junction
The hallmark study by Merritts and Vincent (1989) discerned how fluvial landscapes responded to increasing, then decreasing, uplift rates associated with the approach and passage of the Mendocino triple junction along the northern coast of California. Their inductive approach to tectonic geomorphology required that they first determine the history of uplift-rate changes for small coastal drainage basins.

Spatial variations of tectonic signatures in the hills and streams infer major differences in Late Quaternary rates of local uplift. Rapidly rising fluvial systems near Cape Mendocino have steep, convex hillslopes that plunge down to exceedingly narrow valley floors whose longitudinal profiles are steep (Merritts, et al., 1994). Such fluvial landforms suggest a degree of rock uplift that matches rapid uplift suggested by the presence of Holocene marine terraces where these streams leave steep, rugged coastal hillsides and enter the Pacific Ocean. Several levels of Holocene shore platforms record recent coseismic uplift events. One was a magnitude Ms 7.1 earthquake of 1992 (Merritts, 1996).

Much different landform assemblages greet a visitor to the Fort Bragg area, 140 km to the south along this coast. Convex–concave hillslopes merge with broad valley floors. Neither hillslopes nor stream channels are particularly steep until one gets to the headwaters reaches of these watersheds. Prominent, broad marine terraces are aptly named the Mendocino Staircase. They show us that this coast is rising too, but terminal reaches of streams are estuaries and Holocene marine terraces are absent. Strongly developed soil profiles tell us that the youngest marine terraces pre-date the Holocene by >50 ky.

This is interesting, but we need to quantify spatial variations of uplift rates in order to do more than state generalities about landscape-evolution responses to plate tectonics.

8.3.1 Marine Terraces

Uplift rates vary in both time and space. A shore platform was notched into this rising coast each time a glacio-eustatic sea-level highstand cut a notch into the fluvial landscape. Superb flights of marine terraces along the Mendocino coast (Fig 8.22) provide a framework of reference to determine rates of coastal uplift.

Dating of corals, and analysis of oxygen isotopes, from coralline marine terraces elsewhere provide an essential chronology of matching sea-level changes in the Atlantic and Pacific Oceans. We use the post-340-ka record. A glacio-eustatic still-stand has persisted since 6 ka. A 330 ky time span

is long enough for geomorphic processes to modify the landscapes of these humid-climate watersheds, and for the Mendocino triple junction to shift 18.5 km farther north.

Initial dating of individual marine terraces (Kennedy et al., 1982; Lajoie et al., 1982) revealed much faster uplift in the north – 0.3 m/ky near Fort Bragg and 4 m/ky just south of the triple junction. Analysis of altitudinal spacing of 14 entire flights of marine terraces described a migrating spatial pattern of varying uplift that agrees with the earlier work. Marine-terrace ages at any one place are clearly related to the passage of the Mendocino triple junction (Merritts and Bull, 1989).

Marine terraces were created at times of worldwide sea-level highstands (Chappell, 1983, 2001; Chappell et al., 1996; Edwards et al., 1987a,b; Hamelin et al., 1991; Gallup et al., 1994; Fleming et al., 1998; Israelson and Wohlfarth, 1999; Lajoie, 1986; Lambeck and Chappell, 2001). Global synchroneity of the 124 ka terrace has been demonstrated by essential uranium-series disequilibrium dating of coral from New Guinea, New Hebrides, Barbados, Haiti, the Mediterranean Sea, Hawaii, Japan, and California. This work dated the sea-level highstands and estimated their altitudes of formation (Mesolella et al., 1969; Bloom et al., 1974; Ku et al., 1974; Chappell and Shackleton, 1986).

Newer studies fine-tune and expand our knowledge about Late Quaternary sea-level changes (Ota et al., 1993; Chappell et al., 1996; Chen et al., 1991; Stirling et al., 1998; Yokoyama et al., 2000). New Guinea corals now are used to date sea-level rises of <15 m that lasted only 1–2 ky and match the 6 ky cycles of "Heinrich" episodes of massive North Atlantic ice breakouts (Chappell, 2002). The newer work supports the basic framework of the previous investigators, so this section uses the Merritts and Bull (1989) data to illustrate how to use unique altitudinal spacings of marine terraces to estimate uplift rates along the Mendocino coast since 330 ka.

Changes in surface loading by thick Quaternary ice sheets deformed the surface of the planet and induced flow in the mantle. Such changing surface loads and internal deformation changed sea level too (Chappell et al., 1982; Lambeck, 1993; Dickinson et al., 1994; Dickinson, 2000). This might

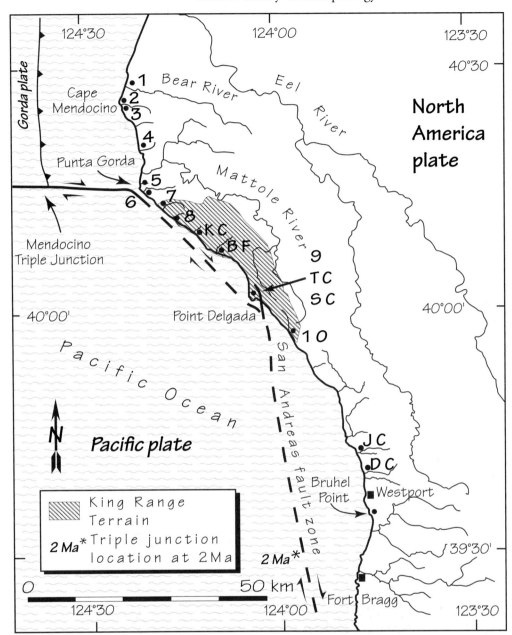

Figure 8.22 Location map of the Fort Bragg–Mendocino triple junction study area showing main fault zones and the locations of streams and surveyed flights of marine terraces. From Merritts and Bull (1989, figure 1). Names of marine terrace sites: 1, Bear River; 2, Cape Ranch; 3, Singley Flat; 4, McNutt Gulch; 5, Smith Gulch; 6, Fourmile Creek; 7, Cooskie Creek; 8, Randall Creek; 9, Kaluna Cliff; 10 Whale Gulch; and Bruhel Point. Names of streams: KC, Kinsey Creek; BF, Big Flat; TC Telegraph Creek; SC Shelter Cove; JC Juan Creek, DC, DeHaven Creek.

be a source of minor differences in sea-level curves in New Guinea as compared to northern California.

Cool-water coasts lack corals so radiocarbon, amino-acid, and thermochronology techniques are used to date deposits laid down on the beveled bedrock of shore platforms. These age estimates constrain approximate times of formation of individual shore-platform remnants. Some shore platform deposits have marine molluscan fauna indicative of sea temperatures cooler than the present (Ota et al., 1996). Warm-water molluscan fauna characterize the major sea-level highstands (330, 212, 124, and 6 ka).

Sea-level highstand shore platforms are not much different in altitude if the uplift rate of coastal hills is a slow 0.1 m/ky, but appear as giant steps on steep mountainsides where uplift is >3 m/ky. The altitude of a terrace is also a function of the altitude of the sea-level-highstand at the time the terrace was created (ranging from +6 m to ~50 m relative to the present level of the sea). Long intervals between times of terrace formation increase the altitudinal spacing between consecutive marine terraces.

This controlling mix of variables creates a unique altitudinal spacing of marine terraces for each uniform uplift rate. Flights of dated global marine terraces rising at 1.3 m/ky on the Huon Peninsula of New Guinea have the same altitudinal spacing as non-coralline marine terraces on another island in the South Pacific Ocean, New Zealand – but only if both coastal uplift rates are uniform 1.3 m/ky.

Merritts needed to assess possible variation of uplift rates for entire flights of marine terraces along the Mendocino coast. Although quite useful, this task that could not be done with the available radiocarbon and amino-acid age estimates. The solution was to estimate ages of every marine terrace in a flight using the approach developed in New Zealand (Bull, 1985; Bull and Cooper, 1986, 1988). A regional uplift-rate curve related to the passage of the Mendocino triple junction was produced when the terrace-spacing approach to dating was combined with dating of individual marine terraces by previous workers.

It is fitting that this chapter about shifting crustal plates and tectonic geomorphology uses a plate-tectonics analogy to estimate the ages of each

former sea level in a flight of marine terraces. The model for marine terrace spacing (Bull, 1985) is conceptually the same as that used for the spacing of sea-floor magnetic stripes. Nonuniform time spans between global magnetic polarity reversals create a variable pattern of magnetic anomalies, even though crustal spreading rates are uniform. Each ocean ridge spreading rate creates a unique spacing of magnetic strips (Vine and Matthews, 1963; Heirtzler et al., 1968). Nonuniform intervals between times of marine-terrace formation and nonuniform altitudes of formation create distinctive altitudinal spacings for flights of marine terraces. Each uniform uplift rate has a unique altitudinal spacing of terraces. Both spacing models – global ocean-floor magnetic anomalies and global marine-terrace altitudes – can be used to infer ages where radiometric age determinations are not available. Both require careful dating in type localities.

Not all global marine terraces may be present for us to use. They may be absent because of fluvial or coastal erosion. A tectonically stable coast will have only marine terraces whose altitude of formation is higher than the present sea level (+6 m at 124 ka and +4 m at 330 ka). Note that, with no uplift or subsidence, the 124 ka highstand would re-occupy the 330 ka terrace location and our data set would consist of the inner edge of only the 124 ka shore platform. A coast that is rising at only 0.4 m/ky will not have post-50-ka terraces because the present unusually high stand of the sea has submerged shorelines created at highstands >40 m below present sea level. Uplift rates of >2.0 m/ky are sufficiently rapid that younger sea-level highstands never submerge older highstands. Such coasts may have numerous marine terraces where rock types favor coastal erosion that bevels shore platforms instead of cliffs that drop into the sea.

Rapidly rising Cape Mendocino shorelines have several Holocene shore platforms (Lajoie et al., 1982; Merritts, 1996), even though sea level has not changed. Such pulses of coseismic uplift show that uplift rates are not perfectly uniform over short time spans. Similar coseismic events occurred during prolonged Late Quaternary sea-level highstands.

Noting both uncertainties and constraints, let us examine the consistency and resolution of altitudinal spacing analyses for slow- and fast-uplift parts of the Mendocino coast. This approach integrates the brief pulses of coseismic uplift into longer-term estimates of uniform rates of coastal uplift. Failure to achieve results suggestive of uniform Late Quaternary uplift requires evaluation for possible spatial or temporal variations of tectonic deformation for a given flight of marine terraces.

Pillans (1990) does not favor use of altitudinal spacing analyses for inferring ages of flights of marine terraces. He correctly points out that erosional notches in ridgecrests may be fluvial, not marine in origin. He wonders if the marine-terrace record at New Guinea is complete for pre-130 ka sea-level highstands. The tests used in the next 7 pages will fail if the diverse New Guinea studies are flawed.

Intuitively, there should be a sea-level highstand and a marine terrace or two between 130 and 170 ka. If so, the Bull and Cooper (1986, 1988) suggested change of uplift rate of the Southern Alps at ~135 ka is apparent, not real. Fortunately they had a dramatic change in landscape characteristics (Fig. 6.5) to support their contention for a recent change of tectonic style. The oxygen-isotope record from deep-sea cores provides a continuous record of the alternating cold and warm climates of the Quaternary (Broecker and van Donk, 1970; Clemens and Tiedemann, 1997). The longest cool spell of the past million years indeed occurred during the 130–170 ka interval (Imbrie et al., 1984; Wright, 1998), so it is unlikely that this was a time of sea-level highstands, warm oceans, and marine terrace formation.

The altitudinal spacing procedure calls for common sense. Continuity of the inferred shoreline along a section of coast is essential. Field surveys of marine-terrace inner edges by different teams have good replication (Table 8.2). A marine origin for a ridgecrest notch is supported by the presence of remnants of a degraded sea cliff, a beveled shore platform surface, and pebbles with shapes and surface textures that match those on nearby present-day beaches. Inferred shorelines on the headwaters divides of drainage basins are at a location where no streams are present to notch a hillslope or leave rounded pebbles (Bull, 1984). Coverbed stratigraphy

Punta Gorda	Smith Gulch
111 m	111 m
123	125
165	162
173	174
185	187
139	241
251	252
274	271

Table 8.2 Continuity of altitudes of marine terrace inner edges surveyed by different survey teams in different years (Merritts, 1987, p. 20).

– loess, colluvium, and soil profiles – needs to match the assigned marine-terrace age.

Fossils, such as marine shell hash, should be examined to see if they record a cold or warm-water fauna, which should match ocean temperatures of a particular sea-level highstand (Ota et al., 1996). For example, the 124 ka sea-level highstand was a time of warm water compared to the lower colder highstand of 53 ka. This allows one to use the ages and heights of marine terraces at sites other than New Guinea to assess the completeness of the global marine terrace record.

We seek and use all numerical age estimates, especially those with small uncertainties. The goal is to have a set of internally consistent age estimates.

All results are regarded as inferred uplift rates. Crosschecks assess if altitudinal spacing analyses give consistent, precise estimates of uplift-rates.
1) Is there only one unique uplift-rate match for the altitudinal spacings of New Guinea and local flights of global marine terraces?
2) Are assigned ages the same for adjacent flights of marine terraces?
3) Does the method fail to produce a uniform rate when a survey transect crosses a known active fault?
4) Do we obtain reasonable differences of uplift rate for flights of marine terraces on opposite sides of an active fault whose style of displacement is known?
5) Do the systematic temporal and spatial changes

Huon Peninsula, New Guinea*		Bruhel Point, northern California**		
Age, ka	Formation altitude, m	Present altitude, m	Inferred uplift, m	Mean uplift rate, m/ky
6	0			
29	−44			
40	−41			
45	−45			
53	−30			
59	−28			
72	−36			
81	−19	10***	29	0.36
96	−26			
100	−9	23	32	0.32
118	0	40	40	0.34
124	6			
172	−50			
240	−10			
286	−46			
305	−27	86	113	0.37
330	4	130	126	0.38

*New Guinea terrace ages and altitudes of formation from Chappell and Shackleton (1986), and Chappell, 1983.

**Inner edge present altitudes surveyed with electronic distance meter, except the 330 ka marine terrace.

*** Terrace age also estimated by amino-acid racemization of molluscs from correlative marine terrace 13 km to south (Kennedy et. al., 1982).

Table 8.3   Data used for correlation of global marine terraces in New Guinea and northern California.

of inferred uplift rate occur along the 140 km of the Mendocino coast, and do they agree with all radiocarbon and amino-acid age estimates of previous workers?

6) Do the inferred ages match the chronosequence for terrace-tread soils and their rates of change of geochemical mass balance?

Inner-edge altitudes of marine terraces in the slowly rising part the Mendocino coast (Table 8.3) illustrate the procedure used by Merritts and Bull (1989). The prediction diagram of Figure 8.23 – after Lajoie (1986, fig. 6.6) – attempts to match New Guinea and California flights of marine terraces. It succeeds only if uplift rates are uniform at both localities. Initial assignment of the age of a particular Mendocino coast terrace is merely a guess based on its strength of soil-profile development and altitude. Another New Guinea age is selected if the first guess results in erratic rates of inferred uplift for individual Mendocino marine terraces.

Crosscheck 1.   Long-term uniform uplift was suggested when the 10 m marine terrace was assigned an age of 81 ka. Mean uplift rates for individual Bruhel Point marine terraces ranged from 0.32 to 0.38 m/ky. The average for this flight is a tight 0.35 ± 0.03 m/ky. This is reassuring considering that potential sources of scatter include New Guinea dating uncertainties, terrace altitude surveying errors at both coasts, and departures from perfectly uniform uplift at both coastal localities. All other age assignments for Bruhel Point result in ~100% variation in inferred uplift rates for individual marine terraces.

This initial result is encouraging but of course leaves one doubting if it works equally well at other sites. The altitudinal spacing analysis model for dating every terrace in a flight consistently provided high-quality results at all 14 transects studied by Merritts and Bull (1989). The vigor of this way to estimate ages of marine terraces is a tribute to the careful fieldwork and geochemical analyses by many diverse earth scientists working on global marine terraces since ~ A.D. 1970.

Crosscheck 2.   An altitudinal spacing analysis for the Bruhel Point flight of marine terraces is supported by all six of the above list of rigorous crosschecks. Shore platform inner-edge altitudes of raised marine terraces are virtually identical at five nearby flights and with similar precision of estimated ages.

Crosschecks 3 and 4.   Altitudinal spacing analyses south of Fort Bragg by MacInnes (2004) also concluded that the lowest raised shore platform was the 81 ka marine terrace. Inferred uplift rates are 0.4 to 0.5 m/ky. The strike-slip San Andreas fault comes onshore here and offsets tilted marine terraces. MacInnes noted a second offset of the marine terrace flight and used it to discover the "Russian

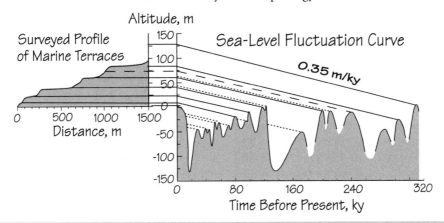

Figure 8.23 Prediction diagram connecting times of New Guinea sea-level high-stands with corresponding inner-edge altitudes of undated marine terraces at Bruhel Point, assuming a uniform rate of uplift. Dashed line is for a missing terrace that presumably was removed by erosion. Dotted lines are for terraces submerged by a more recent sea-level highstand. From figure 2 of Merritts and Bull (1989).

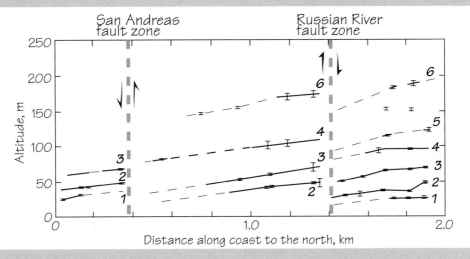

Figure 8.24   Flight of marine terraces offset vertically by the San Andreas and Russian River fault zones. Error bars indicate uncertainties of inner-edge altitudes. Dashed lines indicate correlations where inner edges were not obvious. Inferred ages of numbered terraces (Table 8.3) are based on a local altitudinal spacing analysis; 1, 81 ka; 2, 118 ka; 3, 124 ka; 4, 172 ka; 5, 305 ka; 6, 330 ka. Figure 3 of MacInnes (2004).

River" fault zone (Fig. 8.24). Vertical displacements caused by active faulting changes the altitudinal spacing of the terraces locally for three flights of marine terraces, but inferred uplift rates are uniform for each transect. Accidental mixing of populations of terrace altitudes from opposite sides of an active fault would eliminate the possibility of uniform-uplift-rate results.

The presumption that uplift rates are constant for coasts without structural complications is easily tested. Try it. You will find that coastal uplift rates are surprisingly regular. Time spans for significant change of plate-boundary, or local, uplift rates generally are much longer than the 100 to 300 ky represented by flights of marine terraces. Of course this matching game means that the resulting estimates of uplift rate are inferred.

Suites of inferred uplift rates, when combined with traditional dating results, provided the essential tectonic background information for the many workers who have studied responses of fluvial systems to Mendocino coast uplift. Considering the importance of these estimates of regional rock uplift, let us look more closely at how to determine if a uniform-uplift-rate model characterizes a particular flight of marine terraces.

Crosscheck 1. A survey of six inner edge estimated altitudes at Shell Beach (Fig. 8.25) gave MacInnes (2004) enough data to see if the spacing of these marine terraces might fit a model of uniform uplift. The lowest terrace was assigned the age, and the associated altitude of formation, of a global marine terrace. Three assignments resulted in a 1.6 to 2.1 m/ky variation in average uplift rate; such apparent rapid fluctuations of inferred uplift rate are unlikely to have a tectonic cause. Scatter about a mean value is a minimal ±0.03 m/ky when the youngest terrace is assigned an age of 83 ka. This uniform uplift result makes tectonic sense and is similar to uplift rates at several terrace flights in the adjacent Fort Bragg part of the Mendocino coast..

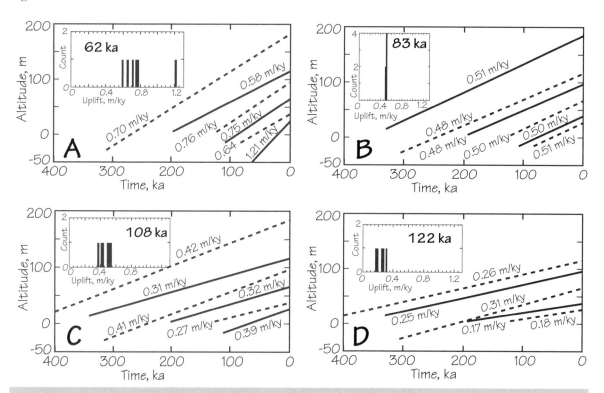

Figure 8.25 Assignment of several sets of ages of global marine terraces to shore platform inner-edge altitudes at Shell Beach on the northern California coast. 38° 26' N, 123° 6' W. Lowest terrace is assigned a global marine terrace age of 62 ka in A, 83 ka in B, 108 ka in C, and 122 ka in D. Inset histograms of inferred uplift rates for individual terraces have a class interval of 0.02 m/ky. Count is the frequency of observations. From figure 7, MacInnes (2004), and using her global-marine terrace ages.

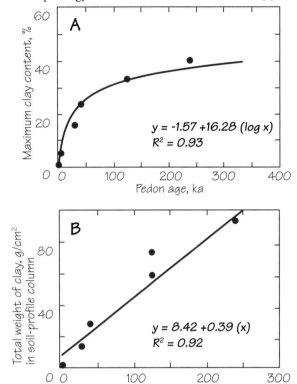

Figure 8.26 Trends in accumulation of pedogenic clay in the soil profiles on marine terrace treads, Cape Mendocino to Punta Gorda, northern California. From Figures 5 and 6 of Merritts et al. (1991).
A. Plot of time versus maximum percent clay.
B. Plot of time versus total weight of clay in a 1 cm² soil column.

<u>Crosscheck 5.</u> The Bruhel Point 10 m marine terrace has an independent age estimate. Evaluation of the amino-acid ratios of cool-water shell types (Kennedy et al., 1982) suggested a correlation with either the middle (5c) or late (5a) oxygen isotope stage 5 (Shackleton and Opdyke, 1973). Uranium series ages of coralline terraces in New Guinea created during these two worldwide sea-level highstands provide ages of 100 ka and 81 ka for these two isotope stages (Bloom et al., 1974; Chappell, 1983). Altitudinal spacing analysis of local terrace flights at nearby Mendocino and Westport indicates that 81 ka is the more likely 10 m terrace age.

<u>Crosscheck 6.</u> Soil-profile development provides a qualitative measure of landform age. Variations of soil-profile characteristics underscore major spatial and temporal changes in coastal uplift suggested by marine-terrace altitudinal spacing analyses. Accumulations of pedogenic clay and iron oxy-hydroxides in soil profiles increase markedly with the passage of time.

Bruhel Point marine terraces at altitudes of only 23 and 40 m have intensity of soil-profile development similar to that of 261–294 m high terrace treads near Cape Mendocino (Merritts et al., 1991, 1992). Maximum percentages of all forms of pedogenic iron increase linearly with time, with high correlation coefficients. Soils developed on marine terraces older than 100 ka have at least 30% maximum clay, and older than 240 ka at least 40–50% maximum clay. Maximum percent clay increases logarithmically with time (Fig. 8.26A) and the total weight of clay in a soil profile increases in a linear fashion with the passage of time (Fig. 8.26B). Data for terraces in the Bruhel Point part of the coast were not included in these regressions because low uplift rates

preclude preservation of terraces younger than 81 ka, and the older terraces, still at low altitudes, have accumulated wind-blown sand. The older terraces also extend 5–10 km inland, support coniferous vegetation instead of bunch grass and wildflowers, and may receive more rainfall than at coastal sites. Maximum percent clay and iron vary consistently with age of each marine terrace. Pedogenic clay and iron is useful for correlating isolated remnants of warped, faulted, and eroded marine terraces along the Mendocino coast where the shore-platform remnants are capped by marine and colluvial deposits.

The Table 8.3 data suggest an arithmetic relation between uplift and terrace age. The Figure 8.27A regression indicates a mean uplift rate of ~0.4 m/ky, but needs closer inspection. The high correlation coefficient is an artifact of regressing cumulative increase of age against cumulative uplift. The inferred uplift rate is merely apparent. It is not truly uniform for the entire 330–0 ka time span because the regression line does not pass through the graph origin. Perhaps some of the small scatter of uplift values noted in Table 8.3 results from non-uniform uplift at Bruhel Point.

A regression for post-200-ka marine terrace uplift (Fig. 8.27B) indeed passes through the graph origin and mean uplift rate appears to be less; 0.34

m/ky. Regressing 330 to 200 ka terrace data provides an inferred uplift rate of 0.44 m/ky. Uplift rates may have decreased by one-third, which is reasonable considering that the Mendocino triple junction passed this part of the coast at ~2.5 Ma. Uplift rates should be gradually decreasing here.

Such Bruhel Point evaluations are only tentative suggestions at this point, because verification is needed from other sites. This is why Merritts and Bull (1989) surveyed many flights of terraces. Each new survey consistently supported a systematic model of spatially changing tectonic deformation caused by shifting crustal plates.

Altitudinal spacing analyses for transects closer to the present position of the Mendocino triple junction document obviously faster uplift rates. Two radiocarbon dated shore platforms continue north to Kaluna Cliff where altitudes of 14 marine-terrace inner edges were surveyed in a flight that rises to an altitude of 400 m. Numerous marine terraces imply a rapid rate of uplift and the Figure 8.28 inferred uplift-rate plot documents a rate of uplift of 1.2 m/ky, three times that of Bruhel Point 55 km to the south. Once again a unique solution was obtained. None of the other correlations with the global marine terraces yields a plot with uniform rates of uplift or a regression line this close to the plot origin.

Uplift rates increase northward still more and at 25 km from Kaluna Cliff they were a maximum of

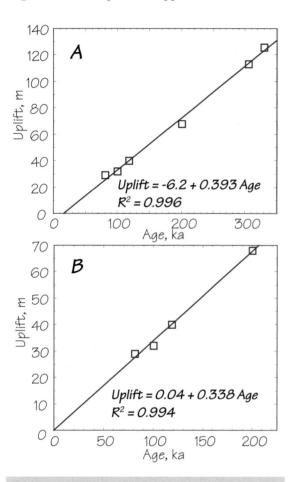

Figure 8.27 Regressions of inferred uplift amounts and terrace ages for the Bruhel Point data set.
A. Regression of marine terrace ages and uplift that does not pass through the graph origin. Inferred uplift rate is 0.39 m/ky.
B. Regression of marine terrace ages and uplift for the four youngest marine terraces. Regression line passes through graph origin. Inferred uplift rate is 0.34 m/ky.

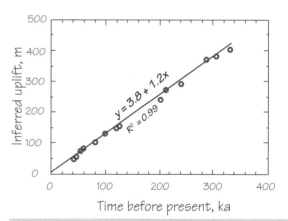

Figure 8.28 Regression of inferred uplift amounts and ages for the Kaluna Cliff marine-terrace transect.

Figure 8.29 Aerial view of Smith Gulch site on the Punta Gorda coast. Prominent break-in-slope at ~ 250 m altitude separates marine terraces formed during times of 0.4 and 3.4 m/ky uplift rates. Arrows point to inner edges of terrace remnants. Prominent Holocene shore platform in foreground. Road for scale. Figure 2.10 of Merritts (1987).

Road

4 m/ky at Randall Creek since 124 ± 6 ka. Multiple Holocene shore platforms as high as 30 m allude to pulses of coseismic uplift and continuing rapid uplift. Radiocarbon dating of these shore platforms (Lajoie et al., 1982) provided uplift-rate estimates that match the results of the altitudinal spacing analyses for the 124 ± 6 ka interval.

The spacing of marine terraces, which stairstep up the hillsides, changes at the Fourmile and Smith Gulch sites just north and south of Punta Gorda. This two-stage uplift history is much different than the uplift style recorded at Kaluna Cliff only 36 km to the southeast. A 250 m high steep coastal escarpment with remnants of former shore platforms changes to a still higher flight of broad, flat marine terraces (Fig. 8.29) that look like Bruhel Point terraces. It is obvious that Late Quaternary rates of uplift increased on this part of the Mendocino coast. Uplift rates since 124 ka were sufficiently rapid that no marine terraces were submerged by younger sea-level highstands. The preferred sequence of assignment of local marine terraces to New Guinea highstands (Fig. 8.30) has uniform styles of early (0.4 to 0.7 m/ky) and recent (3.2 to 4.0 m/ky) uplift rates. This consistency of changing uplift pattern satisfies crosscheck 5.

Uplift styles and rates are forever changing along the Mendocino coast over time spans of 1 My. The agreement between mean uplift rates based on individual terraces dated by radiocarbon and amino-acid methods agrees with uplift rates inferred by analyzing altitudinal spacings of entire flights of marine terraces (Fig. 8.31A). All dating methods are used to define regional trends of Late Quaternary uplift (Fig. 8.31B). Uplift rates increase slowly from Bruhel point towards the north for ~100 km, then increases by tenfold in only 20 km, and then decrease by a third over the next 40 km. These changes in uplift rate were generated by ephemeral crustal thickening (Furlong and Govers, 1999). They suggest that passage of a migrating Mendocino triple junction causes an ephemeral crustal thickening that is the result of viscous coupling between the northward migrating Gorda slab and the base of the North American plate south of the triple junction. Mantle material flows into a slab window that opens as the Gorda plate moves northward.

A corollary to this basic tenet is that each watershed experienced consecutive uplift increase and decrease. Landforms that roughly parallel the San Andreas fault (marine shorelines and tributary valleys to watershed trunk stream channels) should

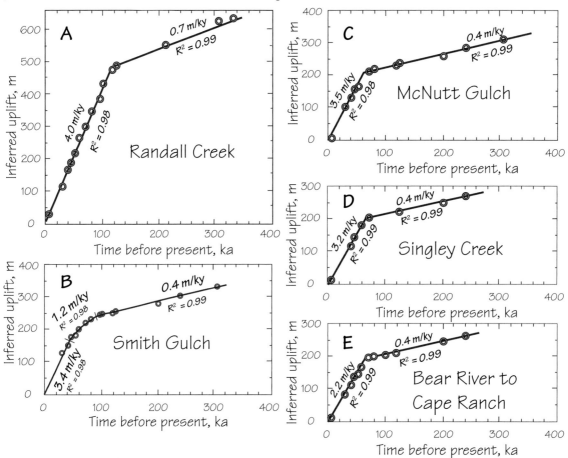

Figure 8.30 Trends of changing inferred uplift rates recorded by marine terrace ages near the Mendocino triple junction. Data and graphs courtesy of Dorothy Merritts.

Approximate distances (+ to North or – to South, in km) are from south edge of the subducted Gorda plate using Bull and Merritts (1989, fig. 1). Randall Creek, –3 km. Smith Gulch, +2 km. McNutt Gulch, +9 km. Singley Creek, +12 km. Bear Creek to Cape Ranch, +16 km.

have been tilted, first to the south and then to the north, as the crustal uplift welt passed by at about 56 m/ky.

Quantification of uplift rates sets the stage for trying to understand the processes and response times of small coastal fluvial systems to plate tectonics generated earth deformation. How dull it would be if uniform uplift had prevailed! Instead, the challenge of deciphering geomorphic responses to variable uplift is our next topic.

8.3.2 Stream Channels
8.3.2.1 Independent Variables for the Coastal Fluvial Systems
Impacts of temporal and spatial variations in uplift rate during the past 330 ky on the hills and streams of many watersheds are easier to evaluate if the other independent variables of the fluvial systems are similar. The Mendocino Coast study area is superlative in this regard.

The present (and Holocene) coastal climate is virtually the same for the Figure 8.32 coastal

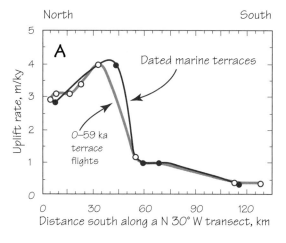

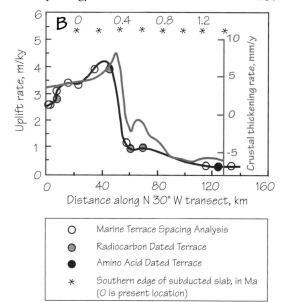

○	Marine Terrace Spacing Analysis
◉	Radiocarbon Dated Terrace
●	Amino Acid Dated Terrace
*	Southern edge of subducted slab, in Ma (0 is present location)

Figure 8.31 Spatial variations of uplift rates along the Mendocino Coast.
A. Comparison of inferred uplift rates based on altitudinal spacing analyses of flights of marine terraces younger than 60 ka, and individual terraces dated by pre-1989 workers.
From Merritts and Bull (1989, figure 5).
B. Overall trend S 30° E of post 72 ka uplift rates (black line) from the Bear River to Mendocino Headlands 15 km south of Fort Bragg. From Merritts and Vincent (1989, figure 4).
Rate of crustal thickening (gray line) is from figure 3 of Furlong and Govers (1999).

watersheds. The Eureka (40° 48.8' N, 124° 10.2' W) and Fort Bragg (39° 26.7' N, 123° 48.2' W) weather stations both have 1000 to 1,200 mm mean annual precipitation and a mean annual temperature of 12°–14°C. Mean August and February temperatures at both are 13°C and 8.5°C. Winter rains begin in October and 93% of the yearly rain falls during the next eight months. Summers are cool, foggy, and dry. Using Appendix A to classify climates, the Mendocino coast is strongly seasonal humid, and weakly seasonal mesic (a cool Mediterranean climate). The 800 to 1200 m high King Range, which extends northwest of Shelter Cove for 60 km, imparts a modest orographic effect and is high enough to have wintertime snow. Hillslope microclimate contrasts are similar for the study-area watersheds because most trunk stream channels flow west-southwest.

Full-glacial August sea-surface temperatures at 18 ka of 13° C were similar to modern sea-surface summer temperatures, but winter temperature gradients were intensified, and sea-surface temperatures were 4°C cooler than present (CLIMAP Project Members, 1981). Being coastal, the cool-moist glacial and warm-dry interglacial Pleistocene climate contrasts were modest because they were stabilized by maritime influences. Johnson (1977) concluded that cold-sensitive species continued to thrive in this coastal refugium.

Franciscan assemblage rock types of the Coastal and King Range terrains (Coney et al., 1980) were created by plate subduction processes between the Cretaceous and Miocene. Both consist of highly sheared, folded, and fractured greywacke sandstone and argillite, rock types that are conducive for high sediment yields and rapid adjustments to tectonic perturbations. This mix of rock types creates variability on a small, local outcrop scale but is considered homogeneous on a regional scale. Colluvial

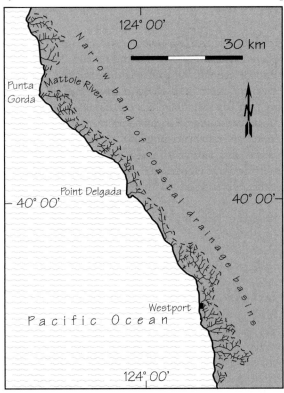

Figure 8.32 Belt of small coastal watersheds of the Mendocino coast study area that is interrupted only by the Mattole River. From Merritts and Vincent (1989).

processes mix detritus derived from a variety of outcrop rock types on the typical hillside. The result is a nice uniformity of soil parent materials in the study region as first noted by Jenny et al.(1969).

Rock mass strength was assessed in two ways; the numerical approach of Selby (1982a,b) and Moon (1984) and qualitative terrain categories of Ellen (written communication to Merritts and Vincent, 1989). Using a scale of 25 to 100, rock mass strength ranges from weak to moderately resistant (values from 42 to 71). Locally resistant sandstones of the two Franciscan terrains are similar. Ellen used the degree of hillslope gullying by small stream channels to classify rock mass strength as hard, intermediate, and soft. Again, both tectonic terrain assemblages were similar, and interestingly

the steepness of hillslopes and stream channels is unrelated to variations in his classes of rock mass strength. Other variable(s), perhaps uplift rate, appear to be the dominant control on watershed steepness and relief.

The Mendocino coast indeed provides a unique setting to assess the geomorphologic impacts of temporally and spatially variable uplift rates because climate, rock mass strength, and drainage-basin areas and orientations are comparable throughout the 140 km long study region. Drainage density of stream-channel networks is controlled in large part by erodibility of hillslope surficial materials (Schumm, 1956, 1997; Strahler 1957, 1958) and provides a test of the conclusion that erodibility is uniform. The drainage density of first- and second-order streams is similar, regardless of uplift rate (Merritts and Vincent, 1989, fig. 6). Other landscape characteristics record plate-tectonics processes.

Northward migration of the Mendocino triple junction at ~56 m/ky generated a broad isostatic crustal welt as fresh hot asthenosphere flowed into the space created by an evolving slab window. The tip of a northward-propagating San Andreas fault creates a prominent restraining bend because of the pronounced turn to the left in order to continue into the east–west Mendocino fracture zone. Migration of the Gorda plate is thought to deform the margins of the slab window and thicken the crust in advance of arrival of the triple junction (Furlong and Schwartz, 2004). Isostatic uplift, oblique-dextral faulting, and crustal thickening are the more important processes that increase rates of uplift to about 4 m/ky and promote a wave of uplift that precedes arrival of the Mendocino triple junction.

Tectonic deformation is much simpler to the south where the Mendocino triple junction has already passed by. Uplift continues, but at decreasing rates as new crustal materials become cooler. Strike-slip faulting becomes the dominant style of the more established (slightly older) straight sections of the San Andreas fault, which is offshore between Shelter Cove and Point Arena (56 km south of Fort Bragg).

Base-level changes for the fluvial systems of the study region include Late Quaternary changes in sea level, regional uplift, and local displacements caused by faulting and folding. How do these perturbations affect landscape evolution? Do they differ in the ways in which they affect hills and streams? Stream channels are the connecting link to headwaters reaches and to all hillslopes. So the focus here is on how streams of various sizes respond to base-level fall. Larger streams have greater annual unit stream power. They respond quickly to tectonically induced base-level fall.

Sea-level changes are caused by climatic (glacio-eustatic) perturbations not related to the migration of the Mendocino triple junction. The sea floor is gently sloping near shore to depths of 30–50 m and then descends the continental slope. Minor declines of sea level will not initiate stream-channel downcutting where seaward extension of river mouths into gentle-slope reaches favors aggradation instead of degradation (Bull, 2007, fig. 2.6). Maximum 130 m sea-level decline was at 18 ka. This was large enough to shift terminal reaches of coastal rivers down the continental slope. The resulting increase in stream power would have tended to shift the "balance" closer to the threshold of critical power and possibly to a degradational mode of operation. Probable increases in sediment yield and bedload transport rate during times of full-glacial climate would increase resisting power, which would tend to offset increases of stream power caused by slope increases.

Bathymetry surveys suggest the net result. They portray a smooth ocean floor instead of entrenched submarine channels leading to the mouths of the larger streams. The Mattole and Delgada submarine canyons are about the only exceptions. Episodes of Late Quaternary falling sea level do not appear to have caused a base-level fall for the small watersheds of the Mendocino coast.

Vertical fault displacements of stream channels generate knickpoints that migrate upstream. The simplest case is a surface rupture that occurs on a range-bounding fault. The entire mountain range is raised and a tectonically induced knickpoint(s) migrates upstream at rates dependent on unit stream power (decreases upstream) and rock mass strength (constant upstream). Each surface rupture increment of uplift migrates upstream as a wave. Reaction times are immediate for stream channels adjacent to the active fault and hillslope response times may be long, especially for headwaters hillslopes.

The way in which regional uplift affects stream-channel gradients is similar to the range-bounding active fault scenario. Faster stream-channel downcutting in the more powerful downstream reaches initiates the steepening of stream gradients. Again the tectonic perturbation is transmitted like a wave to upstream reaches. This tectonically induced perturbation migrates upstream at progressively slower rates, unless prevented by discontinuities (such as waterfalls) in a fluvial system. Reaction times to an increase of regional uplift rate may be <1 ky, and response times for headwaters reaches once again are very long (>100 ky).

More complicated faulting occurs near the propagating tip of the San Andreas fault. Styles of faulting vary spatially and displace stream channels in several reaches. Multiple zones of concurrent uplift shift northward with the Mendocino triple junction and may produce dispersed uplift somewhat akin to regional uplift.

Isostatic uplift, or crustal-thickening processes, are sufficiently regional to raise entire mountain ranges. Uplift may be steady and changes in rate are gradual, unlike short-term pulsatory uplift by an active range-bounding fault.

Higher mountains have increased orographic rainfall from storms moving inland from the Pacific Ocean. Maximum drainage-basin relief increases. Both changes increase denudation rates. Response is quickest in downstream reaches of streams. Terminal reaches of rivers may downcut fast enough to repeatedly return to the base level of erosion as is indicated by the presence of strath terraces in moderately wide valley floors.

8.3.2.2 *Fluvial System Responses to a Shifting Plate Boundary*

The ever-changing tectonic scene affects the landscapes of Mendocino coast watersheds on scales ranging from single hillsides and headwaters rivulets to entire mountain ranges. We start by considering geomorphic factors that influence drainage-basin relief, which is the altitude difference between the watershed mouth and the highest divide summit.

Relief of the belt of coastal drainage basins increases northward. The pattern resembles the trend of Late Quaternary uplift rate (Fig. 8.33). The abrupt increase of relief at a transect distance of 50 km matches the uplift-rate upswing (Fig. 8.31).

Unfortunately, we do not have key information for constructing a model that relates relief to triple junction migration rates. What was the coastal relief profile before arrival of the triple junction? Are ridgecrest denudation rates more or less than 0.4 m/ky uplift rates? So we assess easier questions.

How sensitive is Mendocino coast relief to variations of uplift rate? Maximum unit stream power of large rivers favors stream-channel downcutting equaling the uplift rate. Strath terraces tell us where such equilibrium conditions are achieved (at least for short intervals) in the terminal reaches of intermediate-size trunk streams. Response times after an increase of uplift rate are minimal where unit stream power is greatest.

An opposite scenario prevails on the highest watershed summit. Ridgecrest slope is gentle and the power of collective runoff is in its initial formative stages. This landform will be the last to adjust to tectonically induced stream-channel downcutting at the river mouth. Relief may still be increasing in drainage basins that are now rising only at 0.4 m/ky.

Erosion rates in coastal northern California border on extremely high because frequent flood discharges are effective (Wolman and Miller, 1960) in removing copious landslide detritus generated by a favorable climate and rock type. The 9,400 km² Eel River drainage basin has a mean sediment yield of 1,700 tonnes/km/yr, and some tributary watersheds

of <500 km² have sediment yields of >3,000 tonnes /km/yr (Kelsey, 1980). A sediment yield of 2,000 tonnes /km/yr translates to a mean watershed denudation rate of an impressive 0.7 m/ky.

Such information is not available for the adjacent swath of much smaller (40–24 km²) coastal watersheds studied by Merritts and Vincent (Fig. 8.32), where smaller floods might require more time to remove landslides. Let us assume mean denudation of 0.7 m/ky, and maximum rates of tectonically induced landscape incision that coincide with valley floors and adjacent footslopes. Ridgecrest areas would degrade more slowly, a denudation rate of 0.3 m/ky seems reasonable. Fluvial systems in the slowly rising Bruhel Point part of the coast (0.4 m/ky) would still have rising watershed divides. Headwaters streams and hillslopes have remained rugged and steep even though downstream reaches have subdued characteristics such as broad valley floors and estuaries. Such spatial contrasts in landscape elements are obvious on the Mendocino Coast and are a useful guide to the presence of nonsteady-state conditions in rising mountains.

How significant is the apparent contrast of relief depicted in Figure 8.33? A multivariate correlation analysis matrix (Merritts and Vincent, 1989, table 3) indicates that half the variance of watershed relief can be attributed to variance of uplift rate, r = 0.68. Mean relief is 925 m and mean uplift rate is about 3.3 m/ky for watersheds between transect distances of 0 to 60 km. The southern half of the plot has a mean relief of 645 m and a mean uplift rate of

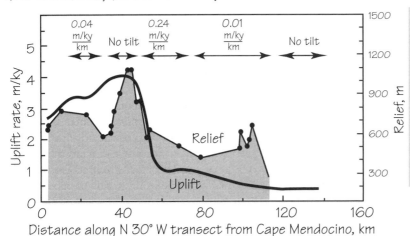

Figure 8.33 Uplift rate, drainage-basin relief, and tilt resulting from spatial variations in uplift rate, along a N30°W transect between Fort Bragg and Cape Mendocino. Merritts and Vincent (1989, figures 6B and 9).

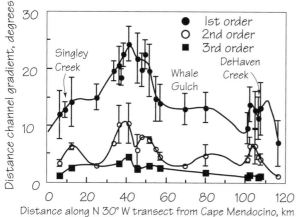

Figure 8.34 Variations of mean gradients of headwaters, transitional, and terminal reaches of streams with increasing distance south from Cape Mendocino. Third-order, and two second-order, streams are graded to sea level. 1 σ error bars. From figure 7B of Merritts and Vincent (1989).

about 0.7 m/ky. The difference in average relief is only 280 m, but the difference in average uplift rate is a high 2.6 m/ky. Rock uplift of a watershed divide could change by an impressive 260 m in only 100 ky. A relief contrast could be generated in a short time span of modest contrast of uplift rates in the Figure 8.33 example.

Relatively brief response times of streams to changing uplift rates favor deciphering their behavior. One way to class streams with different capacities to do work is to use Strahler stream orders (Strahler, 1952). Channel gradients of first- and third-order streams are much different (Fig. 8.34). Headwaters first-order channels have much less unit stream power and may lack the power to efficiently degrade the local bedrock. Their steepness in part records upstream migration of knickpoints generated by more efficient incision of the valley floor further downstream by second- and then third-order streams. Headwaters streams are especially steep where uplift is the most rapid. Second-order streams occupy an intermediate position, and the trend for third-order streams appears relatively flat when included on the scale of this diverse set of graphs. On

closer examination all these streams may have at least a twofold variation in gradient with steeper streams occurring where uplift rates are the highest.

Base-level falls begin as tectonically induced downcutting of the third-order streams so both narrow and inclusive gradient-index (defined in the Glossary) are good ways to study responses of trunk stream channels to different rates of mountain-range uplift. Three examples are used in Figure 8.35 B–D to illustrate clear-cut differences in the rates of adjustment for longitudinal profiles of streams in low, intermediate, and high uplift rate parts of the Mendocino coast.

The DeHaven Creek arithmetic longitudinal profile has a concavity indicative of an equilibrium stream channel. The semi-logarithmic plot is segmented. Larger inclusive gradient-indices for the upstream reaches reveal an inability to achieve the same degree of equilibrium conditions as achieved by downstream reaches. This is to be expected because of the miniscule stream power available to degrade valley floors as one approaches a headwaters divide. DeHaven Creek is not tectonically inactive (0.4 m/ky uplift rate). Small increments of base-level fall induced by uplift continue to propagate upstream. The result contributes to steepening in longitudinal profile characteristics on both the arithmetic and semi-logarithmic plots and to the perpetually rugged headwaters landscapes. The Snyder et al. (2000) analysis reveals similar numerical values for concavity of longitudinal profiles for trunk stream channels of 20 watersheds, although gradients are much steeper in high-uplift rate fluvial systems.

The uplift rate at Telegraph Creek is 1.2 m/ky, which is intermediate for the Mendocino coast but rapid from a global perspective. The arithmetic longitudinal profile is both steeper and more strongly concave than that for DeHaven Creek. Numerous knickpoints, perhaps related to coseismic uplift events, appear as narrow gradient-index anomalies. These could also be generated by lateral erosion induced base-level falls where coastal erosion creates waterfalls that migrate upstream as knickpoints. Local faulting may also be important.

Uplift of 4.0 m/ky at Fourmile Creek is so rapid that the arithmetic longitudinal profile approximates a straight line. The semi-logarithmic plot is

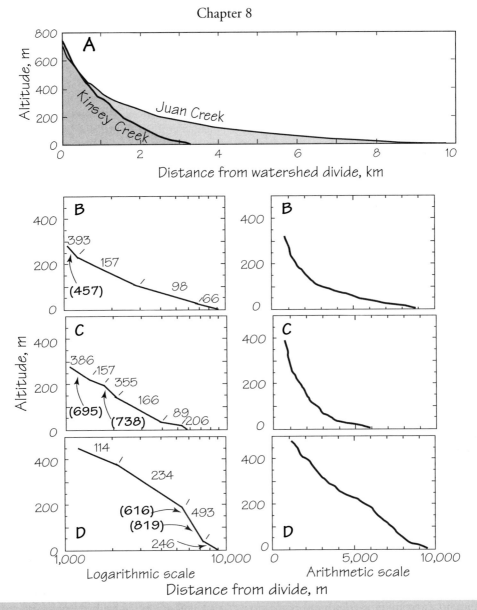

Figure 8.35 Arithmetic and exponential plots of longitudinal profiles of small coastal streams. Values for gradient index (equation 2.13 of Bull, 2007) are noted on the exponential plots. Constant gradient reaches are inclusive gradient indices and values in parentheses are narrow gradient indices. B, C, and D are from Merritts and Vincent (1989, figure 11).

A. Comparison of longitudinal profiles of trunk stream channels in drainage basins with similar total relief. Uplift rate is 0.4 m/ky for Juan Creek and 4.0 m/ky for Kinsey Creek. Profiles were assembled from figure 7 of Snyder et al. (2000).

B. Low ~ 0.4 m/ky uplift rate (DeHaven Creek).

C. Intermediate ~1.2 m/ky uplift rate (Telegraph Creek).

D. High ~ 3.5 m/ky uplift rate (Fourmile Creek).

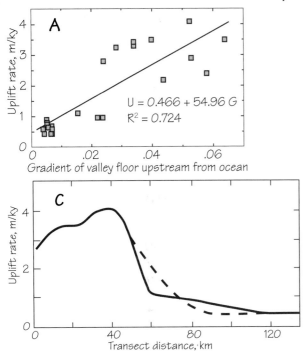

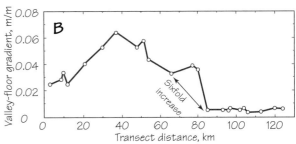

Figure 8.36 Valley-floor gradients for terminal reaches of intermediate-size streams of the Mendocino coast.
A. Regression of Late Quaternary uplift rates and valley-floor gradients.
B. Variation along the 120 km study-area transect.
C. Refined regional uplift rate curve. Dashed line has been replaced by the solid line

convex. Gradient indices for headwaters and terminal reaches are the opposite of those at DeHaven Creek, being least in the headwaters and largest near the basin mouth. This underscores the long time spans needed for tectonically induced downcutting in the third-order reach to be transmitted to the headwaters. The Mendocino triple junction and crustal-uplift welt has arrived so recently (<200 ka) on this part of the coast that headwaters terrain still retains characteristics inherited from times of slow uplift.

Response times are so long for these intermediate-size watersheds that we should consider comparing inclusive gradient-indices only for basin-position coordinates of 50 to 100. The manner in which uplift steepens these actively downcutting stream channels is described by the Figure 8.36A regression. Once again there is a tight clustering of points for the "equilibrium" low-uplift section of the Mendocino coast, and lots of scatter where uplift rates are more than 2 m/ky. Part of the scatter is the result of stream channels displaced by active faults where the San Andreas fault comes ashore and splits into many terminal splays. Such local faulting is an aberration on the overall model of regional uplift.

The terminal reaches of third- and second-order streams in 0.5 to 3 km long reaches just upstream from the ocean provide information about tectonic forcing of stream gradients. They have an average gradient of only 0.0055 m/m at transect distances of 85 to 125 km (Fig. 8.36B). Lack of scatter in the cluster of points in the uplift-gradient regression (Fig. 8.36A) suggests that streams of different sizes have long-term downcutting rates that match the 0.4 m/ky uplift rate. We can class these as type 1 dynamic equilibrium reaches.

Regardless of uplift rate, some response time is needed for terminal reaches to adjust to the most recent increment of continuing uplift (Table 8.4). The time spans needed for landscape adjustment are longer than intervals between uplift events. Upstream reaches and adjacent hillslopes have longer response times. If so, gradients might be still lower than 0.0055 m/m if this coast were tectonically inactive. So a model where they never completely catch up makes more sense than proclaiming these to be examples of steady-state conditions. One alternative is the allometric change conceptual model (Bull, 1975a; 1991, section 1.9) that describes

Geomorphic process or feature	Approximate response time, ky	Comments
Maximum uplift rate related to gradual creation of uplift welt	160	Time needed for crustal processes to change after passage of the south edge of the Gorda plate and creation of new increment of slab window
Tilt of first-order tributaries on the south side of the crustal uplift welt	160	Merritts and Vincent estimate based on mean channel slope differences (Fig. 8.38C)
Tilt of first-order tributaries on the north side of the crustal uplift welt	690	Same as above
Maximum steepness of second-order streams	240	Based on lag of 3 km shown in Figure 8.34
Maximum steepness of first-order streams	290	Based on Figure 8.34
Maximum mountain-range relief	370	Based on Figure 8.33
Time needed to achieve type 1 dynamic equilibrium after brief uplift pulse		
Large streams	<1	Mattole River, 782 km² drainage-basin area; Bear River 332 km²
Intermediate-size watersheds in 0.4 m/ky uplift rate subarea		4 to 24 km² drainage-basin areas
third order	< 5	
second order	>50	
first order	>1000	

Table 8.4 Response times of landscape features to a migrating pulse of rapid uplift.

orderly interactions between dependent variables in a changing landscape before or after the base level of erosion is attained for terminal reaches of the trunk stream channels.

Allometry in biology describes relative systematic changes in different parts of growing organisms. Allometric change in geomorphology describes orderly behavior in nonsteady-state fluvial systems. Hydraulic geometry of stream channels (Leopold and Maddock, 1953) is a good example. Increase of flow width with increasing discharge at a gauging station is static (at a single basin-position coordinate) allometric change. Increase of flow width with downstream increase of discharge (a sequence of basin-position coordinates) is dynamic allometric change. The allometric change model emphasizes

the degree of interconnections between different geomorphic processes and landscape characteristics. For example, on the Mendocino coast it is valuable to know that stream channel concavity has no relation to uplift rate but that gradient does (Snyder et al., 2000). My emphasis is on system behavior instead of on supposed attainment of steady states. "Allometric change" is just a convenient label to alert readers that regressions shown here are based on nonsteady-state assumptions.

Are longer response times needed where uplift is faster for streams of comparable size? Gradients of valley floors are an order of magnitude steeper at transect distances of 30 to 50 km (Fig. 8.36B), and this is where uplift rates are also an order of magnitude greater. Terminal reaches of streams

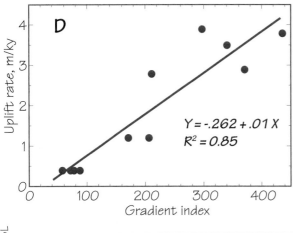

$Y = -.262 + .01 X$
$R^2 = 0.85$

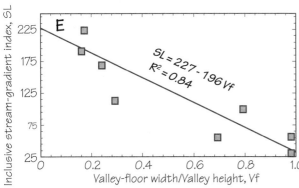

$SL = 227 - 196 Vf$
$R^2 = 0.84$

Figure 8.36 Valley-floor gradients for terminal reaches of intermediate-size streams of the Mendocino coast.
D. Relation between uplift rate and inclusive gradient index for lower reaches (basin-position coordinates of 50 to 100) of stream channels for Mendocino coast intermediate-size watersheds.
E. Relation between valley-floor width/valley height ratios at a basin-position coordinate of 80 and inclusive stream-gradient index between basin-position coordinates of 50 and 80.

attempt to keep pace with uplift that is occurring at an average rate of 2 to 4 m in only 1,000 years, but lag further behind in a systematic manner that is directly proportional to uplift rate (Fig. 8.36A). These reaches are constantly changing and gradient varies with magnitudes of uplift-rate perturbations.

The relation between uplift rates and terminal-reach gradients has two domains of data points, a tight little cluster and a spread-out scatter where uplift exceeds 2 m/ky (Fig. 8.36A). The scatter suggests insufficient time to adjust to variations in rock mass strength and/or pulses of tectonically induced downcutting that migrate upstream as knickpoints. One rock-mass-strength anomaly is the high-uplift-rate Cooskie Creek, which has a gentle gradient of only 0.024 m/m that should be 0.062 m/m if on the regression line. Merritts and Vincent ascribed such Cooskie anomalies as being the result of soft materials of a major shear zone underlying the valley. Like me, they omitted this data point.

The abrupt increase of gradient shown in the Figure 8.36B plot at 85 km does not match the Merritts–Vincent regional uplift-rate trend. However, their plot lacks control points between transect distances of 65 and 115 km. The abrupt sixfold increase of Figure 8.36B suggests a fine-tuning of their model of northward uplift-rate increase (Fig. 8.36C). The low-uplift-rate section of the coast may extend further north to 85 km and the high-uplift domain may extend further south, as indicated by the dashed line. The abrupt crossover at 85 km might reveal the presence of an active fault in the broad Mattole shear zone. If so, the model of uplift caused by deep crustal processes may be complicated locally by shallow faulting.

Uplift rate appears to influence the inclusive gradient index in the downstream half of the Mendocino coast fluvial systems (Fig. 8.36D). Once again, there seems to be more scatter for trunk stream channels that are being raised faster than 2

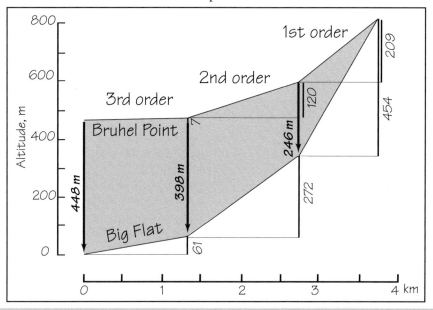

Figure 8.37 Spatial variations of tectonically induced incision of valley floors as a function of stream order. Gray area shows the difference of incision between a representative stream profile for the low uplift rate Bruhel Point area (0.4 m/ky and 0.3°, 5°, 12° segments), as compared to a profile for the high uplift rate Big Flat area (4.0 m/ky and 2.6°, 11°, 25° segments). Numbered arrows show differences in amounts of relative downcutting. Other numbers describe relief for stream-order segments. Portrayed after Merritts and Vincent (1989, figure 8), including the vertical exaggeration of 1:3.64.

m/ky. What other variables are allometrically related to the gradient of these trunk stream channels?

The Figure 8.36E analysis adds the variables of valley-floor width and valley height at a basin-position coordinate of 80 for each stream. Valley floor-width/valley height ratios of less than 1.0 indicate that all of the Mendocino coast watersheds are strongly affected by ongoing uplift – hills as well as streams. Vf ratios (equation 3.14) do indeed increase for the lower-uplift rate systems. The narrower valley floors and higher adjacent ridgecrests of the higher uplift areas fit a model of systematically changing hillslopes, valley floors and stream-channel gradients. Rates of change are different for these diverse suites of nonsteady state landforms.

In tectonically inactive settings, valley-floor widths increase faster than degradational lowering of the adjacent ridgecrests. Valley-floor width/valley height ratios may exceed 5.0. Order-of-magnitude

variations make this tool valuable for comparisons of drainage basins in different tectonic settings.

Nonsteady-state landscape evolution results in different rates of change for different parts of a fluvial system. The valley-floor width/valley height ratio varies as a function of tectonically induced downcutting. This is an allometric index of response times to tectonic base-level fall. Stream channels respond quickly and their rate of downcutting is matched by increase of relief between the stream channel and the adjacent ridgecrests. In rapidly rising landscapes, valley-floor widths remain narrow while relief of valley cross sections increases. Valley floors are just wide enough to convey flood streamflow events and have widths that are less than 0.5 of the mean height of the valley (Fig. 8.36E).

Tectonically induced downcutting steepens drainage networks, but not uniformly. Rapid incision in downstream reaches minimizes gradient increase and creates the base-level fall that steepens upstream reaches. Upstream migration of base-level fall decreases exponentially, but here (Fig. 8.37) is assumed to be uniform so downcutting amounts are the same throughout each stream-order reach. This analysis is only for valley floors. Changes in hillslope relief upstream from first-order stream channels are not included because different processes degrade hillslopes.

Merritts and Vincent quantified the consequences of spatially variable mountain-range uplift by comparing gradient decreases for consecutive stream orders (Fig. 8.37). Two representative profiles were constructed from suites of watersheds in low (Bruhel Point) and high uplift (Big Flat) parts of the Mendocino coast. Heads of first-order reaches are the common starting point for both profiles.

The third-order streams are closest to maintaining the same gradient with the passage of time and at Bruhel Point can be considered as type 1 dynamic equilibrium reaches. The upstream stream orders are less able to downcut at rates that equal the base-level fall imposed by incision of the third-order reach. The 120 m of relief for second-order reaches records two aspects. First is the normal upstream increase of gradient that would occur in an equilibrium stream of a tectonically inactive mountain range. Second is an unknown amount of gradient increase generated by the smaller unit stream power being unable to equal the base-level fall imposed by tectonically induced downcutting in the third-order reach (tectonic-relief increase). Time lags are so long for first-order streams that even at Bruhel Point much of the gradient represents accumulated tectonic-relief increases.

The Big Flat representative profile is assumed to be for a hydrologically similar watershed, but uplift rates are ten times faster than at Bruhel Point. The additional 54 m of third-order relief is a tectonic-relief increase that terminal-reach incision has yet to assimilate. At least 152 m of the second-order relief and at least 245 m of the third-order relief can also be classed as tectonic-relief increases.

Stream channels may be steepened as the crustal uplift welt passes by during time spans of >100 ky. The uplift-rate curve (Fig. 8.31B, 8.36C) can be used to assess magnitudes of differential uplift between two parts of the coast. It depicts decreasing uplift rates between Smith Gulch and Bear River so the landscape should have been raised more at Smith Gulch. Marine terraces along this part of the coast are tilted northward 0.09°, a minimum value because the azimuth of maximum tilt is not known.

Merritts and Vincent considered three models. The transform boundary model is the obvious migration direction of the Mendocino triple junction as defined by the San Andreas fault, and a perpendicular uplift axis would be N 60° E. Asthenospheric upwelling and accretion processes would result in isostatic uplift and should parallel the south edge, S 80° E, of the northward migrating Gorda plate. Their third model uses both of these plate tectonics processes, which results in a domal style of uplift with a larger span of possible tilt maxima.

Trunk stream channels flow west-southwest to the coast, so it is the first-order tributaries that are likely to be best oriented to evaluate possible differences in tilt direction on opposite sides of the crustal uplift welt. A schematic view up a trunk valley is sketched in Figure 8.38A. The profile across part of the hypothetical uplift indicates decreasing amounts of uplift towards the NNE during the past 50 to 100 ky. Uplift affects both ridgecrests but the SSW side of the valley is raised more. Such differential uplift steepens the SSW side of the valley and tends to increase the slope of first-order tributaries. Tilt has the opposite effect on the other side of the valley.

The magnitude of the presumed perturbation is fairly small but may be discernible in a topographic analysis. The upvalley diagram in Figure 8.38B of two trunk valleys on opposite sides of the crustal uplift welt has first-order tributary gradients of 15°. Uplift rates are large and asymmetric (Fig. 8.31B). A maxima of 4.0 m/ky decreases gradually to the north passing through 2.9 m/ky. Uplift deceases quickly to the SSW passing through 1.2 m/ky.

Tilt varies spatially. The watershed divide between the two trunk streams is raised the most but is not tilted because local uplift is spatially uniform. Tilt of 1.4° to the NNE causes both increase and decrease of gradient of the first-order tributaries. Slope-increase channels now are inclined 16.4° and slope-decrease tributaries now are inclined 13.6°.

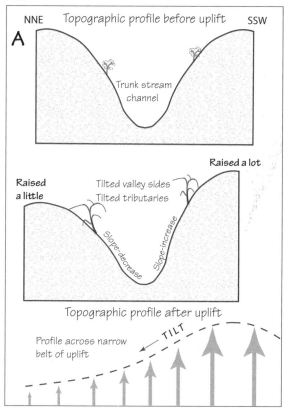

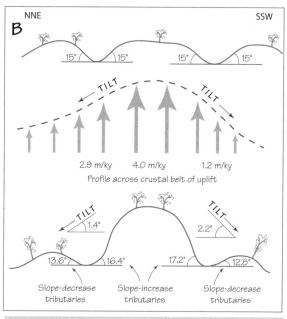

Figure 8.38 Sketches showing how tilting north and south of the area of maximum uplift, 40 km S 30° E of Cape Mendocino, tends to make streams flowing in the direction of tilt steeper, or vice-versa.
A. Diagrammatic sketch showing how differential uplift created by hypothetical narrow anticlinal folding changes hillslope and tributary stream gradients on opposite sides of a broad valley. Tributary streams become steeper or gentler on opposite sides of the valley.

Figure 8.38 Sketches showing how tilting north and south of the area of maximum uplift, 40 km S 30° E of Cape Mendocino, tends to make streams flowing in the direction of tilt steeper, or vice-versa.
B. Estimates of tilt amounts of 2.2° to the south and 1.4° to the north on opposite sides of the asymmetrical crustal uplift welt created by migration of the Mendocino triple junction. Channel-gradient changes are mean values for the watersheds analyzed by Merritts and Vincent (1989, figure 9).

Potential impacts are even larger on the more strongly tilted SSW side of the crustal uplift welt, 12.8° and 18.2°.

Can such departures be observed? Only some streams flow in the range of directions needed to assess possible tectonic controls. Gradients of small tributaries are likely to be strongly affected by variations in rock mass strength. Base-level fall caused by downstream valley-floor incision is only partly transmitted upstream far enough to steepen the headwaters reaches (Fig. 8.37) but surely has to be a factor in their present slope. The best we can hope for is that tectonically induced base-level fall is about the same for suites of first-order streams used in this analysis. Ideally we need pairs of observations from opposite sides of many valleys.

The tilt analysis for first-order channel gradients indeed suggests separate classes of channel steepness as dictated by differential uplift rates. Slope-increase and slope-decrease stream channels

Figure 8.38 Sketches showing how tilting north and south of the area of maximum uplift, 40 km S 30° E of Cape Mendocino, tends to make streams flowing in the direction of tilt steeper, or vice-versa.
C. Change from steeper north-draining first-order tributaries to steeper south-draining first-order tributaries coincides with approximate axis of maximum uplift rate. Solid circles are mean gradients of channels flowing south, and open circles flowing north. Plate tectonics model 3 (see text) is used here.
Merritts and Vincent (1989, figure 10C).

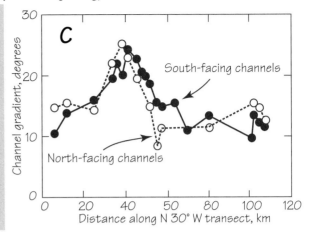

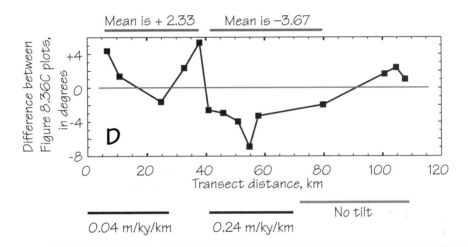

Figure 8.38 Sketches showing how tilting north and south of the area of maximum uplift, 40 km S 30° E of Cape Mendocino, tends to make streams flowing in the direction of tilt steeper, or vice-versa.
D. Departures of the two plots of stream gradients shown in Figure 8.38C. Minimal tilt occurs near summit of presumed domal uplift or along constant-uplift-rate parts of the coast.

plot as different curves (Fig. 8.38C) and means of departures of the two populations are significant (Fig. 8.38D). Tilting has steepened some streams by 2°, which is 40% of the 5° difference with untilted first-order channels in the low uplift rate Bruhel Point area. The remaining 3° of steepness contrast can be attributed to upstream propagation of base-level falls.

8.4 Summary

Northward propagation of the Walker Lane–Eastern California dextral shear zone occurred after Baja California split off from the North American plate at ~5 Ma. The broad plate-boundary transition operates as a system of conjugate fault blocks with sinistral as well as dextral displacement characteristics.

Upwelling of hot asthenosphere after delamination of batholithic crustal roots raised both the Panamint Range and the Sierra Nevada more than 1 km. Both have a conspicuous break-in-slope that is ~1 km above the front of a rugged escarpment. Relict landscapes in the highest part of the range consist of low hills and broad valleys and may pre-date the Cenozoic.

Panamint Range stream channels were tilted downstream on the east side and subject to mountain front base-level fall on the west side. Tilt produced exceptionally long straths with minimal concavity, whereas convex, disequilibrium, longitudinal profiles characterize the west-side streams.

Sierra Nevada uplift enhanced Pleistocene glaciation that complicated responses of rivers to uplift. The ~4 to 3 Ma uplift event in the southern Sierra Nevada caused the large rivers to incise inner gorges with narrow valley floors and few straths. Batholithic rocks are strong. Only the larger tributary streams had sufficient unit stream power to erode at the same rate as trunk rivers. Rates of Feather River stream-channel downcutting in the northern Sierra Nevada are increasing, an opposite trend from that of the Kings River in the southern Sierra Nevada.

Continued tectonic subsidence of the center of Owens Valley is the result of transtensional faulting that is part of the Walker Lane–Eastern California dextral shear zone. It is not related to the Sierra Nevada range-bounding fault, which is virtually inactive. This impressive mountain front has valley and piedmont characteristics indicative of formerly rapid uplift followed by 2 My of tectonic quiescence. Encroachment of Owens Valley normal faulting and subsidence has created discontinuous, short, normal faults along the mountain–piedmont junction since ~100 ka.

Transitory upwelling of asthenosphere followed migration of the Mendocino triple junction up the California coast. Rock uplift rates increased as the triple junction approached, achieved a maximum as it passed, then gradually slowed as injected new crust cooled. Holocene uplift rates range from 0.3 to 4 m/ky in a 150 km long study area with similar climate and rock type. A similar progression of uplift rates occurred as the triple junction passed a given locality.

Analyses of global marine terraces allows estimation of local uplift-rate histories since 330 ka. Each uplift rate has a unique altitudinal spacing of terraces, and local crosschecks of terrace ages are provided by radiocarbon and amino-acid age estimates, and by soil-profile characteristics.

Increasing uplift rates over a 150 ky timespan had a profound influence on streams and hillslopes. Geomorphic response times can be assessed where each watershed experienced an increase, and then a decrease, of uplift rates.

Terminal reaches of these small coastal watersheds have the greatest unit stream power. Passage of the crustal uplift welt steepened stream channel longitudinal profiles, and rapid valley-floor incision decreased the valley-floor width/valley height ratio. It is only when uplift rates became a slow 0.4 m/ky that the valley floors of these terminal reaches were able to achieve equilibrium gradients (strath formation and downcutting rates equal the uplift rate). Tectonically induced downcutting by third-order terminal streams is a base-level fall that eventually affects the entire watershed. Unit stream power decreases sufficiently in the upstream direction to prevent achievement of equilibrium valley floors of the second- and first-order streams. Hillslopes, with their huge volumes of material available for erosion, responded even more slowly than the valley floors. Headwaters terrains may retain the landscape characteristics inherited from times of fast uplift even after 200 ky have passed.

Glossary

Accuracy The correctness of a measurement or numerical result. Accuracy is determined by comparing a measurement with its true, actual value. It is the comparison of measurement value (such as an estimated age) with the true value (known age). *See* precision.

Active channel The bed, banks, and floodplain of a stream changed by streamflows.

Adsorption Adhesion of molecules of ions and molecules to electrically charged surfaces of clay minerals, especially very fine platy montmorillonite.

Aggradation A disequilibrium mode of operation that raises the altitude of an active stream channel by depositing bedload.

Allometric change The tendency for orderly adjustment among interdependent materials, processes, and landforms in a geomorphic open system.

Alluvial fan A fluvial deposit whose surface forms the segment of a cone that radiates downslope from the point where the stream leaves a source area.

Altitude The specific term for height above present sea level. Preferred to the engineering term "elevation", which can have several geologic connotations, including rock uplift.

Antecedent stream A stream that was present before the formation of a fold, fault scarp, or other tectonic landform, and which maintains its course despite subsequent earth deformation.

Antislope scarp *See* ridge rent.

Antithetic fault A secondary fault in a set of faults with a sense of slip is opposite that of the primary fault from which it originates.

Apparent throw Estimates of magnitude of structural throw based on simple scarp height (may be too large), or the vertical separation of the topographic profile for the landform that was faulted (may be too small). Such estimates of throw are only apparent.

Aquiclude An areally extensive body of saturated but relatively impermeable material that does not yield appreciable quantities of water to wells. Aquicludes constitute boundaries of aquifer flow systems; term is synonymous with confining bed.

Aquifer An areally extensive body of confined or unconfined saturated permeable material (such as sand, gravel, porous sandstone, fractured limestone or greywacke) that readily yields water to wells and springs. Unconsolidated alluvium is a common water-bearing formation.

Aquifer system A heterogeneous body of interbedded permeable and poorly permeable layers that functions regionally as a water-yielding hydraulic unit. It comprises two or more aquifers (permeable formations) separated by laterally discontinuous aquitards that locally impede ground-water movement but do not greatly affect the overall hydraulic continuity of the system.

Aquitard A saturated, but poorly permeable, bed that locally impedes ground-water movement and does not yield water freely to wells.

Asthenosphere A layer of soft, hot, plastic rock in the Earths mantle. It is the lower part of the upper mantle at depths of about 100 to 300 km and is weaker than the solid lithosphere above it or rocks below it.

Base level The altitude below which a reach of a stream cannot degrade its bed. Sea level is the ultimate base level.

Base-level fall Decrease of the altitude of a reach of stream as a result of faulting, folding, or erosion.

Base level of erosion A concept that integrates the system, equilibrium, and base-level concepts. It is the equilibrium longitudinal profile below which a stream cannot degrade and neither net vertical erosion nor deposition occurs. See type 1 and type 2 dynamic equilibrium.

Base-level processes Fluvial aggradation or degradation, or tectonic deformation that changes the altitude of a reach of a stream.

Base-level rise Elevation of a reach of a stream by tectonic or depositional processes relative to the adjacent upstream reach.

Beheaded stream channel Two sections of a stream channel that have been separated by strike-slip faulting so that the downstream reach no longer receives flow from its former headwaters reach.

Basin-position coordinate See hydraulic coordinates.

Blind thrust fault A thrust fault that has yet to rupture up to the land surface. Such young, active thrust faults generate earthquakes, but lack fault scarps for paleoseismologists to study.

Cataclasite A metamorphic rock consisting of angular fragments created by shearing in a fault zone, commonly in a weakly cohesive matrix.

Classes of relative tectonic activity A hierarchy of five classes – highly active to inactive – of mountain-front faults and folds. They are based on fluvial landscape responses to rates of tectonic base-level fall at mountain fronts.

Colluvium Loose detritus deposited on hillslopes by fluvial and mass-movement processes.

Concentration The type of geomorphic processes in areas of converging water flow lines that notch hillslopes, localize sediment flux, and result in mass movements such as debris flows.

Consequent stream A streamcourse that develops on a slope created by tectonic deformation.

Convex–straight transition The hillside location where a topographic profile changes from convex to straight hillslope morphology. Straight elements may be short or long and are uniform to irregular.

Degradation event Stream-channel downcutting into Quaternary alluvium and bedrock between the times of formation of an aggradation surface and attainment of the lowest possible base level of erosion for the stream. This would be a type 1 dynamic equilibrium condition in tectonically active reaches.

Denudation rate The average long-term erosional lowering of the land surface within a watershed. It ranges from < 0.01 to > 10 m/ky.

Diffusion Types of geomorphic processes in areas of divergent flow lines that tend to disperse water and sediment on hills.

Disequilibrium A reach of a stream that is aggrading, or is downcutting, and has yet to attain the base level of erosion.

Dominant watershed A drainage basin that is slowly, or intermittently, expanding its area by capturing hillslopes and/or streams of adjacent subordinate watersheds.

Drainage basin The watershed whose divides direct flow of water derived from precipitation down hillslopes into channels of a drainage net that conveys streamflow to the drainage-basin mouth.

Drainage density The total length of stream channels per square kilometer of a watershed, as measured in the field, or on digital terrain models or aerial photographs. (also see valley density)

Dynamic allometric change The interrelations of measurements made of a landform or process at different times. It is time dependent. For example, the changes in width, depth, and velocity of flow that occur as a rising stream fills its channel. *See also* static allometric change.

Dynamic equilibrium An adjustment between streamflow variables that occurs when a reach of a stream has downcut to the minimum gradient needed to transport the imposed sediment load. *See also* base level of erosion; type 1 and type 2 dynamic equilibrium.

Earthquake recurrence interval The mean time span between characteristic surface-rupture events for a segment of a fault.

Elongation ratio A dimensionless index of the degree of circularity of a drainage-basin planimetric shape. For example, the ratio of the diameter of a circle with the same area as the basin and the map length between the two most distant points in the basin.

Ephemeral stream A stream that flows only briefly in direct response to rainfall or snowmelt. Its streambed remains above the groundwater table.

Equilibrium landform An element of the fluvial landscape that does not change its topographic configuration with time as detritus is washed through it. Examples include stream-channel straths, pediments, and some concave footslopes of hills.

Exhumation Exposure of deep-seated rocks by erosion, which if rapid can bring hot, high-grade metamorphic rocks to the land surface.

Flexural isostasy model Adjustment of land-surface height when extensional stresses break the elastic plate (Earth's strong lithosphere) allowing upward flow of the weaker asthenosphere to elevate a mountain range and balance the gravitational forces between crustal masses of different density. Isostatic adjustments continue as erosion removes mass from the land surface.

Floodplain A planar lowland bordering a stream channel that is flooded at high stages of streamflow.

Fluvial system The interacting hills and streams of a drainage basin. Hillslope and adjacent stream subsystems comprise the watershed, or input reaches, and piedmont depositional subsystems comprise the output reaches of fluvial systems.

Glacial equilibrium line altitude Or ELA. It is the altitude on a glacier where annual accumulation equals annual ablation and the net mass balance is zero.

Gradient-index *See* inclusive gradient-index and narrow gradient-index.

Gruss Decay of plutonic rocks into a loose accumulation of crystals such as feldspar, quartz, amphiboles, and mica.

Hillslope-position coordinate *See* hydraulic coordinates.

Hillslope sediment reservoir Accumulated colluvium and soils on hillslopes that are sources of sediment yielded to streams. Net reservoir gain or loss is sensitive to Quaternary climatic change.

Hydraulic coordinates Dimensionless planimetric coordinates within drainage basins that are ratios of horizontal distances of flow direction down hills and streams. The *hillslope-position coordinate* is the ratio of horizontal distance from a ridgecrest to a hillside point to total distance to the edge of a valley floor. The *basin-position coordinate* is the planimetric length from the headwaters divide of a drainage basin along the trunk valley to a streambed point, divided by the total length of the valley.

Hydraulic gradient The change in static head per unit of distance in the direction of the maximum rate of decrease in head.

Inclusive gradient-index Long reaches where a longitudinal profile of a stream has a constant rate of decrease in gradient with progressively larger streamflows with increasing distance from the headwaters divide. Inclusive gradient-indices describe situations of type 1 or type 2 dynamic equilibrium. *See also* narrow gradient index.

Inselberg Bedrock outlier that rises above a pediment; commonly a remnant of a former spur ridge.

Intermittent stream A stream that flows during part of the year. Its water is derived from rainfall, snowmelt, or occasional groundwater discharge into the stream channel.

Knickpoint An abrupt increase in stream gradient represented by a waterfall that migrates upstream by the process of headcutting.

Knickzone A short reach of anomalously steep stream gradient, such as rapids or riffles between pools. It commonly is a degraded knickpoint that has migrated upstream.

Land subsidence Sinking or settlement of the land surface, due to diverse causes and generally occurring on a large scale. Usually the term refers to the vertical downward movement of the land surface although small-scale horizontal movements may be present. The term does not include landslides which have large-scale horizontal displacements, or settlement of artificial fills.

Lateral erosion base-level fall Nontectonic lowering of base level of hillside footslopes caused by lateral erosion by streams or waves.

Lateral slip The relative displacement of a topographic or geologic feature that crosses a strike-slip fault. Termed right-lateral if it appears to have shifted to the right, and left-lateral if displaced toward the left.

Listric fault Curved normal faults. Hanging-wall block rotates as it slides along a concave fault plane whose dip decreases with depth.

Local base levels Resistant outcrops, lakes, and dams that remain at the same altitude at the downstream ends of reaches of streams.

Longitudinal profile A topographic profile along a valley or stream.

Mode of operation The behavior of a reach of stream where interactions among variables cause one of three conditions: aggradation, degradation, or equilibrium.

Mountain–piedmont junction The planimetric junction between a mountainous escarpment and an adjacent piedmont. It tends to be straight in tectonically active settings and sinuous in inactive settings.

Mylonite A metamorphic rock formed by recrystallization during plastic flow below the brittle zone of the Earths crust.

Narrow gradient-index An anomalous stream-gradient index for a longitudinal profile. It describes magnitudes of longitudinal-profile abnormalities for short distances and only one or several contour intervals. *See also* inclusive gradient index.

Neutral stress Hydrostatic (pore) pressure that does not change grain-to-grain stresses in an aquifer system, or in a saturated hillslope.

Oxygen $^{18}O/^{16}O$ ratio ^{18}O, having 10 instead of 8 neutrons, is heavier than ^{16}O which is 500 times more common. Fractionation results in less ^{18}O in water evaporated from the ocean and in cold ice and snow. The ratio $^{18}O/^{16}O$ is compared with a standard ratio of ocean water. Ratio values are more negative with colder temperatures, allowing temperature histories to be reconstructed from the Antarctica and Greenland ice sheets.

Orbital parameters Three changing features of the Earth's orbit that tend to cause climatic changes: (1) variations of eccentricity (the slightly elliptical path of the Earth's orbit around the sun), (2) obliquity of axial tilt of the Earth to the plane of the orbit, and (3) precession of the seasonal equinoxes (progressive changes in seasonal timing of the earth's orbital perihelion and aphelion that result from a wobble in the Earth's axis of rotation).

Paraglacial Geomorphic processes that adjust to the change from full glacial to interglacial climate; glacier-outburst floods are an example.

Parallel partitioning Thrust, or normal, faulting and strike-slip faulting in two nearby, but separate, parallel fault zones. *See also* serial partitioning.

Partial area contribution That portion of a watershed that supplies runoff of water and large amounts of sediment to a stream during a rainfall event.

Pediment An erosion surface formed by retreat of an escarpment and by continued planation and degradation of rocks and basin fill downslope from the escarpment.

Perennial stream A stream that flows continuously as a result of surface runoff from rain and snowmelt, and influx of groundwater.

Piedmont A plain that slopes away from the base of a mountainous escarpment.

Piercing point Matching parts of geology (a dike, strata) or landscape (a stream channel, glacial moraine, cinder cone) are called piercing points when used to estimate the amount of separation caused by faulting.

Piggyback basin Tectonic depression above or in back of active thrust fault(s).

Potentiometric surface A surface which represents the static ground-water head. As related to an aquifer, it is defined by the levels to which water will rise in tightly cased wells. Also called piezometric surface.

Precision The instrumental error of a measured value, generally indicated as a ± range about the mean value that includes instrumental measurement errors, and ideally includes the variation in measured values that are a function of choice of procedural assumptions. *See also* accuracy.

Reach That part of a stream and its adjacent hillslopes with characteristics that differ from the adjacent upstream and downstream parts of a valley.

Reaction time The interval between the time of initiation of a perturbation to a fluvial system and the time that a change in system behavior is first observed.

Rainy-day normal (RDN) A climatic index defined as the ratio between mean annual precipitation and the average annual number of rainy-days.

Relaxation time The time needed to establish a new equilibrium or threshold condition after a fluvial system has reacted to a perturbation (Thornes and Brunsden, 1977).

Response time The sum of reaction and relaxation times.

Ridge rent (sackungen, antislope scarp) A ridge top or hillside uphill-facing scarp caused by gravitational settlement of a slab into a valley that has been steepened by glacial or fluvial erosion.

Rock mass strength The total resistance of a rock type to the applied stresses of flowing water, taking into account cohesive strength, presence of fractures, joints, shears, and groundwater levels and seepage forces (Selby, 1982a,b).

Rock uplift The sum of tectonic and isostatic uplift of earth materials, which is greater than uplift of a land surface that is being eroded.

Sackungen See ridge rent.

Sarsen stones Large sandstone blocks originally found in south-central England eroded from a bed of dense, hard silica-cemented sandstone.

Scarp height The difference in altitude between the base and crest of a fault scarp.

Secular equilibrium Occurs where the quantity of a radioactive terrestrial cosmogenic nuclide remains constant – its production rate equals its decay rate.

Sediment yield Amount of earth materials yielded per square kilometer of drainage basin per year.

Sediment-yield index Numerical appraisal of long-term rates of sediment yield based on comparison of sizes of depositional and erosional landforms. For example, the amount of alluvial-fan area per unit drainage-basin area represents the sediment yield from the source watershed.

Self-arresting feedback Interactions between variables that tend to eliminate changes that are occurring in geomorphic processes or landforms.

Self-enhancing feedback Interactions between variables that tend to perpetuate changes in geomorphic processes or landforms.

Seepage force The force exerted in the direction of groundwater flow as hydrodynamic viscous drag transfers energy from water to the solid particles of an aquifer system. A seepage force is equal to the decrease of hydraulic head.

Serial partitioning Alternating oblique thrust and strike-slip fault segments. *See also* parallel partitioning.

Slope distance The distance along the land surface between two points on a hillslope.

Slope fall The vertical distance between two points on a hillslope.

Slope length The horizontal distance between two points on a hillslope.

Soils chronosequence A temporal sequence of soils formed on alluvial geomorphic surfaces of different ages and similar parent materials based on how soil-profile horizons change with time.

Static allometric change Refers to the interrelations of measurements made of a landform or process at a single time; it is time independent. An example, is the changes in width, depth, and velocity of flow that occur in the downstream direction for a stream at bankfull stage. *See also* dynamic allometric change.

Steady state (geomorphic) A landscape where every slope and every form is adjusted to every other. It occurs when denudation maintains hillslope length and relief through time (Burbank and Anderson, 2001).

Steady state (terrestrial cosmogenic nuclides) *See* secular equilibrium.

Strath A beveled bedrock surface formed by streamflow during an interval when the valley floor is neither aggrading nor degrading.

Stream-gradient index Describes spatial variations in the character of the longitudinal profile. It is the product of slope of a reach (change of the altitude/length ratio of a reach) and the horizontal length to the midpoint of the reach from the watershed divide. See definitions for inclusive and narrow gradient-indices.

Stream order The relative position of stream segments in a channel network where stream channel join systematically downstream.

Stream power The power available to transport bedload. It consists of those variables (such as discharge and gradient) that if increased, may favor bedload transport and fluvial degradation.

Stream terrace Former level of a broad valley floor that was created by aggradation or by lateral erosion, but now is above the present valley floor because of incision by the stream. A terrace consists of a tread and a riser, which separates the tread from the stream or a lower terrace. Paired stream terraces have continuity along a valley but unpaired terraces occur only in one reach.

Stress, applied The downward stress imposed at an aquifer boundary. It differs from effective stress in that it defines only the external stress tending to compact a deposit rather than the grain-to-grain stress at any depth within a compacting deposit.

Stress, effective Stress (pressure) that is borne by and transmitted through the grain-to-grain contacts of a deposit, and thus affects its porosity or void ratio and other physical properties. In one-dimensional compression, effective stress is the average grain-to-grain load per unit area in a plane normal to the applied stress. At any given depth, the effective stress is the weight (per unit area) of sediments and moisture above the water table, plus the submerged weight (per unit area) of sediments between the water table and the specified depth, plus or minus the seepage stress (hydrodynamic drag) produced by water movement through the saturated sediments above the specified depth. Algebraic sum of the gravitational stress and seepage stress.

Stress, neutral Fluid pressure exerted equally in all directions at a point in a saturated deposit by the head of water. Neutral pressure is transmitted to the base of the deposit through the pore water, and does not have a measurable influence on the void ratio or on any other mechanical property of the deposits.

Stress, seepage *See* seepage force.

Summer monsoon Seasonal influx of rain drawn into a large continental area by sustained atmospheric low pressure caused by prolonged intense solar heating.

Surface uplift The algebraic sum of rock uplift and geomorphic processes (erosion or deposition).

Subordinate watershed A watershed that is becoming smaller because of gradual or abrupt expansion of adjacent dominant watershed(s).

Synthetic fault A minor fault whose sense of displacement is similar to its associated strike-slip, normal, or thrust fault.

Tectonic aneurism Advection of deep crustal material into a relatively weak zone characterized by rapid rock uplift, and rapid erosion. In the Himalayas rapid uplift leads to continued rapid erosion by large rivers such as the Indus and Tsangpo, producing more isostatic uplift.

Tectonic perturbation A tectonic increase in relief, or horizontal disruption of fluvial landforms, of sufficient strength to affect rates of aggradation or degradation in a fluvial system.

Tectonic translocated watershed Where strike-slip faulting bisects a watershed and shifts the upstream portion, eventually to where the stream flows into a different watershed.

Terrestrial cosmogenic nuclide The flux of cosmic rays at the Earths surface is radiation that creates nuclides within rock minerals. The abundance of cosmogenic nuclides present in rocks is useful for dating length of surface exposure or burial, and erosion rates.

Throw The amount of vertical displacement caused by faulting. *See* apparent throw.

Transpressive duplex A lens between two strike-slip faults with a reverse component that migrates when the reverse faults propagate. The trailing edge of the duplex undergoes a reversal of style collapsing back into the now dilating part of the wedge.

Threshold-intersection point A point on a longitudinal profile that separates aggrading and degrading reaches of a stream.

Threshold of critical power A transition between the modes of net deposition and net erosion by a stream. Opposing tendencies for aggradation and degradation are described by a ratio whose numerator consists of variables that if increased favor degradation (gradient, discharge) and whose denominator consists of variables that if increased favor aggradation (bedload, hydraulic roughness).

Type 1 dynamic equilibrium A condition of adjustment between processes operating in a reach of a stream that occurs when rates of tectonically induced downcutting equal rates of uplift, thus allowing the longitudinal profile of a stream to attain a succession of base levels of erosion in a tectonically active environment. Flights of strath terraces, broad straths, and longitudinal profiles of valley floors that plot as straight lines on semi-logarithmic graphs are common in such reaches.

Type 2 dynamic equilibrium A condition of adjustment between processes in streams that have a strong tendency toward, but lack of clear attainment of, the base level of erosion. Diagnostic landforms include valley floors that are narrow and lack strath terraces, but whose longitudinal profiles plot as concave lines on arithmetic graphs and as straight lines on semi-logarithmic graphs.

Transport-limited slope A scarp or other hillslope that lacks a cliffy slope element. It tends to change morphology by processes of slope decline at rates that are largely controlled by the rates at which erosional processes remove weakly cohesive earth materials.

Transrotational tectonics Crustal rotations about vertical axes that contribute to the translation of regional shear across the diverse tectonic blocks in a broad transform plate boundary.

Triple junction the small area where the boundaries of three tectonic plates meet. The Mendocino Triple Junction is where the transform San Andreas and Mendocino faults meet the Cascadia subduction zone.

Unconfined aquifer A geologic formation of permeable material that has a water table as the upper surface.

Underfit valley A valley whose stream is much to small to have eroded the valley.

Valley density The total length of valleys per square kilometer of a watershed, as measured using the crenulations of contour lines on topographic maps of 1:24,000 scale or larger. (*See also* drainage density.)

Water gap A valley where erosion by persistent streamflow has been able to maintain its course through a resistant ridge or rising fold. *See also* wind gap.

Water table The upper surface of the zone of saturation in a phreatic aquifer in which the pressure is atmospheric.

Weathering-limited slope A cliffy slope element on a scarp or other hillslope. It tends to change morphology by the processes of parallel erosional retreat, whose rates are largely controlled by the rates at which rock materials are made available for geomorphic processes to remove.

Wind gap A former stream channel notched into a ridgecrest that has been abandoned because of stream capture. *See also* water gap.

Wrench faults Strike-slip faults that involve basement rocks, and where a regional shear couple rotates basement rocks and overlying sediments to create complex and changing sets of faults and folds.

Xenolith A fragment of country rock enclosed in plutonic or volcanic rocks.

Appendix A

Precipitation		Temperature	
Mean annual (mm)		**Mean annual (°C)**	
Extremely arid	<50	Pergelic	< 0
Arid	50–250	Frigid	0–8
Semiarid	250–500	Mesic	8 -15
Semihumid	500–1,000	Thermic	15–22
Humid	1,000–2,000	Hyperthermic	>22
Extremely humid	>2,000		
Seasonality index (Sp)*		**Seasonality index (St)****	
Nonseasonal	1–1.6	Nonseasonal	< 2
Weakly seasonal	1.6–2.5	Weakly seasonal	2–5
Moderately seasonal	2.5–10	Moderately seasonal	5–15
Strongly seasonal	>10	Strongly seasonal	>15

* Precipitation seasonality index (Sp) is the ratio of the average total precipitation for the three wettest consecutive months (Pw) divided by the average total precipitation for the three consecutive driest months (Pd).

$$Sp = Pw/Pd$$

** Temperature seasonality index (St) is the mean temperature of the hottest month (Th) minus the mean temperature of the coldest month (Tc) in °C.

$$St = Th - Tc$$

Classification of climates.

References Cited

Abbott, L.D., Silver, E. A., Anderson, R.S.; Smith, R., Ingle, J.C., Kling, S.A., Haig, D., Small, E., Galewsky, J., Sliter, W.S, 1997, Measurement of tectonic surface uplift rate in a young collisional mountain belt: Nature, v. 385, p. 501–507. DOI: 10.1038/385501a0

Adams, J., 1980, Contemporary uplift and erosion of the Southern Alps, New Zealand: Geological Society of America Bulletin, v. 91, p. 1–114.

Nature 385, 501 - 507 (06 February 1997); doi:10.1038/385501a0

Adams, J., 1981, Earthquake-dammed lakes in New Zealand: Geology, v. 9, p. 215–219.

Adams, J., 1985, Large-scale tectonic geomorphology of the Southern Alps, New Zealand, in, Morisawa, M., and Hack, J.T., editors, Tectonic Geomorphology: Allen and Unwin, Winchester, Massachusetts, p. 105–128.

Adams, K., Briggs, R., Bull, W., Brune, J., Granger, D., Ramelli A, Riebe, C, Sawyer, T. , Wakabayashi, J., and Wills, C., 2001, Guidebook to the northern Walker Lane and northeast Sierra Nevada: Friends of the Pleistocene Pacific Cell field trip, October 12–14, 2001.

Ahnert, F., 1967, The role of the equilibrium concept in the interpretation of landforms of fluvial erosion and deposition, P. Macar, editor: in, L'evolution des Versants, Univ. of Liege, Liege, France, p. 23–41.

Ahnert, F., 1970, Functional relationships between denudation, relief, and uplift in large mid-latitude drainage basins: American Journal of Science, v. 268, p. 243–263.

Ahnert, F., 1973, COSLOP 2, a comprehensive model program for simulating slope profile development: Geocom Bulletin, v. 6, p. 99–122.

Ahnert, F., 1976, Brief description of a comprehensive three-dimensional process–response model of landform development: Zeitschrift fur Geomorphologie, v. 25, p. 29–49.

Aitken, S.R., and Leigh, C.H., 1992, Vanishing Rain Forests, the Ecological Transition in Malaysia: Oxford: Oxford University Press, 194 p.

Almond, P., Roering, J., and Hales, T.C., 2007, Using soil residence time to delineate spatial and temporal patterns of transient landscape response: Journal of Geophysical Research, v. 112, F03S17, doi:10.1029/2006JF000568.

Ambraseys, N.N., and Jackson, J.A. 1990, Seismicity and associated strain of central Greece between 1890 and 1988: Geophysical Journal International, v. 101, p. 663-708.

Amit, R., Harrison, J.B.J., and Enzel, Y., 1995, Use of soils and colluvial deposits in analyzing tectonic events — The southern Arava Rift, Israel: Geomorphology, v. 12, p. 91–107.

Anders, M.H., Geissman, J.W., and Sleep, N.H., 1990, Comment on northeastern Basin and Range Province active tectonics, An alternative view: Geology, v. 18, p. 914–917.

Anderson, L.W., and Piety L.A., 2001, Geologic seismic source characterization of the San Luis–O'Neill area, eastern Diablo Range, California for B.F. Fisk and O'Neill forebay dams, San Luis unit, Central Valley project, California: U.S. Bureau of Reclamation Seismotectonics Report 276 p.

Anderson, R. S., 1994, Evolution of the Santa Cruz Mountains, California, through tectonic growth and geomorphic decay, Journal of Geophysical Research, v. 99, 20,161–20,174.

Andrews, D.J., and Bucknam, R.C., 1987, Fitting degradation of shoreline scarps by a nonlinear diffusion model: Journal of Geophysical Research, v. 92, p. 12,857–12,867.

Angevine, C.L., Heller, P.L., and Paola, C,. 1990, Quantitative Sedimentary Basin Modelling: American Association of Petroleum Geologists Continuing Education Course Note Series, No. 32., 133 p.

Argus, D.F., and Gordon, R.G., 2001, Present tectonic motion across the Coast Ranges and San Andreas fault system in central California: Geological Society of America Bulletin, v. 113, p. 1580–1592.

Arkley, R.J., 1962, The geology, geomorphology, and soils of the San Joaquin Valley in the vicinity of the Merced River, California: California Division of Mines and Geology Bulletin 182, p. 25–31.

Armijo, R., Meyer, B., King, G.C.P., Rigo, A., Papanastassiou, D., 1996, Quaternary evolution of the Corinth Rift and its implications for the Late Cenozoic evolution of the Aegean: Geophysical Journal International, v. 126, p. 11–53.

Armijo, R., Meyer, B., Hubert, A., and Barka, A, 1999, Westward propagation of the North Anatolian Fault into the northern Aegean; timing and kinematics: Geology, v. 27, p. 267–270.

Atwater, B.F., Adam, D.P., Bradbury, J.P., Forester, R.M., Mark, R.K., Lettis, W.R., Fisher, G.R., Gobalet, K.W., and Robinson, S.W., 1986, A fan dam for Tulare Lake, California, and implications for the Wisconsin glacial history of the Sierra Nevada: Geological Society of America Bulletin, v. 97, p. 97–109.

Atwater, T., 1970, Implications of plate tectonics for

the Cenozoic tectonic evolution of western North America: Geological Society of America Bulletin, v. 81, p. 3513–3536.

Atwater, T., and Stock, J., 1998, Pacific–North America plate tectonics of the Neogene southwestern United States, An update: International Geology Review, v. 40, p. 375–402.

Augustinus, P.C., 1992, The influence of rock mass strength on glacial valley cross-profile morphometry: A case study from the Southern Alps, New Zealand: Earth Surface Processes and Landforms, v. 17, p. 39–51.

Augustinus, P.C., 1995, Glacial valley cross-profile development: the influence of *in situ* rock stress and rock mass strength, with examples from Southern Alps, New Zealand: Geomorphology, v. 11, p. 87–97.

Azor A., Keller, E.A., and Yeats, R.S., 2002, Geomorphic indicators of active fold growth: South Mountain–Oak Ridge anticline, Ventura basin, southern California: Geological Society of America Bulletin, v. 114, p. 745–753.

Bahat, D., Grossenbacher, K., and Karasaki, K., 1999, Mechanism of exfoliation joint formation in granitic rocks at Yosemite National Park: Journal of Structural Geology, v. 21, p. 85–96.

Bakker, J.P., and Le Heux, J.W.N., 1952, A remarkable new geomorphological law: Koninklijke Nederlandsche Akademie van Wetenshappen Series B, v. 55, p. 399–410 and p. 554–571.

Ballantyne, C.K., 2002, Paraglacial geomorphology: Quaternary Science Reviews, v. 21, p. 1935–2017.

Ballantyne, C.K., and Benn, D.I., 1996. Paraglacial slope adjustment during recent deglaciation, implications for slope evolution in formerly glaciated terrain: in, Brooks, S., and Anderson, M.G., editors, Advances in Hillslope Processes, Wiley, Chichester.

Bard, E., Hamelin, B. Fairbanks, R.G., and Zindler, A. 1990, Calibration of the ^{14}C timescale over the past 30,000 years using mass spectrometric U–Th ages from Barbados corals: Nature, v. 345, p. 405–410.

Batt, G.E., Braun, J., Kohn, B.P., and McDougal, I., 2000, Thermochronological analysis of the dynamics of the Southern Alps, New Zealand: Geological Society of America Bulletin, v. 112, p. 250–266.

Batt, G.E., Brandon, M.T., Farley, K.A., and Roden-Tice, M., 2001, Tectonic synthesis of the Olympic Mountains segment of the Cascadia wedge, using 2-D thermal and kinematic modeling of isotopic ages: Journal of Geophysical Research, v. 106, p. 26,731–26,746.

Beanland, S., and Barrow-Hurlbert, S.A., 1988, The Nevis–Cardrona fault system, Central Otago, New Zealand, Late Quaternary tectonics and structural development: New Zealand Journal of Geology and Geophysics, v. 31, p. 337–352.

Beanland, S., and Berryman, K.R., 1989, Style and episodicity of Late Quaternary activity on the Pisa-Grandview fault zone, central Otago, New Zealand: New Zealand Journal of Geology and Geophysics, v. 32, p. 45–461.

Beanland, S., and Clark, M.M., 1994, The Owens Valley fault zone, eastern California, and surface faulting associated with the 1872 earthquake: U.S. Geological Survey Bulletin, 29 p., 4 sheets.

Beaumont, C., Fullsack, P., and Hamilton, J., 1992, Erosional control of active compressional orogens: in, McClay, K.R., editor, Thrust Tectonics, Chapman and Hall, London, p. 1–18.

Beck, A.C., 1968, Gravity faulting as a mechanism of topographic adjustment: New Zealand Journal of Geology and Geophysics, v. 11, p. 191–199.

Belloni, L.G., and Stefani, R., 1987, The Vaiont slide; instrumentation, past experience and the modern approach: in, Leonards, G.A., editor, Dam Failures: Engineering Geology, v. 24, p. 445–474.

Benedetti, L., Tapponnier, P., King, G., Meyer, B., and Manighetti, I., 2000, Growth folding and active thrusting in the Montello region, Veneto, northern Italy: Journal of Geophysical Research, v. 105, p. 739–766.

Benn, D.I., Owen, L.A., Finkel, R.C., and Clemmens, S., 2006, Pleistocene lake outburst floods and fan formation along the eastern Sierra Nevada, California: implications for the interpretation of intermontane lacustrine records: Quaternary Science Reviews, v. 25, p. 2729–2748.

Bennett, E.R., Youngson, J.H., Jackson, J.A., Norris, R.J., Raisbeck, G.M., Yiou, F., and Fielding, E., 2005, Growth of South Rough Ridge, Central Otago, New Zealand, using *in-situ* cosmogenic isotopes and geomorphology to study an active, blind reverse fault: Journal of Geophysical Research, v. 110, B02404, doi:10.1029/2004JB003184.

Bennett, E.R., Youngson, J.H., Jackson, J.A., Norris, R.J., Raisbeck, G.M., and Yiou, F., 2006, Combining geomorphic observations with *in-situ* cosmogenic isotope measurements to study anticline growth and fault propagation in Central Otago, New Zealand: New Zealand Journal of Geology and Geophysics, v. 49, p. 217–231.

Bennett, R.A., Davis, J.L., Wernicke, B.P., 1999: Present-day pattern of Cordilleran deformation in the western United States: Geology, v. 27, p. 371–374.

Berger, A., 1988, Milankovitch theory and climate: Reviews of Geophysics, v. 26, p. 624–657.

Berggren, W.A., Hilgen, F.J, Langereis, C.G., Kent, D.V., Obradovich, J.D., Raffi, I., Raymo, M.E, and Shackleton, N. J., 1995, Late Neogene chronology, New perspectives in high-resolution stratigraphy: Geological Society of America Bulletin, v.107, p. 1272–1287.

Berryman, K.R., Beanland, S., Cooper, A.F., Cutten, H.N., Norris, R.J., and Wood, P.R., 1992, The Alpine Fault, New Zealand, variation in Quaternary structural style and geomorphic expression: Annals Tectonica, v. 6, p. 126–163.

Beschta, R.L., 1978, Long-term patterns of sediment production following road construction and logging in the Oregon Coast Range: Water Resources Research, v. 14, p. 1011–1016.

Betson, R.P., and Marius, J.B., 1969, Source areas of storm runoff: Water Resources Research, v. 5, p. 574–582.

Bielecki, A.E., and Mueller, K.J., 2002, Origin of terraced hillslopes on active folds in the southern San Joaquin Valley, California: Geomorphology, v. 42, p. 131–152.

Bierman, P.R., and Caffee, M.W., 2001, Slow rates of rock surface erosion and sediment production across the Namib Desert and escarpment, Southern Africa: American Journal of Science, v. 301, p. 326–358.

Bierman, P.R., and Caffee, M.W., 2002, Cosmogenic exposure and erosion history of Australian bedrock landforms: Geological Society of America Bulletin, v. 114, p. 787–803.

Bierman, P.R., Gillespie, A.R., and Caffee, M.W., 1995, Cosmogenic ages for earthquake recurrence intervals and debris flow fan deposition, Owens Valley, California: Science, v. 270, p. 447–450.

Bierman, P., Zen, E., Pavich, M., and Reusser, L.J., 2004, The incision history of a passive margin river, the Potomac near Great Falls, in, Southworth, S., and Buron, W., editors, Geology of the National Capital Region, Field Trip Guidebook for NE/SE Geological Society of America Meeting, p. 191–122.

Billingsley, G.H., and Workman, J.B., 2000, Geologic map of the Littlefield 309 3 609 Quadrangle, Mohave County, northwestern Arizona: U.S. Geological Survey Geologic Investigations Map I-2628, 1 sheet, scale 1:48 000, 25 p.

Billiris, H., Paradissis, D., Veis, G., England, P., Featherstone, W., Parsons, B., Cross, P., Rands, P., Rayson, M., Sellers, P., Ashkenazi, V., Davison, M., Jackson, J. and Ambraseys, N., 1991, Geodetic determination of tectonic deformation in central Greece from 1900 to 1988: Nature, v. 350, p. 124–129.

Bischoff, J.L., Cummins, K., 2001, Wisconsin glaciation of the Sierra Nevada (79,000–15,000 yr BP) as recorded by rock flour in sediments of Owens Lake, California: Quaternary Research, v. 55, p. 14–24.

Bishop, D G, 1985, Inferred uplift rates from raised marine surfaces, southern Fiordland, New Zealand: New Zealand Journal of Geology and Geophysics, v. 28, p. 243–251.

Bishop, D.G., 1991, High-level marine terraces in western and southern New Zealand, indicators of the tectonic tempo of an active continental margin: in, Macdonald, D.I.M., editor, Sedimentation, Tectonics, and Eustasy, Special Publication of the International Association of Sedimentology, v. 12, p. 69–78.

Bishop, D.G. 1994, Extent and regional deformation of the Otago peneplain: New Zealand Institute of Geological and Nuclear Sciences Report 94/1.

Bishop, P., and Brown, R. 1992, Denudational isostatic rebound of intraplate highlands: The Lachlan River Valley, Australia: Earth Surface Processes and Landforms, v. 17, p. 345–360.

Bjorse, G. and Bradshaw, R. 1998, 2000 years of forest dynamics in southern Sweden; suggestions for forest management: Forest Ecology and Management, v. 104, p. 15–26.

Blackwelder, E., 1928, Mudflow as a geologic agent in semiarid mountains: Geological Society of America Bulletin, v. 39, p. 465–483.

Blair, T.C., 2001, Outburst flood sedimentation on the proglacial Tuttle Canyon alluvial fan, Owens Valley, California, USA: Journal of Sedimentary Research, v. 71, p. 657–679.

Blair, T.C., 2002, Alluvial-fan sedimentation from a glacial-outburst flood, Lone Pine, California, and contrasts with meteorological flood deposits. in, Martini, I.P., Baker, V.R., and Garzon, G., editors, Flood and Megaflood Processes and Deposits: Recent and Ancient Examples. Special Publications of International Association of Sedimentology, 32, p. 113–140.

Blakely, R.J., Jachens, R.C., Calzia, J.P., and Langenheim, V.E., 1999, Cenozoic basins of the Death Valley extended terrane as reflected in regional-scale gravity anomalies (p. 1–16), in, Wright, L.A., and

B.W. Troxel, editors, Cenozoic Basins of the Death Valley Region: Geological Society of America Special Paper 333, 381 p.

Bloom, A.L., Broecker, W.S., Chappell, J., Matthews, R.K., Mesolella, K.J., 1974, Quaternary sea level fluctuations on a tectonic coast: new ^{230}Th/^{234}U dates from Huon Peninsula, New Guinea: Quaternary Research, v. 4, p.185–205.

Bookhagen, B., Thiede, R.C., and Strecker, M.R., 2005, Abnormal monsoon years and their control on erosion and sediment flux in the high, arid northwest Himalaya: Earth and Planetary Science Letters: v. 231, p. 131–146.

Bookhagen, B., Fleitmann, D., Nishiizumi, K., Strecker, M.R., and Thiede, R.C., 2006a, Holocene monsoonal dynamics and fluvial terrace formation in the northwest Himalaya, India: Geology, v. 34, p. 601–604.

Bookhagen, B., Burbank, D.W., Strecker, M.R., Thiede, R.C., Nishiizumi, K, 2006b, Contrasts between short- and long-term erosion rates in the NW Himalaya; Disequilibrium at 10^3 to 10^6-yr time scales: American Geophysical Union, Fall Meeting, abstract #T13E-03.

Borchers, J.W., editor, 1998, Land subsidence case studies and current research: Proceedings of the Dr. Joseph F. Poland Symposium on Land Subsidence: Land Subsidence Case Studies and Current Research, Star Publishing Company, Belmont, California, 576 p.

Broecker, W. S., and van Donk, J., 1970, Insolation changes, ice volumes, and ^{18}O record in deep-sea cores, Reviews of Geophysics and Space Physics, v. 8, p. 169–198, doi:10.1029/RG008i001p00169.

Brook, E.J., Brown, E.T. , Kurz, M.D., Ackert, R.P. , Raisbeck, G.M., and Yiou, F. , 1995, Constraints on age, erosion, and uplift of Neogene glacial deposits in the Transantarctic Mountains determined from *in situ* cosmogenic ^{10}Be and ^{26}Al: Geology, v. 23, p.1063– 1066.

Brook, M.S., Kirkbride, M.P., and Brock, B.W., 2004, Rock strength and development of glacial valley morphology in the Scottish Highlands and northwest Iceland: Geografiska Annaler, v. 86, p. 225–237.

Brozovic, N., Burbank, D.W., and Meigs, A.J., 1997, Climatic limits on landscape development in the northwestern Himalaya: Science, v. 276, p. 571–574.

Bruun, F., Brodie, J.W., and Fleming, C.A., 1955, Submarine geology of Milford Sound, New Zealand: New Zealand Journal of Science and Technology, v. 36B, p. 397–410.

Bull, L.J., and Kirkby, M.J., 1997, Gulley processes and modelling: Progress in Physical Geography, v. 21, p. 354–374.

Bull, W.B., 1964, History and causes of channel trenching in western Fresno County, California: American Journal of Science, v. 262, p. 249–258.

Bull, W.B., 1973, Geologic factors affecting compaction of deposits in a land subsidence area: Geological Society of America Bulletin, v. 84, p. 3783–3802.

Bull, W.B., 1975a, Allometric change of landforms: Geological Society of America Bulletin, v. 85, p. 1489–1498.

Bull, W.B., 1975b, Land subsidence due to groundwater withdrawal in the Los Banos–Kettleman City area – Part 2 – Subsidence and compaction of deposits: U.S. Geological Survey Professional Paper 437F, 90 p.

Bull, W.B., 1977, Landforms that do not tend toward a steady state: in, W.N. Melhorn and R. Flemal, editors, Theories of Landform Development, Proceedings 6th Annual Geomorphology Symposium, State University New York at Binghamton, p. 111–128.

Bull, W.B., 1984, Tectonic geomorphology: Journal of Geological Education, v. 32, p. 310–324.

Bull, W.B., 1985, Correlation of flights of global marine terraces: in, M. Morisawa and J. Hack, editors, Tectonic Geomorphology, Proceedings 15th Annual Geomorphology Symposium, State University of New York at Binghamton, George Allen and Unwin, Boston.

Bull, W.B., 1991, Geomorphic Responses to Climatic Change: New York, Oxford University Press, 326 p.

Bull, W.B., 1996, Prehistorical earthquakes on the Alpine fault, New Zealand: Journal of Geophysical Research, Solid Earth, Special Section "Paleoseismology", v. 101, p. 6037–6050.

Bull, W.B., 1997, Discontinuous ephemeral streams: Geomorphology, v. 19, p. 227–276.

Bull, W.B., 1998, Tectonic controls of geomorphic processes in a land subsidence area: in, J.W. Borchers, editor, Land Subsidence Case Studies and Current Research, Star Publishing Company, Belmont, California, p. 29–36.

Bull, W.B., 2004, Sierra Nevada earthquake history from lichens on rockfall blocks: Sierra Nature Notes, v. 4, 20 p. This can be downloaded from (http://www.yosemite.org/naturenotes/LichenIntro.htm)

Bull, W.B., 2007, Tectonic Geomorphology of Mountains, a New Approach to Paleoseismology: Blackwell Publishing, Oxford, 320 p.

Bull, W.B., and Cooper, A.F., 1986, Uplifted marine terraces along the Alpine fault, New Zealand: Science, v. 234, p. 1225–1228.

Bull, W.B., and Cooper, A.F., 1988, Uplifted marine terraces; uplift rates: Response to Technical Comment by C.M. Ward, Science, v. 236, p. 803–805.

Bull, W.B., and Poland, J.F., 1975, Land subsidence due to groundwater withdrawal in the Los Banos–Kettleman City area – Part 3 – Interrelations of water-level change, change in aquifer-system thickness, and subsidence: U.S. Geological Survey Professional Paper 437G, 62 p.

Bull, W.B., Schlyter, P., and Brogaard, S., 1995, Lichenometric analysis of the Kaerkerieppe slush-avalanche fan, Kaerkevagge, Sweden: Geografiska Annaler, v. 77A, p. 231–240.

Burbank, D.W., and Anderson, R.S., 2001, Tectonic Geomorphology: Blackwell Science, Oxford, 274 p.

Burbank, D.W., and Whistler, D.P., 1987, Temporally constrained tectonic relations derived from magnetostratigraphic data: Implications for the initiation of the Garlock fault, California: Geology, v. 15, p. 1172–1175, doi: 10.1130/0091-7613(1987)15<1172:TCTRDF>2.0.CO;2.

Burbank, D.W., Leland, J., Fielding, E., Anderson, R.S., Brozovic, N., Reid, M.R., and Duncan, C., 1996a, Bedrock incision, rock uplift, and threshold hillslopes in the northwestern Himalaya: Nature, v. 379, p. 505–510.

Burbank, D., Meigs, A., and Brozovic, N., 1996b, Interactions of growing folds and coeval depositional systems: Basin Research, v. 8, p., 199–223.

Burbank, D.W., McLean, L. J., Bullen, M.E., Abdrakmatov, K.Y., and Miller M., 1999, Partitioning of intermontane basins by thrust-related folding, Tien Shan, Kyrgyzstan: Basin Research, v. 11, p. 75– 92.

Burbank, D.W., Blythe, A.E., Putkonen, J.L., Pratt-Situala, B.A., Gabet, E.J., Oskin, M.E., Barros, A.P., and Ohja, T.P., 2003, Decoupling of erosion and climate in the Himalaya: Nature, v. 426, p. 652–655.

Burchfiel, B.C., and Davis, G.A., 1981, Mojave Desert and environs, in, Ernst, W.G., editor, The geotectonic development of California (Rubey volume 1): Englewood Cliffs, N.J., Prentice-Hall, p. 217–252.

Burchfiel, B.C., and Stewart, J.H., 1966, "Pull-apart" origins of the central segment of Death Valley, California: Geological Society of America Bulletin, v. 77, p. 439–442.

Burchfiel, B.C., Hodges, K.V., and Royden, L.H., 1987, Geology of Panamint Valley–Saline Valley pull-apart system, California: Palinspastic evidence for low-angle geometry of a Neogene range-bounding fault: Journal of Geophysical Research, v. 92, no. B10, p. 10422–10426.

Burchfiel, B.C., Molnar, P., Zhang, P., Deng, Q., Zhang, W., and Wang, Y., 1995, Example of a supradetachment basin within a pull-apart tectonic setting: Mormon Point, Death Valley, California: Basin Research, v. 7, p. 199–214.

Busby, C.J., and Ingersoll, R.V., editors, 1995, Tectonics of Sedimentary Basins: Blackwell Science, 579 p.

Caine, N, 1980, The rainfall intensity-duration control of shallow landslides and debris flows: Geografiska Annaler, Series A: Physical Geography, v. 62, p. 23–27.

Campbell, R.H., 1975, Soil slips, debris flows, and rainstorms in the Santa Monica Mountains and vicinity, southern California: U.S. Geological Survey Professional Paper 851, 51 p.

Cannon, P.J., 1976, Generation of explicit parameters for a quantitative geomorphic study of the Mill Creek drainage basin: Oklahoma Geology Notes, v. 36, p. 3–16.

Cannon, S.H., 1988, Regional rainfall-threshold conditions for abundant debris-flow activity, in, Ellen, S.D., and Wieczorek, G.F., editors., Landslides, Floods, and Marine Effects of the Storm of January 3–5, 1982, in the San Francisco Bay Region, California: U.S. Geological Survey Professional Paper 1434, p. 35–42.

Cannon, S.H., 2001, Debris-flow generation from recently burned watersheds: Environmental and Engineering Geoscience, v. 7, no. 4, p. 321–341.

Cannon, S.H., and Ellen, S., 1985, Rainfall conditions for abundant debris avalanches, San Francisco Bay region, California: California Geology, v. 38, p. 267–272.

Cannon, S.H., and Ellen, S.D., 1988, Rainfall that resulted in abundant debris- flow activity during the storm, in, Ellen, S.D., and Wieczorek, G.F., editors, Landslides, Floods, and Marine Effects of the Storm of January 3–5, 1982, in the San Francisco Bay Region, California: U.S. Geological Survey Professional Paper 1434, p. 27–34.

Carson, M.A., 1976, Mass-wasting, slope development and climate: in, Derbyshire, E., editor, Geomorphology and climate: New York, John Wiley and Sons, p. 101–136.

Carson, M.A., and Petley, D.J., 1970, The existence of threshold hillslopes in the denudation of the landscape: Institute of British Geographers Transactions, v. 49, p. 71–95.

Carver, G.A., 1987, Late Cenozoic tectonics of the Eel river basin, coastal northern California: in, Schymiczek, H., and Suchsland, R., editors, Tectonics, Sedimentation and Evolution of the Eel River and Associated Coastal Basins of Northern California, San Joaquin Geological Society Miscellaneous Publication No. 37, Bakersfield, California, p. 61–72.

Carver, G.A., and Burke R.M., 1992, Late Cenozoic deformation on the Cascadia subduction zone in the region of the Mendocino triple junction, in, Burke, R.M., and Carver, G.A., editors, A Look at the Southern End of the Cascadia Subduction Zone and Mendocino Triple Junction: Pacific Cell, Friends of the Pleistocene Guidebook for the Field Trip to Northern Coastal California, p. 31–63.

Castle, R. O., 1999, Work in Progress—A Connection Through the Brittle Crust Between Death Valley and the Salton Basin; in, Slate, J.L., editor, Proceedings of Conference on Status of Geologic Research and Mapping, Death Valley National Park: U.S. Geological Survey Open-File Report 99–153, p. 163–164.

Cenderelli, D.A., and Wohl, E.E., 2003, Flow hydraulics and geomorphic effects of glacial-lake outburst floods in the Mount Everest region, Nepal: Earth Surface Processes and Landforms, v. 28, p. 385–407.

Chappell, J., 1983, A revised sea-level curve for the last 300,000 years from Papua New Guinea: Search, v. 14, p. 99–101.

Chappell, J., 2001, Sea level change through the last glacial cycle: Science, v. 292, p. 679–686.

Chappell, J., 2002, Sea level changes forced ice breakouts in the Last Glacial cycle: new results from coral terraces: Quaternary Science Reviews, v. 21, p. 1229–1240.

Chappell, J., Omura, A., Esat, T., McCulloch, M., Pandolfi, J., Ota, Y., and Pillans, B., 1996, Reconciliation of Late Quaternary sea levels derived from coral terraces at Huon Peninsula with deep sea oxygen isotope records: Earth and Planetary Science Letters: v. 141, p. 227–236.

Chappell, J., Rhodes, E.G., Thorn, B.G., and Wallensky, E., 1982. Hydro-isotasy and the sea-level isobase of 5500 BP in north Queensland, Australia: Marine Geology, v. 49, p. 81–90.

Chappell, J., and Shackleton, N.J., 1986, Oxygen isotopes and sea level: Nature, v. 324, p. 137–140.

Chase, C.G., 1978, Plate kinematics; The Americas, East Africa, and the rest of the world: Earth and Planetary Science Letters, v. 37, p. 353–368.

Chase, C.G., 1992. Fluvial landsculpting and the fractal dimension of topography: Geomorphology, v. 5, p. 39–57.

Chase, C.G., and Wallace, T.C., 1986, Uplift of the Sierra Nevada, California: Geology, v. 14, p. 730–733.

Chen J.H., Curran H.A., White B., and Wasserburg G.J., 1991, Precise chronology of the last interglacial period; ^{234}U-^{230}Th data from fossil reefs in the Bahamas: Geological Society of America Bulletin, v. 103, p. 82–97.

Cheng, G., and Wu, T., 2007, Responses of permafrost to climate change and their environmental significance, Qinghai-Tibet Plateau: Journal of Geophysical Research, v. 112, F02S03, doi:10.1029/2006JF000631.

Chinn, T.J., and Whitehouse, I.E., 1980, Glacier snow line variations in the Southern Alps, New Zealand: in World Glacier Inventory, Riederalp (Switzerland) Workshop, Proceedings, International Association of Hydrological Sciences, IAHS-AISH Publication No. 126, p. 219–228.

Chinn, T.J., McSavney, M.J., and McSaveney, E.R., 1992, The Mount Cook Rock Avalanche of 14 December, 1991: Institute of Geological and Nuclear Sciences Ltd., Lower Hutt, New Zealand, Information Series 20.

Chow, V.T., editor, 1964, Handbook of Applied Hydrology: McGraw-Hill Book Co., New York.

Christensen, M.N., 1966, Late Cenozoic crustal movements in the Sierra Nevada of California: Geological Society of America Bulletin, v. 77, p. 163–182.

Church, M., 2005, Continental drift: Earth Surface Processes and Landforms, v. 30, p. 129–130.

Cichanski, M., 2000, Low-angle, range-flank faults in the Panamint, Inyo, and Slate ranges, California; Implications for recent tectonics of the Death Valley region: Geological Society of America Bulletin, v. 112, p. 871–883.

Clarke, P.J., Davies, R.R., England, P.C., Parsons, B.E., Billiris, H., Paradissis, D., Veis, G., Denys, P.H., Cross, P.A., Ashkenazi, V., and Bingley, R. 1997. Geodetic estimate of seismic hazard in the Gulf of Korinthos. Geophysical Research Letters, v. 24, p. 1303–1306.

Clemens, S.C., and Prell, W.L., 2003, A 350,000 year summer-monsoon multi-proxy stack from the Owen Ridge, northern Arabian Sea: Marine Geology, v. 201, p. 35–51.

Clemens, S.C., and R. Tiedemann, 1997, Eccentricity forcing of Pliocene-Early Pleistocene climate revealed in a marine oxygen-isotope record, Nature, v. 385, 801–804.

CLIMAP, 1981, Seasonal reconstruction of the Earth's surface at the last glacial maximum: Geological Society of America, Map and Chart Series MC-36: p. 1–18.

Cloetingh S., Boldreel L.O., Larsen B.T., Heinesen M., and Mortensen L., 1998, Tectonics of sedimentary basin formation, models and constraints: Tectonophysics, v 300, p. 1–11.

Cockburn, H.A.P., Siedl, M.A., and Summerfield, M.A., 1999, Quantifying denudation rates on inselbergs in the Central Namib Desert using *in situ*-produced cosmogenic ^{10}Be and ^{26}Al: Geology, v. 27, p. 399–402.

Collier, R.E.L., Pantosi, D., D'Addezio, G., De Martini, P.M., and Masana, E., 1998, Paleoseismicity of the 1981 Corinth earthquake fault, Seismic contribution to extensional strain in central Greece and implications for seismic hazard: Journal of Geophysical Research, v. 103, p. 30,001–30,019.

Coney, P.J., Jones, D.L., and Monger, J.W.H., 1980, Cordilleran suspect terranes: Nature, v. 288, p. 329–333.

Cooper, A.F., 1980, Retrograde alteration of chromium kyanite in metachert and amphibolite whiteschist from the Southern Alps, New Zealand, with implications for uplift on the Alpine Fault: Contributions to Mineralogy and Petrology, v. 75, p. 153–164.

Cooper, A.F., and Bishop, D.G., 1979, Uplift rates and high level marine platforms associated with the Alpine fault at Okuru River, south Westland, in, Walcott, R.I., and Cresswell, M.M., editors, The origin of the Southern Alps: Royal Society of New Zealand Bulletin 18, p. 35–43.

Costa, J.E., 1984, Physical geomorphology of debris flows, in, J.E. Costa and P.J. Fleischer, editors, Developments and Applications of Geomorphology: Springer-Verlag, Berlin, p. 269–317.

Costa, J.E., 1994, Evolution of Sediment Yield from Mount St. Helens, Washington, 1980-1993 Water Fact Sheet: U.S. Geological Survey Open-File Report 94-313.

Cotton, C. A., 1917, Block Mountains in New Zealand: American Journal of Science, v. 44, p. 249–293.

Cowan, H.A., 1991, The North Canterbury Earthquake of September 1, 1888, Journal of the Royal Society of New Zealand, v. 21, p. 1–12.

Cowan, H.A., Nicol, A., and Tonkin, P., 1996, A comparison of historical and paleoseismicity in a newly formed fault zone and a mature fault zone, North Canterbury, New Zealand: Journal of Geophysical Research, v. 101, p. 6021–6036.

Cowie, P.A., and Roberts, G.P., 2001, Constraining slip rates and spacings for active normal faults: Journal of Structural Geology, v. 23, p. 1901–1915.

Crampton, J., Laird, M., Nicol, A., Townsend, D., and Van Dissen, R. 2003, Palinspastic reconstructions of southeastern Marlborough, New Zealand, for mid-Cretaceous-Eocene times: New Zealand Journal of Geology and Geophysics, v. 46, p. 153–175.

Croft, M.G., 1968, Geology and radiocarbon ages of Late Pleistocene lacustrine clay deposits, southern part of the San Joaquin Valley, California: in Geological Survey Research 1968, U.S. Geological Survey Professional Paper 600B, p. B151–156.

Croft, M.G., 1972, Subsurface geology of the late Tertiary and Quaternary water-bearing deposits of the southern part of the San Joaquin Valley, California: U.S. Geological Survey Water-Supply Paper 1999-H, 29 p.

Crowell, J.C., 1974, Origin of late Cenozoic basins in southern California: in Tectonics and Sedimentation, W.R. Dickinson, editor, Special Publication 22, Society of Economic Paleontologists and Mineralogists, p. 190–204.

Crowell, J.C., 1982, Tectonics of Ridge Basin, southern California: in, Crowell, J.C., and Link, M.H., editors, Geologic History of Ridge Basin, Southern California,: Pacific Section of the Society of Economic Paleontologists and Mineralogists, Los Angeles, CA, p. 25–42.

Crozier, M.J., 1999, Prediction of rainfall-triggered landslides, a review of the antecedent water status model: Earth Surface Processes and Landforms, v. 24, p. 825–833.

Crutzen, P.I., and Stoermer, E.F., 2000, The "Anthropocene": International Geosphere–Biosphere Programme Newsletter 41, p. 12.

Culling, W.E.H., 1960, Analytical theory of erosion: Journal of Geology, v. 68, p. 336–344.

Culling, W.E.H., 1963, Soil creep and the development of hillside slopes: Journal of Geology, v. 71, p. 127–161.

Culling, W.E.H., 1965, Theory of erosion on soil-covered slopes: Journal of Geology, v. 73, p. 230–254.

Dadson, S.J., and 10 others, 2004, Earthquake-triggered increase in sediment delivery from an active mountain belt: Geology, v. 32, p. 733–736.

Dalrymple, G.B., and Hamblin, W.K., 1998, K–Ar of Pleistocene lava dams in the Grand Canyon in Arizona: National Academy of Science Proceedings, v. 95, p. 9744–9749.

Davis, D., Suppe, J., and Dahlen, F. A., 1983, Mechanics of fold-and-thrust belts and accretionary wedges: Journal of Geophysical Research, v. 88, p. 1153–1172.

Davis, G.A., and Burchfiel, B.C., 1973, Garlock fault: An intracontinental transform structure, southern California: Geological Society of America Bulletin, v. 84, p. 1407–1422.

Davis, G.H., and Coplen, T.B., 1989, Late Cenozoic paleohydrology of the western San Joaquin Valley, California, as related to structural movements in the central Coast Ranges: Geological Society of America Special Paper 234, 40 p.

Davis, G.H., and Green, J.H., 1962, Structural control of the interior drainage, southern San Joaquin Valley, California: U.S. Geological Survey Professional Paper 450D, p. D89–D91.

Davis, W.M., 1892, The convex profile of bad-land divides: Science, v. 200, p. 245.

Davis, W.M., 1898, The grading of mountain slopes: Science, v. 7, p. 81.

Davis, W.M., 1899, The geographical cycle: Geographical Journal, v. 14, p. 481–504.

Davis, W.M., 1909, The systematic description of land forms: Geographical Journal of the Royal Geography Society, v. 34, p. 300–326.

Dawers, N.H., and Anders, M.H., 1995, Displacement-length scaling and fault linkage. Journal of structural Geology: v. 17, p. 60–614.

Dawson, T.E., McGill, S.F., and Rockwell, T.K., 2003, Irregular recurrence of paleoearthquakes along the central Garlock fault near El Paso Peaks, California: Journal of Geophysical Research, v. 108, doi: 10.1029/2001JB001744.

Dawson, T.E., Rockwell, T., McGill, S., and Black, N., 2000, A 7000 year record of paleoearthquakes on the central Garlock fault, near El Paso Peaks, California: in, Okumura, K., Takada, K., and Goto, H., editors, Active Fault Research for the New Millennium, Proceedings of the Hokudan International Symposium and School on Active Faulting, Letter Press Ltd., Hiroshima, Japan, p. 65–68.

D'Odorico, P. and Porporato, A, 2004, Preferential states in soil moisture and climate dynamics: Proceedings of the National Academy of Sciences, USA, v. 101, p. 8848-8851.

D'Odorico, P., Laio, F., and Ridolfi, L., 2005, Noise-induced stability in dryland plant ecosystems: Proceedings of the National Academy of Sciences, USA, v. 102, p. 10819–10822.

DeMets, C., Gordon, R.G., Argus, D.F., and Stein, S., 1990, Current plate motions: Geophysics Journal International, v. 101, p. 425–478.

De Polo, C.M., and Anderson, J.G., 2000, Estimating the slip rates of normal faults in the Great Basin, USA: Basin Research, v. 12, p. 227–240.

De Vita, P., and Reichenbach P., editors, 1998, Rainfall-triggered landslides, a reference list: Environmental Geology, v. 35, p. 219–233.

Densmore, A.L., and Anderson, R.S., 1997, Tectonic geomorphology of the Ash Hill Fault, Panamint Valley, California: Basin Research, v. 9, p. 53–63.

Densmore, A.L., Anderson, R.S., McAdoo, B.G., and Ellis, M.A., 1997, Hillslope evolution by bedrock landslides: Science, v. 275, p. 369–372.

Densmore, A.L., Allen, P.A., and Simpson, G., 2007, Development and response of a coupled catchment fan system under changing tectonic and climatic forcing: Journal of Geophysical Research, v. 112, F01002, doi:10.1029/2006JF000474.

Densmore, A.L., and Hovius, N., 2000, Topographic fingerprints of bedrock landslides: Geology, v. 28, p. 371–374, doi: 10.1130/0091-7613(2000)0282.3.CO;2.

Dibblee, T. W., Jr., 1973, Regional Geologic Map of San Andreas and related Faults in Carrizo Plain, Temblor, Caliente, and La Panza Ranges and Vicinity, California, U. S. Geological Survey, Miscellaneous Geologic Investigations Map I-757.

Dickinson, W.R., 1974, Tectonics and Sedimentation: Society of Economic Paleontologists and Mineralogists Special Publication 22, p. 1–27.

Dickinson, W.R., 1981, Plate tectonics and the continental margin of California, in, Ernst, W.G., editor, The Geotectonic Development of California (Rubey volume 1): Prentice-Hall, Englewood Cliffs, N.J., p. 1–28.

Dickinson, W.R., 1996, Kinematics of transrotational tectonism in the California Transverse Ranges and its contribution to cumulative slip along the San Andreas transform fault system: Geological Society of America Special Paper 305, 46 p.

Dickinson, W.R., 1997, Tectonic implications of Cenozoic volcanism in coastal California: Geological

Society of America Bulletin v. 109, p. 936–954.

Dickinson, W.R., 2000, Hydro-isostatic and tectonic influences on emergent Holocene paleoshorelines in the Mariana Islands, western Pacific Ocean: Journal of Coastal Research, v. 16, p. 735–746.

Dickinson, W.R., and Snyder, W.S., 1978, Plate tectonics of the Laramide orogeny: in, Matthews, V., III, editor, Laramide Folding Associated with Basement Block Faulting in the Western United States: Geological Society of America Memoir 151, p. 355–366.

Dickinson, W.R., and Wernicke, B.P., 1997, Reconciliation of San Andreas slip discrepancy by a combination of interior Basin and Range extension and transrotation near the coast: Geology, v. 25, p. 663–665.

Dickinson, W.R., Burley, D. V., and Shutler, R., 1994, Impact of hydro-isostatic Holocene sea-level change on the geologic context of island archeological sites, northern Ha'apai group, Kingdom of Tonga: Geoarchaeology, v. 9, p. 85–111.

Dietrich, W.E., Reneau, Steven.L, and Wilson, C.J, 1987, Overview; "zero-order basins" and problems of drainage density, sediment transport and hillslope morphology: in, Beschta, R.L, Blinn, T, Grant, G.E, Ice, G.G, Swanson, F.J., editors, Proceedings of an International Symposium on Erosion and Sedimentation in the Pacific Rim: International Association of Hydrological Sciences IAHS-AISH Publication 165, p. 27–37.

Dietrich, W.E., Wilson, C.J., Montgomery, D.R., McKean, J., and Bauer, R., 1992, Erosion thresholds and land surface morphology, Geology, v. 20, p. 675–679.

Dietrich, W.E., Wilson, C.J., Montgomery, D.R., and McKean, J., 1993, Analysis of erosion thresholds, channel networks, and landscape morphology using a digital terrain model: Journal of Geology, v. 101, p. 259–278.

Dietrich, W.E., Reiss, R., Hsu, M.L., and Montgomery, D., 1995, A process-based model for colluvial soil depth and shallow landslides using digital elevation data: Hydrological Processes, v. 9, p. 383–400.

Dilles, J.H., and Gans, P.B., 1995, The chronology of Cenozoic volcanism and deformation in the Yerington area, western Basin and Range and Walker Lane: Geological Society of America Bulletin, v. 107, p. 474–486.

Dinklage, W.S., 1991, The evolution of drainage density on Wheeler Ridge, an active fault-bend fold anticline, southern San Joaquin Valley, M.S. thesis: Santa Barbara, University of California, 18 p.

Dixon, T.H., Robaudo, S., Lee, J., and Reheis, M.C., 1995, Constraints on present-day Basin and Range deformation from space geodesy: Tectonics, v. 14, p. 755–772, doi: 10.1029/95TC00931.

Dixon, T.H., Miller, M., Farina, F., Wang, H., and Johnson, D., 2000, Present day motion of the Sierra Nevada block and some tectonic implications for the basin and Range province: North American Cordillera: Tectonics, v. 19, p. 1–24, doi: 10.1029/1998TC001088.

Dixon, T.H., Norabuena, E., and Hotaling, L., 2003, Paleoseismology and Global Positioning System: Earthquake-cycle effects and geodetic versus geologic fault slip rates in the Eastern California shear zone: Geology, v. 31, p. 55–58, doi: 10.1130/0091-7613(2003)031<0055:PAGPSE>2.0.CO;2.

Dokka, R.K., and Travis, C.J., 1990a, Late Cenozoic strike slip faulting in the Mojave Desert, California: Tectonics, v. 9, p. 311–340.

Dokka, R.K., and Travis, C.J., 1990b, Role of the eastern California shear zone in accommodating Pacific– North American plate motion: Geophysical Research Letters, v. 17, p. 1323–1326.

Du, Y., and Aydin, A., 1996, Is the San Andreas big bend responsible for the Landers earthquake and the eastern California shear zone?: Geology, v. 24, p. 219–222.

Ducea, M.N., and Saleeby, J.B., 1996, Buoyancy sources for a large, unrooted mountain range, the Sierra Nevada, California: evidence from xenolith thermobarometry: Journal of Geophysical Research, v. 101, p. 8229–8244.

Ducea, M.N., and Saleeby, J.B., 1998, A case for delamination of the deep batholithic crust beneath the Sierra Nevada: International Geology Review, v. 40, p. 78–93.

Dumitru, T.A., 2000, Fission-track geochronology: in, Noller, J.S., Sowers, J.M., Lettis, W.R. editors, Quaternary Geochronology; Methods and Applications, American Geophysical Union, Washington, D.C., p. 131–155.

Dunne, T., and Black, R.D., 1970, Partial area contributions to storm runoff in a small New England watershed: Water Resources Research, v. 6, p. 1296–1311.

Dunne, T., Moore, T.R., and Taylor, C.H., 1975, Recognition and prediction of runoff-producing zones in humid regions: Hydrological Sciences Bulletin, v. 20, p. 305–327.

Dutton, C.E., 1882, The Tertiary History of the Grand Canyon: U.S. Geological Survey Monograph, v. 2, U.S. Government Printing Office, Washington, D.C., 260 p.

Dykoski, C.A., Edwards, R.L., Cheng, H., Yuan, D.X., Cai, Y.J., Zhang, M.L., Lin, Y.S., Qing, J.M., An, Z.S., and Revenaugh, J., 2005, A high-resolution, absolute-dated Holocene and deglacial Asian monsoon record from Dongge Cave, China: Earth and Planetary Science Letters, v. 233, p. 71–86.

Dzurisin, D., 1975, Channel responses to artificial stream capture, Death Valley, California: Geology, v. 3, p. 309–312.

Eberhart-Phillips, D., and 28 others, 2003, The 2002 Denali Fault earthquake, Alaska, A large magnitude, slip-partitioned event: Science, v. 300, p. 1113–1118. doi: 10.1126/science.1082703.

Eden, D.E., and Hammond, A.P., 2003, Dust accumulation in the New Zealand region since the last glacial maximum: Quaternary Science Reviews, v. 22, p. 2037–2052.

Edwards, R.L., Chen, J.H., and Wasserburg, G.J., 1987a, ^{238}U/^{234}U^{230}Th/^{232}Th systematics and the precise measurement of time over the past 500,000 years: Earth and Planetary Science Letters, v. 81, 175–192.

Edwards, R.L., Chen, J.H., Ku, T.L., and Wasserburg, G.J., 1987b, Precise timing of the last interglacial period from mass spectrometric determination of thorium-230 in corals: Science, v. 236, p. 1547–1552.

Ehlig, P.L., Ehlert, K.W., 1978, Engineering geology of a Pleistocene landslide in Palos Verdes: in, D.L. Lamar, chairperson, Geologic Guide and Engineering Geology Case Histories, Los Angeles metropolitan area, Association of Engineering Geologists, California Section, p. 159–166.

Ehlig, P.L., and Ehlert, K.W, 1979, South Shores landslide, Palos Verdes Hills: in, W. Cotton, editor, A Guidebook for Visiting Selected Southern California Landslides, 18 p.

Ellen, S.D., and Wieczorek, G.F., editors, 1988, Landslides, floods, and marine effects of the storm of January 3–5, 1982, in the San Francisco Bay region: U.S. Geological Survey Professional Paper 1434, 310 p.

Emery, D., and Myers, K.J., 1996, Sequence stratigraphy: Blackwell Science, Oxford, 297 p.

Emmett, W.W., 1970, The hydraulics of overland flow on hillslopes: U.S. Geological Survey Professional Paper 662–A, 68 p.

England, P., and Thompson, A.B., 1984, Pressure–temperature time paths of regional metamorphism, I; Heat transfer during the evolution of regions of thickened continental crust: Journal of Petrology, v. 25, p. 894–928.

Eusden, J.D. Jr., Pettinga, J.R., and Campbell, J.K., 2000, Structural evolution and landscape development of a collapsed transpressive duplex on the Hope Fault, North Canterbury, New Zealand: New Zealand Journal of Geology and Geophysics, v. 43, p. 391–404.

Eusden, J.D. Jr., Pettinga, J.R., Campbell, J.K., 2005a, Structural collapse of a transpressive hanging-wall fault wedge, Charwell region of the Hope Fault, South Island, New Zealand: New Zealand Journal of Geology and Geophysics, v. 48, p. 295–309.

Eusden, J.D., Jr., Koons, P.O., Pettinga, J.R., and Upton, P., 2005b, Structural geology and geodynamic modeling of linking faults in the Marlborough fault zone, South Island, New Zealand: New Zealand Geological Society Annual Meeting abstracts, p. 26.

Farley, K.A., 2000, Helium diffusion from apatite: general behavior as illustrated by Durango fluorapatite: Journal of Geophysical Research, v. 105, p. 2903–2914.

Farmer, G.L., Glazner, A. F., Manley, C.R., 2002, Did lithospheric delamination trigger late Cenozoic potassic volcanism in the southern Sierra Nevada, California?: Geological Society of America Bulletin, v. 114, p. 754–768.

Fenton, C.R., Webb, R.H., Cerling, T.E., Poreda, R.J., and Nash, B.P., 2001, Cosmogenic ^{3}He ages and geochemical discrimination of lava-dam outburst-flood deposits in western Grand Canyon, Arizona: in, House, K., et al., editors, Ancient Floods and Modern Hazards, Principles and Applications of Paleoflood Hydrology: Washington, D.C., American Geophysical Union, Water Science and Application Series, v. 4.

Fernandes, N.F., and Dietrich, W.E., 1997, Hillslope evolution by diffusive processes: The timescale for equilibrium adjustments: Water Resources Research, v. 33, p. 1307–1318.

Ferretti, D.F., and ten other authors, 2005, Unexpected changes to the global methane budget over the past 2,000 years: Science, v. 309, p. 1714–1717. doi: 10.1126/science.1115193.

Fitzgerald, P.G., and Gleadow, A.J.W., 1988, Fission track geochronology, tectonics and structure of the Transantarctic Mountains in northern Victoria Land, Antarctica: Isotope Geoscience, v. 73, p. 169–198.

Fleming K., Johnston P., Zwartz D., Yokoyama Y., Lambeck K., and Chappell J., 1998, Refining the eustatic sea-level curve since the Last Glacial Maximum using far- and intermediate-field sites: Earth and Planetary Science Letters, v. 163, p. 327–342.

Fliedner, M., Klemperer, S.L., and Chrisyensen, N.I., 2000, Three-dimensional seismic model of the Sierra Nevada arc, California, and its implications for crustal and upper mantle composition: Journal of Geophysical Research, v. 105, p. 10,899–10,921.

Frankel, K.L., 2007, Fault slip rates, constancy of seismic strain release, and landscape evolution in the eastern California shear zone: Ph.D. dissertation, Los Angeles, University of Southern California, 179 p.

Frankel, K.L., Brantley, K.S., Dolan, J.F., Finkel, R.C., Klinger, R., Knott, J., Machette, M., Owen, L.A., Phillips, F.M., Slate, J.L., and Wernicke, B.P., 2007a, Cosmogenic ^{10}Be and ^{36}Cl geochronology of offset alluvial fans along the northern Death Valley fault zone: Implications for transient strain in the eastern California shear zone: Journal of Geophysical Research, v. 112, B06407, doi: 10.1029/2006JB004350.

Frankel, K.L., Dolan J.F., Finkel R.C., Owen L.A., and Hoeft, J.S., 2007b, Spatial variations in slip rate along the Death Valley–Fish Lake Valley fault system determined from LiDAR topographic data and cosmogenic ^{10}Be geochronology: Geophysical Research Letters, v. 34, doi:10.1029/2007GL030549.

Frankel, K.L., and 17 other authors, 2008, Active tectonics of the eastern California shear zone: Geological Society of America Field Guide 11, p. 43–81.

Frostick, L. and Steel, R., editors, 1994, Tectonic Controls and Signatures in Sedimentary Successions: International Association of Sedimentologists Special Publication 20, p. 259–276.

Furlong, K.P., and Govers, R., 1999, Ephemeral crustal thickening at a triple junction: The Mendocino crustal conveyor: Geology, v. 27, p. 127–130.

Furlong, K.P., and Schwartz, S.Y., 2004, Influence of the Mendocino triple junction on the tectonics of coastal California: Annual Reviews of Earth and Planetary Sciences, v. 32, p. 403–433.

Gabet, E.J., Pratt-Situala, B.A., and Burbank, D.W., 2004, Climatic controls on hillslope angle and relief in the Himalayas: Geology, v. 32, p. 629–632.

Galloway, D.L., Jones, D.R., and Ingebritsen, S.E., 1999, Land subsidence in the United States: U.S. Geological Survey Circular 1182, 175 p.

Gallup, C.D., Edwards, R.L., and Johnson, R.G., 1994, The timing of high sea levels over the past 200,000 years: Science, v. 263, p. 796–800.

Gansser, A., 1991, Facts and theories on the Himalayas: Eclogae Geologicae Helvetiae, v. 84, p. 33–59.

Garvin, C.D., Hanks, T.C., Finkel, R.C., and Heimsath, A.M., 2005, Episodic incision of the Colorado River in Glen Canyon, Utah: Earth Surface Processes and Landforms, v. 30 p. 973–984. doi:10.1002/esp.

Gasparini, N.M., Whipple, K.X., and Bras, R.L., 2007, Predictions of steady state and transient landscape morphology using sediment-flux-dependent river incision models: Journal of Geophysical Research, v. 112, F03S09, doi:10.1029/2006JF000567.

Gasse, F., Arnold, M., Fontes, J.C., Fort, M., Gibert, E., Huc, A., Li, B.Y., Li, Y.F., Lju, Q., Melieres, F., Vancampo, E., Wang, F.B., and Zhang, Q.S., 1991, A 13,000-year climate record from western Tibet: Nature, v. 353, p. 742–745.

Gilbert, G.K., 1877, Geology of the Henry Mountains (Utah): Geographical and Geological Survey of the Rocky Mountain Region, U.S. Government Printing Office, Washington, D.C, 170 p.

Gilbert, G.K., 1904, Domes and dome structures of the high Sierra: Geological Society of America Bulletin, v. 15, p. 29–36.

Gilbert, G.K., 1909, The convexity of hilltops: Journal of Geology, v. 17, p. 344–350.

Gillespie, A.R., 1991, Quaternary subsidence of Owens Valley, California, in, Hall, C.A., Jr., Doyle-Jones, V., and Widawski, B., editors, Natural History of Eastern California and High-Altitude Research: Los Angeles, University of California, White Mountain Research Station Symposium, v. 3, p. 365–382.

Gillon, M.D., 1992, Landslide stabilisation at the Clyde Power Project; a major geotechnical undertaking: New Zealand Engineering, v 47, p. 27–29.

Glazner, A., 2003, Volcanic evidence for Pliocene delamination of the Sierra Nevada batholithic root: Appendix 5, in, Stock, G., editor; 2003 Pacific Cell, Friends of the Pleistocene field trip, Sequoia and Kings Canyon, p. 79–83.

Gleadow, A.J.W., Kohn, B.P., Brown, R.W., O'Sullivan, P.B., and Raza, A., 2002, Fission track thermotectonic imaging of the Australian continent: Tectonophysics, v. 349, p. 5–21.

Goldrick, G., and Bishop, P., 1995, Differentiating the roles of lithology and uplift in the steepening of bedrock river long profiles; an example from southeastern Australia: Journal of Geology, v. 103, p. 227–231.

Gosse, J.C., and Phillips, F. M., 2001, Terrestrial *in situ* cosmogenic nuclides: theory and application: Quaternary Science Reviews, v. 20, p. 1475–1560.

Govi, M., and Sorzana, P. F., 1980, Landslide susceptibility as a function of critical rainfall amount in Piedmont Basin (NW Italy): Studia Geomorphologica Carpatho-Balcanica, v. 14, p. 43–61.

Graham, S.A., Stanley, R.G., Bent, J.V., and Carter, J.R., 1989, Oligocene and Miocene paleogeography of central California and displacement along the San Andreas fault: Geological Society of America Bulletin, v. 101, p. 711–730.

Granger, D.E., and Muzikar, P.F., 2001, Dating sediment burial with *in situ*-produced cosmogenic nuclides: Theory, techniques and limitations: Earth and Planetary Science Letters, v. 188, p. 269–281, doi: 10.1016/S0012- 821X(01)00309-0.

Grant, L.B., Mueller, K.J., Gath, E.M., Cheng, H., Edwards, R.L., Munro, R., and Kennedy, G.L., 1999, Late Quaternary uplift and earthquake potential of the San Joaquin Hills, southern Los Angeles basin, California: Geology, v. 27, p. 1031–1034.

Griffiths, G.A., and McSaveney, M.J., 1983, Hydrology of a basin with extreme rainfalls, Cropp River, New Zealand: New Zealand Journal of Science, v. 26, p. 293–306.

Gruber, S., and Haeberli, W., 2007, Permafrost in steep bedrock slopes and its temperature-related destabilization following climate change: Journal of Geophysical Research, v.112, F02S18, doi:10.1029/2006JF000547.

Gunnell, Y., 2000, Apatite fission thermochronology: an overview of its potential and limitations in geomorphology: Basin Research, v. 12, p. 115–132.

Gutenberg, B., 1955, Epicenter and origin time of the main shock on July 21 and travel times of major phases, in earthquakes in Kern County, California during 1952: California Division of Mines Bulletin 181, p. 165–175.

Guth, P.L., 1997, Tectonic geomorphology of the White Mountains, eastern California: Geological Society of America Abstracts with Programs, v. 29, p. 235.

Guzzetti, F., Peruccacci, S., Rossi, M., and Stark, C.P., 2007, Rainfall thresholds for the initiation of landslides in central and southern Europe: Meteorology and Atmospheric Physics, v. 98, p. 239–267.

Hack, J.T., 1957, Studies of longitudinal stream profiles in Virginia and Maryland: U.S. Geological Survey Professional Paper 294-B, p. 45–97.

Hack, J.T., 1960, Interpretation of erosional topography in humid temperate regions: American Journal of Science (Bradley Volume), v. 258-A, p. 80–97.

Hack, J.T., 1965, Geomorphology of the Shenandoah Valley, Virginia and West Virginia, an origin of the residual ore deposits: U.S. Geological Survey Professional Paper 484, 84 p.

Hack, J.T., 1973a, Stream-profile analysis and stream-gradient index: U.S. Geological Survey Journal of Research, v. 1, p. 421–429.

Hack, J.T., 1973b, Drainage adjustment in the Appalachians: in, Morisawa, M., editor, Fluvial Geomorphology, Proceedings, 4th Annual Geomorphology Symposia, Binghamton, New York, p. 51–74.

Hack, J.T., 1982, Physiographic divisions and differential uplift in the piedmont and Blue Ridge: U.S. Geological Survey Professional Paper 1265, 49 p.

Hack, J.T., and Goodlett, J.C., 1960, Geomorphology and forest ecology of a mountain region in the central Appalachians: U.S. Geological Survey Professional Paper 347, 66 p.

Hack, J.T., and Young, R.S., 1959, Intrenched meanders of the North Fork of the Shenandoah River, 4 Virginia: U. S. Geological Survey Professional Paper 354-A, 10 p.

Hall, C.E., 1965, Las Aguilas land surface: in International Association for Quaternary Research, VII Congress, Guidebook for Field Conference I, northern Great Basin and California, p. 145.

Hallet, B., and Molnar, P., 2001, Distorted drainage basins as markers of crustal strain east of the Himalaya: Journal of Geophysical Research, v. 106, p. 13697–13710.

Hamblin, W.K., 1994, Late Cenozoic lava dams in the western Grand Canyon: Geological Society of America Memoir 183, 135 p.

Hamelin B., Bard, E., Zindler, A., and Fairbanks, R.G., 1991, $^{234}U/^{238}U$ mass spectrometry of corals : How accurate is the U-Th age of the last interglacial period?: Earth and Planetary Science Letters, v. 106, p.169–180.

Hancox, G.T., McSaveney, M.J., Manville, V.R., and Davies, T.R., 2005, The October 1999 Mt. Adams rock avalanche and subsequent landslide dambreak flood and effects in Poerua River, Westland, New Zealand: New Zealand Journal of Geology and Geophysics, v. 48,p. 683–705.

Haneberg, W.C., 1991, Observation and analysis of pore pressure fluctuations in a thin colluvium landslide complex near Cincinnati, Ohio: Engineering Geology, v. 31, p. 159–184.

Hanks, T.C., and Finkel, R.C., 2006, Early Pleistocene incision of the San Juan River, Utah, dated with [26]Al and 10Be: Comment and Reply: Geology Forum, p. 77, 78.

Hanks, T.C., and Webb, R.H., 2006, Effects of tributary debris on the longitudinal profile of the Colorado River in Grand Canyon: Journal of Geophysical Research, v. 111, doi:10.1029/2004JF000257.

Hanks, T.C., Lucchitta, I., Davis, S.W., Davis, M.E., Finkel, R.C., Lefton, S.A., and Garvin, C.D., 2003, The Colorado River and the Age of Glen Canyon: in, Young, R.A. and Spamer, E.E., Editors, The Colorado River, Origin and Evolution, Grand Canyon Association Monograph, p. 129–133.

Harbert, W., 1991, Late Neogene relative motions of the Pacific and North America plates: Tectonics, v. 10, p. 1–15.

Harden, J.W., 1987, Soil development in granitic alluvium near Merced, California, U.S. Geological Survey Bulletin 1590-A, 65p.

Harding, T.P., 1974, Petroleum traps associated with wrench faults: American Association of Petroleum Geologists Bulletin, v. 58, p. 1290–1304.

Hasbargen, L.E., and Paola, C., 2000, Landscape instability in an experimental drainage basin: Geology, v. 28, p. 1067-1070. doi: 10.1130/00917613(2 000)028<1067:LIAED:>2.3.CO;2.

Haefeli, R., 1948, The stability of slopes acted upon by parallel seepage: Proceedings of the 2nd International Conference on Soil Mechanics, Rotterdam, v. 1, p. 57–62.

Heimsath, A.M., Dietrich, W.E., Nishiizumi, K. and Finkel, R.C., 1997. The soil production function and landscape equilibrium: Nature, v. 388, p. 358–361.

Heimsath, A.M., Chappell, J., Dietrich, W.E., Nishiizumi, K. and Finkel, R.C., 2001, Soil production on a retreating escarpment in southeastern Australia: Geology, v. 28, p. 787–790.

Heimsath, A.M., Dietrich, W.E., Nishiizumi, K., and Finkel, R.C., 2001, Stochastic processes of soil production and transport: Erosion rates, topographic variation and cosmogenic nuclides in the Oregon Coast Range: Earth Surface Processes and Landforms, v. 26, no. 5, p. 531–552, doi: 10.1002/esp.209.

Heirtzler, J.R., Dickson, G.O., Herron, E.M., Pitman, W.C., and Le Pichon, X., 1968, Marine magnetic anomalies, geomagnetic field reversals, and motions of the ocean floor and continents: Journal of Geophysical Research, v. 73, p. 2,119–2,139.

Hendron, A.J., and Patton, F.D., 1985, The Vaiont slide, a geotechnical analysis based on new geologic observations of the failure surface: US Army Corps of Engineers Technical Report GL-85-5.

Hendron, A.J., and Patton, F.D., 1987, The Vaiont slide – a geotechnical analysis based on new geologic observations of the failure surface: Leonards, G.A. editor, in Dam Failures, Engineering Geology, v. 24, p. 475–491.

Henry, C.D., and Perkins, M.E., 2001, Sierra Nevada–Basin and Range transition near Reno, Nevada, Two stage development at 12 and 3 Ma: Geology, v. 29, p. 719–722.

Hill, M.L., 1984, Earthquakes and folding, Coalinga, California: Geology, v. 12, p. 711–712.

Hillel, D. 1998, Environmental Soil Physics: Academic Press, San Diego, 771 p.

Hirano, M., 1968, A mathematical model of slope development – an approach to the analytical theory of erosional topography: Journal of Geosciences Osaka City University, v. 11, p. 13–52.

Hodges, K.V., McKenna, L.W., and Harding, M.B., 1990, Structural unroofing of the central Panamint Mountains, Death Valley region, southeastern California: Wernicke, B. P., editor, in Basin and Range Extensional Tectonics Near the Latitude of Las Vegas, Nevada: Geological Society of America Memoir 176, p. 377–390.

Hodges, K.V., Hurtado, J.M., and Whipple, K.X., 2001, Southward extrusion of Tibetan crust and its effect on Himalayan tectonics: Tectonics, v. 20, 799–809.

Hollett, K.J., Danskin, W.R., McCaffrey, W.F., and Walti, C.L., 1991, Geology and water resources of Owens Valley, California: U.S. Geological Survey Water-Supply Paper 2370B, 77 p.

Holeman, J. N. 1968. The sediment yield of major rivers of the world: Water Resources Research, v. 4, p. 737–747.

Holm, D.K., Geissman, J.W., and Wernicke, B., 1993, Tilt and rotation of the footwall of a major normal fault system, Paleomagnetism of the Black Mountains, Death Valley extended terrane, California: Geological Society of America Bulletin, v. 105, p. 1373–1387.

Holzer, T.L., 1998, The history of the aquitard-drainage model: in, J.W. Borchers, editor, Land Subsidence Case Studies and Current Research, Star Publishing Company, Belmont, California, p. 7–13.

Holzhausen, G.R., 1989, Origin of sheet fracture, 1, Morphology and boundary conditions: Engineering Geology, v. 27, p. 225–278.

Hooke, R.LeB., 1967, Processes on arid-region alluvial fans: Journal of Geology, v. 75, p. 438–460.

Hooke, R.LeB., 1968, Steady-state relationships on arid-region alluvial fans in closed basins: American Journal of Science, v. 266, p. 609–629.

Hooke, R.LeB., 1972, Geomorphic evidence for late-Wisconsin and Holocene tectonic deformation, Death Valley, California: Geological Society of America Bulletin, v. 83, p. 2073–2098.

Hooke, R.LeB., 2003, Time constant for equilibration of erosion with tectonic uplift: Geology, v. 31, p. 621–624, doi: 10.1130/0091-7613(2003)0312.0.CO;2.

Horton, R.E., 1932, Drainage basin characteristics: Transactions of the American Geophysical Union, v. 13, p. 350–361.

Horton, R.E., 1945, Erosional development of streams and their drainage basins: hydrophysical approach to quantitative morphology: Geological Society of America Bulletin, v. 56, p. 275–370.

House, M.A., Wernicke, B.P., and Farley, K.A., 1998, Dating topography of the Sierra Nevada, California, using apatite (U-Th)/He ages: Nature, v. 396, p. 66–69.

House, M.A., Wernicke, B.P., and Farley, K.A., 2001, Paleo-geomorphology of the Sierra Nevada, California, from (U–Th)/He ages in apatite: American Journal of Science: v. 301, p. 77–102.

Hovius, N., and Stark, C.P. , 2006, Landslide-driven erosion and topographic evolution of active mountain belts: in, Evans, S.G. et al., editors, Landslides From Massive Rock Slope Failure, Springer, p. 573–590.

Hovius, N, Stark, C.P., and Allen, P.A., 1997, Sediment flux from a mountain belt derived from landslide mapping: Geology, v. 25, p. 231–234.

Hovius, N., Stark, C.P., Tutton, M.A., and Abbot, L.D., 1998, Landslide-driven drainage network evolution in a pre-steady-state mountain belt: Finisterre Mountains, Papua New Guinea: Geology, v. 26, p. 1071–1074.

Howard, A.D., 1994, A detachment limited model of drainage basin evolution: Water Resources Research, v. 30, p. 2261–2285.

Howard, A.D., 1997, Badland morphology and evolution, Interpretation using a simulation model: Earth Surface Processes and Landforms, v. 22, p. 211–227.

Huber, N.K., 1981, Amount and timing of late Cenozoic uplift and tilt of the central Sierra Nevada, California-evidence from the upper San Joaquin river basin: U.S. Geological Survey Professional Paper 1197, 28 p.

Huber, N.K., 1990, The late Cenozoic evolution of the Tuolumne River, central Sierra Nevada, California: Geological Society of America Bulletin, v. 102, p. 102–115.

Huftile, G.J., and Yeats, R.S., 1995, Convergence rates across a displacement transfer zone in the western Transverse Ranges, Ventura basin, California: Journal of Geophysical Research, v. 100, p. 2043–2067.

Hughes, M.W., 2008, Late Quaternary landscape evolution and environmental change in Charwell Basin, South Island, New Zealand: Lincoln University (New Zealand) Ph.D. dissertation.

Hughes, M.W., Schmidt, J. and Almond, P.C., 2009, Automatic landform stratification and environmental correlation for modelling loess landscapes in North Otago, South Island, New Zealand: Geoderma (in Press).

Humphreys, E.D., and Weldon, R.J., 1994, Deformation across the western United States: A local estimate of Pacific–North America transform deformation: Journal of Geophysical Research, v. 99, p. 19,975–20,010, doi: 10.1029/94JB00899.

Hunt, C.B., and Mabey, D.R., 1966, Stratigraphy and structure, Death Valley, California: U.S. Geological Survey Professional Paper 494-A, 162 p.

Imbrie, J., Hays, J.D., Martinson, D.G., McIntyre, A., Mix, A.C., Morley, J.J., Pisias, N.G., Prell, W.L., and Shackleton, N.J., 1984, The orbital theory of Pleistocene climate: support from a revised chronology of the marine $\delta^{18}O$: in, Berger, A., Imbrie, J., Hays, J., Kukla, G., and Saltzman, B., editors, Milankovitch and Climate, Part 1, Dordrecht, Reidel Publishing Co., p. 269–305.

Israelson, C., and Wohlfarth, B., 1999. Timing of the Last Interglacial high sea level on the Seychelles Islands, Indian Ocean: Quaternary Research, v. 51, p. 306–316.

Iverson, R.M., 1997, The physics of debris flows: Reviews of Geophysics, v. 35, p. 245–296.

Iverson, R.M., 2000, Landslide triggering by rain infiltration, Water Resources Research, v. 36, p. 1897–1910.

Iverson, R.M., and Major, J.J., 1987, Rainfall, groundwater flow, and seasonal motion at Minor Creek landslide, northwestern California; Physical interpretation of empirical relations: Geological Society of America Bulletin, v. 99, 579–594.

Jackson, G.W., 1990, Tectonic geomorphology of the Toroweap fault, western Grand Canyon, Arizona: Implications for transgression of faulting on the Colorado Plateau: Arizona Geological Survey Open-File Report, v. 90-4, p. 1–66.

Jackson, J.A., and Leeder, M.R., 1994, Drainage systems and the development of normal faults, an example from Pleasant Valley, Nevada: Journal of Structural Geology, v. 16, p. 1041–1059.

Jackson, J., and McKenzie, D., 1988, The relationship between plate motions and seismic moment tensors, and the rates of active deformation in the Mediterranean and the Middle East: Geophysical Journal, v. 93, p. 45–73.

Jackson, J.A., Gagnepain, J., Houseman, G., King, G.C.P., Papadimitriou, P., Soueris, C., Virieux, J., 1982, Seismicity, normal faulting, and the geomorphological development of the Gulf of Corinth, Greece, The Corinth earthquakes of February and March 1981: Earth and Planetary Science Letters, v. 57, p. 377–397.

Jackson, J., Norris R., and Youngson J., 1996, The structural evolution of active fault and fold systems in central Otago, New Zealand: Evidence revealed by drainage patterns: Journal of Structural Geology, v. 18, p. 217–234.

Jackson, J.R, Ritz, J.F, Siame, L., Raisbeck, G., Yiou, F., Norris, R.J., Youngson, J.H., and Bennett, E. 2002. Fault growth and landscape development rates in Otago, New Zealand, using in situ cosmogenic ^{10}Be: Earth and Planetary Science Letters, v. 195, p.185–193.

Jade, S., Bhatt, B.C., Yang, Z., Bendick, R., Gaur, V.K., Molnar, P., Anand, M.B., Kumar, Dilip, 2004, GPS measurements from the Ladakh Himalaya, India; Preliminary tests of plate-like or continuous deformation in Tibet: Geological Society of America Bulletin, v. 116, p. 1385–1391.

James, L.A., Harbor, J., Fabel, D., Dahms, D., and Elmore, D., 2002, Late Pleistocene glaciations in the northwestern Sierra Nevada, California: Quaternary Research, v. 57, p. 409–419.

Janda, R.J., 1965, Quaternary alluvium near Friant, California, in Northern Great Basin and California, 7th International Association of Quaternary Research Congress, USA, Guidebook for Field Conference 1, p. 128–133.

Jennings, C.W., 1994, Fault Activity Map of California and Adjacent Areas: California Department of Conservation, Division of Mines and Geology, Geologic Data Map No. 6, Scale 1:750,000.

Jenny, H., Arkley, R.J., and Schultz, A.M., 1969, The pygmy forest-podzol ecosystem and its dune associates of the Mendocino coast: Madrona, v. 20, p. 60–74.

Johnson, A.M., 1970, Physical Processes in Geology: Freeman, Cooper and Company, San Francisco, California, 577 p.

Johnson, D.L., 1977, The Late Quaternary climate of coastal California, evidence for an ice age refugium: Quaternary Research, v. 8, p. 154–179.

Johnson, R.G., 1982, Brunhes–Matuyama magnetic reversal dated at 790,000 years B.P. by marine-astronomical correlations: Quaternary Research, v. 17, p. 135–147.

Jones, C.H., Kanamori, H., and Roecker, S.W., 1994, Missing roots and mantle "drips": Regional Pn and teleseismic arrival times in the southern Sierra Nevada and vicinity, California: Journal of Geophysical Research, v. 99, p. 4567–4601.

Jones, C.H., Unruh, J., and Sonder, L.J., 1996, The role of gravitational potential energy in active deformation in the southwestern United States: Nature, v. 381, p. 37–41.

Jones, C.H., Farmer, G.L., and Unruh, J. R., 2004, Tectonics of Pliocene removal of lithosphere of the Sierra Nevada, California: Geological Society of America Bulletin, v. 116, p. 1408–1422.

Juracek, K. E., 1999, Estimation of potential runoff-contributing areas in the Kansas–Lower Republican River Basin, Kansas: U.S. Geological Survey Water-Resources Investigations Report 99-4089, 24 p.

Judson, S., and Kauffman, M.E., 1990, Physical Geology: 8th edition, Prentice Hall, Englewood Cliffs, NJ, 534 p.

Kamai, T., Wang, W.N., and Shuzui, H., 2000, The landslide disaster induced by the Taiwan Chi-Chi earthquake of 21 September 1999: Landslide News, v. 13, p. 8–12.

Kamp, P.J.J., and Tippett, J.M., 1993, Dynamics of Pacific plate crust in the South Island (New Zealand) zone of oblique continent-continent convergence: Journal of Geophysical Research, v. 98, p. 16105–16118.

Kamp, P.J.J., Green, P. F., and White, S.H., 1989, Fission track analysis reveals character of collisional tectonics in New Zealand: Tectonics, v. 8., p. 169–195.

Keefer, D.K., editor, 1998, The Loma Prieta, California, earthquake of October 17, 1989 – Landslides: U.S. Geological Survey Professional Paper 1551-C, 185 p.

Keefer, D.K., 1999, Earthquake-induced landslides and their effects on alluvial fans: Journal of Sedimentary Research, v. 69, 84–104.

Keefer, D.K., 2000, Statistical analysis of an earthquake-induced landslide distribution – the 1989 Loma Prieta, California Event: Engineering Geology, v. 58, p. 213–249.

Keefer, D.K., 2002, Investigating landslides caused by earthquakes—A historical review: Surveys in Geophysics, v. 23, p. 473–510.

Keefer, D.K., and Tannaci, N.E., 1981, Bibliography on landslides, soil liquefaction, and related ground failures in selected historic earthquakes: U.S. Geological Survey Open-File Report 81-572, 38 p.

Keefer, D.K., Wilson, R.C., Mark, R.K., Brabb, E.E., Brown, W.M., Ellen, S.D., Harp, E.L., Wieczorek, G.F., Alger, C.S., and Zatkin, R.S., 1987, Real-time landslide warning during heavy rainfall: Science, v 238, p. 921–925.

Keller, E.A., Zapeda, R.L., Rockwell, T.K., Ku, and T.L., Dinklage, W.S., 1998, Active tectonics at Wheeler Ridge southern San Joaquin Valley, California: Geological Society of America Bulletin, v. 110, p. 298–310.

Keller, E.A., Gurrola, L., and Tierney, T.E., 1999, Geomorphic criteria to determine direction of lateral propagation of reverse faulting and folding: Geology, v. 27, p. 515–518.

Keller, E.A., Seaver, D.B., Laduzinsky, D.L., Johnson, D.L., and Ku, T.L., 2000, Tectonic geomorphology of active folding over buried reverse faults: San Emigdio Mountain front, southern San Joaquin Valley, California: Geological Society of America Bulletin v. 112, p. 86–97.

Keller, E.A. and Pinter., N., 1996, Active Tectonics; Earthquakes, Uplift and Landscape: Prentice-Hall, 338 p.

Kelsey, H.M., 1980, A sediment budget and an analysis of geomorphic process in the Van Duzen River basin, north coastal California, 1941–1975: Geological Society of America Bulletin, v. 91, p. 190–195.

Kelsey, H.M, 1985, Hillslope evolution in zero-order headwater catchments sculpted by periodic debris avalanching, Northern California: Eos, Transactions, American Geophysical Union, v. 66, p. 46.

Kelsey, H.M., 1987, Controls on the relation of streamside landsliding to channel sediment storage in a region of active uplift: in, Beschta, R.L., Blinn, T., Grant, G.E. Ice, G.G., and Swanson, F.J., editors, Proceedings of International Symposium on Erosion and Sedimentation in the Pacific Rim, F.J.International Association of Hydrological Sciences, Publication 165, p. 505–506.

Kelsey, H.M., 1988, Formation of inner gorges: Catena, v. 15, no.5, p. 433–458.

Kelsey, H.M., and Bockheim, J.G., 1994, Coastal landscape evolution as a function of eustasy and surface uplift rate, Cascadia margin, southern Oregon, Geological Society of America Bulletin, v. 106, 840–854.

Kelsey, H.M., Engebretson, D.C., Mitchell, C.E., and Ticknor, R. L., 1994, Topographic form of the Coast Ranges of the Cascadia margin in relation to coastal uplift rates and plate subduction: Journal of Geophysical Research, v. 99, 12,245–12,255.

Kelsey, H.M., Coghlan, M., Pitlick, J., and Best, D., 1995, Geomorphic analysis of streamside landslides in the Redwood Creek basin, Northwestern California, in, Nolan, K.M., Kelsey, H.M., and Marron, D.C., editors, Geomorphic Processes and Aquatic Habitat in the Redwood Creek Basin, Northwest California: U.S. Geological Survey Professional Paper 1454-J, p. J1–J12.

Kelsey, H.M., Ticknor, R.L., Bockheim, J.G., and Mitchell, C.E., 1996, Quaternary upper plate deformation in coastal Oregon: Geological Society of America Bulletin, v. 108, p. 843–860, doi: 10.1130/0016-7606(1996)1082.3.CO;2.

Kennedy, G.L., Lajoie, K.R., and Wehmiller, J.F., 1982, Aminostratigraphy and faunal correlations of Late Quaternary marine terraces, Pacific Coast, U.S.A.: Nature, v. 299, p. 545–547.

Kiersch, G.A., 1964, Vaiont Reservoir disaster: Civil Engineering, v. 34 , p. 32–39.

Kiersch, G.A., 1988, Lessons from notable events; Vaiont reservoir disaster: in, Jansen, R.B., editor, Advanced Dam Engineering for Design, Construction, and Rehabilitation, Van Nostrand Reinhold, New York, p. 41–53.

Kilburn C.R.J., and Petley, D.N., 2003, Forecasting giant, catastrophic slope collapse, lessons from Vajont, Northern Italy: Geomorphology, v. 54, p. 21–32.

King, G., and Stein, R., 1983, Surface folding, river terrace deformation rate and earthquake repeat time in a reverse faulting environment: the Coalinga, California, earthquake of May 1983, in, Bennett, J.H. and Sherburne R.W., editors, The 1983 Coalinga, California, Earthquake Bennett, Sacramento, California, California Division of Mines and Geology, Special Publication 66, p. 261–274.

Kirkby, M.J., 1971, Hillslope process–response models based on the continuity equation: Transactions of the Institute of British Geographers Special Publication No. 3, p. 15–30.

Kirkby, M.J., 1985, A model for the evolution of regolith-mantled slopes, in, Woldenberg M.J. editor, Models in Geomorphology, Allen and Unwin, Winchester, Mass, p. 213–237.

Klinger, R.E., 1999, Tectonic geomorphology along the Death Valley fault system—Evidence for recurrent Late Quaternary activity in Death Valley National Park, in, Slate, J.L., editor., Proceedings of Conference on Status of Geologic Research and Mapping, Death Valley National Park: U.S. Geological Survey Open-File Report 99-153, p. 132–140.

Klinger, R.E., and Piety, L.A., 1996, Final Report—Evaluation and Characterization of Quaternary faulting on the Death Valley and Furnace Creek faults, Death Valley, California: Yucca Mountain Project, Activity 8.3.1.17.4.3.2: Denver, Colorado, Bureau of Reclamation Seismotectonic Report 96-10, prepared for the U.S. Geological Survey in cooperation with the U.S. Department of Energy, 98 p.

Knopf, A, 1918, A Geologic Reconnaissance of the Inyo Range and The Eastern Slope of the Southern Sierra Nevada, California: U.S. Geological Survey Professional Paper 110, 130 p.

Knuepfer, P.L.K., 1992, Temporal variations in latest quaternary slip across the Australian–Pacific Plate boundary, northeastern South Island, New Zealand: Tectonics, v. 11, p. 449–464.

Knuepfer, P.L.K, 2004, Extracting tectonic and climatic signals from river and terrace long profiles in active orogens, Taiwan and New Zealand: Geological Society of America Abstracts with Programs, v. 36, No. 5, 307.

Koons, P. O., 1989, The topographic evolution of collisional mountain belts; A numerical look at the Southern Alps, New Zealand: American Journal of Science, v. 289, p. 1041–1069.

Koons, P.O., 1990, The two sided wedge in orogeny; erosion and collision from the sand box to the Southern Alps, New Zealand: Geology, v. 18, p. 679–682.

Koons, P.O., 1995, Modeling the topographic evolution of collisional mountain belts: Annual Reviews of Earth and Planetary Sciences, v. 23, p. 375–408.

Koons, P.O., 1998, Big mountains, big rivers, and hot rocks; beyond isostasy: Eos (Transactions, American Geophysical Union), v. 79, no. 45, supplement, p. F908.

Koons, P.O., Norris, R.J., Craw, D., and Cooper, A.F., 2003, Influence of exhumation on the structural evolution of transpressional plate boundaries; An example from the Southern Alps, New Zealand: Geology: v. 31, p. 3–6.

Korup, O., 2004, Geomorphic implications of fault zone weakening: slope instability along the Alpine Fault, South Westland to Fiordland: New Zealand Journal of Geology and Geophysics, v. 47, p. 257–267.

Korup, O., 2005a, Large landslides and their effect on sediment flux in South Westland, New Zealand: Earth Surface Processes and Landforms, v. 30, p. 305–323.

Korup, O., 2005b, Distribution of landslides in southwestern New Zealand: Landslides, v. 2, p. 43–51.

Korup, O., 2006, Effects of large deep-seated landslides on hillslope morphology, western Southern Alps, New Zealand: Journal of Geophysical Research, v. 111, F01018, doi:10.1029/2004JF000242.

Korup, O., 2008, Rock type leaves topographic signature in landslide-dominated mountain ranges: Geophysical Research Letters, v. 35, L11402, doi:10.1029/ 2008GL034157.

Korup, O., and Crozier, M., 2002, Landslide types and geomorphic impact on river channels, Southern Alps, New Zealand: in, Rybar, J., Stemberk, J., and Wagner, P., editors, Proceedings of the First European Conference on landslides, Prague, p. 233–238.

Korup, O., McSaveney, M.J., and Davies, T.R.H., 2004, Sediment generation and delivery from large historic landslides in the Southern Alps, New Zealand: Geomorphology v. 61, p. 189–207.

Kramer, S., and Marder, M., 1992, Evolution of river networks: Physical Review Letters: v. 68, p. 205–208.

Ku, T.L, Kimmel, M.A., Easton, W.H., and O'Neil, T.J., 1974, Eustatic sea level 120,000 years ago on Oahu, Hawaii: Science, v. 183, p. 959–962.

Lachenbruch A.H., and Sass J.H., 1980. Heat flow and energetics of the San Andreas fault zone: Journal of Geophysical Research, v. 85, p. 6185–222.

Laduzinsky, D. M., 1989, Late Pleistocene–Holocene chronology and tectonics, San Emigdio Mountains: M.S. thesis, Santa Barbara, University of California, 95 p.

LaForge, R., and Lee, W.H.K., 1982, Seismicity and tectonics of the Ortigalita fault and southeast Diablo Range, California: California Division of Mines and Geology Special Publication, v. 62, p. 93–101.

Lajoie, K.R., 1986, Coastal tectonics: Chapter 6 in Active tectonics, National Academy Press, Washington, p. 95–124.

Lajoie, K.R., Sarna-Wojcicki, A.M., and Ota, Y., 1982, Emergent marine terraces at Ventura and Cape Mendocino, California—Indicators of high uplift rates: Geological Society of America Abstracts with Programs, v. 14, p. 178.

Lambeck, K., 1993, Glacial rebound and sea-level change; an example of a relationship between mantle and surface processes: Tectonophysics, v. 223, p. 15–37.

Lambeck, K., and Chappell, J., 2001, Sea level change through the last glacial cycle: Science, v. 292, p. 679– 686.

Lan, H., Zhou, C., Lee, C.F., Wang, S., and Wu, F., 2003, Rainfall-induced landslide stability analysis in response to transient pore pressure, A case study of natural terrain landslide in Hong Kong: Science in China Series E, Technological Sciences, v. 46, p. 52–68.

Landis, C.A., and Youngson, J.H., 1996, Waipounamu erosion surface, the Otago Peneplain: Geological Society of New Zealand Annual Conference 1996, field trip FT2; Geological Society of New Zealand Miscellaneous Publication 91B: FT2-1 to FT2-9.

Landis, C.A., Campbell, C.A., Begg, J.G., Mildenhall, A.M., Patterson, A.M., and Trewick, A.M., 2008, The Waipounamu Erosion Surface, questioning the antiquity of the New Zealand land surface and terrestrial fauna and flora: Geological Magazine; v. 145; no. 2; p. 173-197; doi: 10.1017/S0016756807004268.

Langbein, W.B., and Schumm, S.A., 1958, Yield of sediment in relation to mean annual precipitation: American Geophysical Union Transactions, v. 39, p. 1076–1084.

Lavé, J., Avouac, J.P., Laccassin, R., Tapponnier, P., Montagner, J.P., 1996, Seismic anisotropy beneath Tibet: evidence for eastward extrusion of the Tibetan lithosphere?: Earth and Planetary Science Letters, v. 140, p. 83–96.

Le, K., Lee, J., Owen, L.A., and Finkel, R., 2007, Late Quaternary slip rates along the Sierra Nevada frontal fault zone, California: Slip partitioning across the western margin of the Eastern California Shear Zone–Basin and Range Province: Geological Society of America Bulletin, v. 119, p. 240–256, doi: 10.1130/B25960.1.

Leeder, M.R., Seger, M.J., and Stark, C.P., 1991, Sedimentology and tectonic geomorphology adjacent to active and inactive normal faults in the Megara basin and Alkyonides Gulf, central Greece: Journal of the Geological Society, London, v. 148, p. 331–343.

Leheny, R.L., and Nagel, S.R., 1993, Model for the evolution of river networks: Physical Review Letters, v. 71, p.1470–1473.

Le Pichon, X., and Angelier, J., 1981. The Aegean Sea: Philosophical Transactions of the Royal Society London, v. 300, p. 357–372.

Leopold, L.B., 1969, The rapids and pools – Grand Canyon, in The Colorado River Region and John Wesley Powell: U.S. Geological Survey Professional Paper 669, p. 131–145.

Leopold, L.B., and Bull, W.B., 1979, Base level, aggradation, and grade: Proceedings of American Philosophical Society, v. 123, p. 168–202.

Leopold, L.B., and Maddock, T., Jr., 1953, The hydraulic geometry of stream channels and some physiographic implications: U.S. Geological Survey Professional Paper 252, 56 p.

Leopold L.B., and Wolman, M.G., 1957, River channel patterns, braided, meandering, and straight: U.S. Geological Survey Professional Paper 282B, 51 p.

Lettis, W.R., 1982, Late Cenozoic stratigraphy and structure of the western margin of the central San Joaquin Valley, California: U.S. Geological Survey Open-File Report 82-526, 203 p.

Lettis, W.R., 1985, Late Cenozoic stratigraphy and structure of the west margin of the central San Joaquin Valley, California: in, Weide D.L., editor, Soils and Quaternary geology of the southwestern United States, Geological Society of America Special Paper 203, p. 97–114.

Lettis, W.R, 1988, Quaternary geology of the Northern San Joaquin Valley: in, Graham, S.A., editor, Studies of the Geology of the San Joaquin Basin, Society of Economic Paleontologists and Mineralogists Bulletin, v. 60, p. 333–351.

Lettis, W.R., and Hanson, K. L., 1991, Crustal strain partitioning; implications for seismic-hazard assessment in western California: Geology, v. 19, p. 559–562.

Lettis, W.R., and Unruh, J.R., 1991, Quaternary geology of the Great Valley, California: in, Morrison, R.B., editor, Quaternary Non-glacial Geology of the Western United States; Decade of North American Geology, v. K-2: Geological Society of America, p. 164–176.

Lin, J., and Stein, R.S., 2006, Seismic constraints and Coulomb stress changes of a blind thrust fault system, 1: Coalinga and Kettleman Hills, California: U.S. Geological Survey Open-File Report 2006-1149, 17 p. Available at URL http://pubs.usgs.gov/of/2006/1149/

Lindgren, W., 1911, The Tertiary gravels of the Sierra Nevada of California: U.S. Geological Survey Professional Paper 73, 226 p.

Lindqvist, J.K. 1990, Deposition and diagenesis of Landslip Hill silcrete, Gore Lignite Measures (Miocene), New Zealand: New Zealand Journal of Geology and Geophysics, v. 33, p. 137–150.

Litchfield, N.J., 2001, The Titri Fault System; Quaternary-active faults near the leading edge of the Otago reverse fault province: New Zealand Journal of Geology and Geophysics, v. 44, p. 517–534.

Litchfield, N.J., and Norris, R.J., 2000, Holocene motion on the Akatore Fault, south Otago coast, New Zealand: New Zealand Journal of Geology and Geophysics, v. 43, p. 405–418.

Liu, M., and Shen, Y.Q., 1998, Sierra Nevada uplift: A ductile link to mantle upwelling under the basin and range province: Geology, v. 26, p. 299–302.

Lofgren, B.E., 1979, Changes in aquifer-system properties with ground-water depletion: in, Saxena, S.K., editor., Evaluation and prediction of subsidence, New York, American Society of Civil Engineers, p. 26–46.

Loomis, D.P., and Burbank, D.W., 1988, The stratigraphic evolution of the El Paso basin, southern California; Implications for the Miocene development of the Garlock fault and uplift of the Sierra Nevada: Geological Society of America Bulletin: v. 100, p. 12–28.

Lu, H., Fulthorpe, C.S., Mann, P. and Kominz, M.A., 2005, Miocene–recent tectonic and climatic controls on sediment supply and sequence stratigraphy, Canterbury basin, New Zealand: Basin Research, v. 17, p. 311–328.

Lucchitta, I., 1979, Late Cenozoic uplift of the southwestern Colorado Plateau and adjacent lower Colorado River region: in, McGetchin, T.R., and Merrill, R.B., editors, Plateau Uplift; Mode and Mechanism, Tectonophysics v. 61, p. 63–95.

Lucchitta, I., Curtis, G.H., Davis, M.E., Davis, S.W., and Turrin, B., 2000, Cyclic aggradation and downcutting, fluvial response to volcanic activity, and calibration of soil-carbonate stages in the western Grand Canyon, Arizona: Quaternary Research, v. 53, p. 23–33.

Machette, M.N, 1985, Calcic soils of the southwestern United States: in, Weide, D.L., editor, Soils and Quaternary Geology of the Southwestern United States: Geological Society of America Special Paper 203, p. 1–21.

MacInnes, B., 2004, Uplift and deformation of marine terraces along the San Andreas fault; Duncan's Landing to Fort Ross, California: Senior Integrative Exercise, Carleton College, Northfield, Minnesota, 26 p. http://keck.wooster.edu/archives/symposium/04/California_pdfs/bremacinnesabs.pdf

Mackay, D.A., 1984, Kowhai management area detritus survey: New Zealand Department of Lands and Survey, Marlborough Catchment and Regional Water Board Report, 38 p.

Magleby, D.C., and Klein I.E., 1965, Ground-water conditions and potential pumping resources above the Corcoran Clay—An addendum to the groundwater geology and resources definite plan appendix, 1963: U.S. Bureau of Reclamation Open-File Report, 21 plates.

Mahood, G.A., Nibler, G.E., and Halliday, A.N., 1996, Zoning patterns and petrologic processes in peraluminous magma chambers: Hall Canyon pluton, Panamint Mountains, California: Geological Society of America Bulletin, v. 108, p. 437–453.

Manley, C.R., Glazner, A.F., and Farmer, G.L., 2000, Timing of volcanism in the Sierra Nevada of California: evidence for Pliocene delamination of the batholithic root?: Geology, v. 28, p. 811–814.

Marchand, D.E., 1977, The Cenozoic history of the San Joaquin Valley and the adjacent Sierra Nevada as inferred from the geology and soils of the eastern San Joaquin Valley, in, Singer, M.J., editor, Soil Development, Geomorphology, and Cenozoic History of the Northeastern San Joaquin Valley and Adjacent Areas, California: University of California Press. Guidebook for Joint Field Session, Soil Science Society of America and Geological Society of America, p. 39–50.

Marchand, D.E., and Allwardt, A., 1981, Late Cenozoic Stratigraphic Units, Northeastern San Joaquin Valley, California: U.S. Geological Survey Bulletin 1470, 70 p.

Markley, M., and Norris, R. J., 1999, Structure and neotectonics of the Blackstone Hill antiform, Central Otago, New Zealand: New Zealand Journal of Geology and Geophysics, v. 42: p. 205–218.

Matthews, W.H., 1979, Landslides of Central Vancouver Island and the 1946 Earthquake: Seismological Society of America Bulletin, v. 69, p.445–450.

Maxson, J.H., 1950, Physiographic features of the Panamint Range, California: Geological Society of America Bulletin, v. 61, p. 99–114.

Mayer, L., 1979, The evolution of the Mogollon Rim in central Arizona: in, McKetchin T.R., Merill R.B., Kisslinger, C., and Z. Suzuki, Z., editors, Plateau Uplift Mode and Mechanism, Tectonophysics, v. 61, p. 49–62.

McCaffrey, R., 1992, Oblique plate convergence, slip vectors, and forearc deformation: Journal of Geophysical Research, v. 97, p. 8905– 8915.

McCalpin, J.P., 1996, Field Techniques in Paleoseismology, in, McCalpin, J.P., editor, Paleoseismology: Academic Press, New York, p. 33–84.

McCalpin, J.P., and Hart, E.W., 2004, Ridge-top spreading features and relationship to earthquakes, San Gabriel Mountains region, Southern California: in, Hart, E.W., editor, Ridge-Top Spreading in California, California Geological Survey, CD-ROM.

McCalpin, J.P., and Irvine, J.R., 1995, Sackungen at Aspen Highlands Ski Area, Pitkin County, Colorado: Environmental and Engineering Geoscience, v. 1, p. 277–290.

McClusky, S.C., Bjornstad, S.C., Hager, B.H., King, R.W., Meade, B.J., Miller, M.M., Monastero, F.C., and Souter, B.J., 2001, Present day kinematics of the eastern California shear zone from a geodetically constrained block model: Geophysical Research Letters, v. 28, p. 3369–3372, doi: 10.1029/2001GL013091.

McKean, J.A., and Roering, J.J., 2004, Landslide detection and surface morphology mapping with airborne laser altimetry: Geomorphology, v. 57, p. 331–351. doi:10.1016/S0169-555X(03)00164-8.

McKean, J.A., Dietrich, W.E., Finkel, R.C., Southon, J.R., and Caffee, M.W., 1993, Quantification of soil production and downslope creep rates from cosmogenic ^{10}Be accumulations on a hillslope profile: Geology, v. 21, p. 343–346.

McKenzie, D.P., 1972, Active tectonics of the Mediterranean region: Geophysical Journal of the Royal Astronomical Society, v. 30, p. 109–185.

McKenzie, D.P., 1978, Active tectonics of the Alpine–Himalayan belt; the Aegean Sea and surrounding regions: Geophysical Journal of the Royal Astronomical Society, v. 55, p. 217–254.

McKenzie, D.P., and Morgan, W.J., 1969, Evolution of triple junctions: Nature, v. 224, no. 5215, p. 125–133.

McLaughlin, R.J., Sliter, W.V., Frederikson, N.O., Harbert, W.P., and McCulloch, D.S., 1994, Plate motions recorded in tectonostratigraphic terranes of the Franciscan Complex and evolution of the Mendocino triple junction, northwestern California: U.S. Geological Survey Bulletin 1997, 60 p.

McQuarrie, N., and Wernicke, B. P., 2005, An animated tectonic reconstruction of southwestern North America since 36 Ma: Geosphere, v.1, p. 147–172.

McSaveney, M.J. and Davies, T.R., 2005, Dynamics of large high-speed blockslides: in, Senneset, K., Flaate, K. and Larsen, J.O., editors, Landslides and Avalanches, ICFL 2005, Norway, Taylor and Francis Group, London, p. 257–264.

Medwedeff, D.A., 1992, Geometry and kinematics of an active, laterally propagating wedge thrust, Wheeler Ridge, California: in, Mitra, S., and Fisher, G. W., editors, Structural Geology of Fold-and-Thrust belts, Johns Hopkins University Press, Baltimore, Maryland, p. 3–28.

Merritts, D.J. 1987 Geomorphic responses to Late Quaternary tectonism, coastal northern California, Mendocino triple junction region: Ph.D. dissertation, University of Arizona, 190 p.

Merritts, D.J., 1996, The Mendocino triple junction: Active faults, episodic coastal emergence, and rapid uplift: Journal of Geophysical Research, v. 101, p. 6051–6070.

Merritts, D.J., and Bull, W.B., 1989, Interpreting Quaternary uplift rates at the Mendocino triple junction, northern California, from uplifted marine terraces: Geology, v. 17, p. 1020–1025.

Merritts, D.J., and Ellis, M., 1994, Introduction to special section on tectonics and topography: Journal of Geophysical Research, v. 99, p. 12135–12141.

Merritts, D.J., and Vincent, K.R., 1989, Geomorphic response of coastal streams to low, intermediate, and high rates of uplift, Mendocino triple junction region, northern California: Geological Society of America Bulletin, v. 117, p. 1373–1388.

Merritts, D.J., Chadwick, O.A., and Hendricks, D.M., 1991, Rates and processes of soil evolution on uplifted marine terraces, northern California: in, Pavich, M.J., editor, Weathering and Soils, Geoderma, v. 51, p. 241–275.

Merritts, D.J., Dunklin, T.B., Vincent, K.R., Wohl, E.E., and Bull, W.B., 1992, Quaternary tectonics and topography, Mendocino triple junction: in, Burke, R.M. and Carver, G.A. , editors, A Look at the Southern End of the Cascadia Subduction Zone and the Mendocino Triple Junction; Field Trip Guidebook, Pacific Cell, Friends of the Pleistocene,

Northern Coastal California, Humboldt State University, Arcata, California, p. 119–169.

Merritts, D.J., Vincent, K.R., and Wohl, E.E., 1994, Long river profiles, tectonism, and eustasy: A guide to interpreting fluvial terraces: Journal of Geophysical Research (Special Issue on Tectonics and Topography), v. 99 (B7), p. 14031–14050.

Mesollela, K.J., Matthews, R.K., Broecker, W.S., and Thurber, D.L., 1969, The astronomical theory of climatic change, Barbados data: Journal of Geology, v. 77, p. 250–274.

Miller, R.E., Green, J.H., and Davis, G.H., 1971, Geology of the compacting deposits in the Los Banos–Kettleman City subsidence area, California: U.S. Geological Survey Professional Paper 497-E, 46 p.

Miller, S.R., Slingerland, R.L., and Kirby, E., 2007, Characteristics of steady state fluvial topography above fault-bend folds: Journal of Geophysical Research, v. 112, F04004, doi:10.1029/2007JF000772.

Mitchell, C.E., Vincent, P., Weldon, R.J., and Richards M., 1994, Present-day vertical deformation of the Cascadia margin, Pacific Northwest, United States: Journal of Geophysical Research, v. 99, 12,257–12,277.

Molnar, P., and England, P.C., 1990, Late Cenozoic uplift of mountain ranges and global climate change; chicken or egg?: Nature, v. 346, no. 6279, p. 29–34, doi: 10.1038/ 346029a0.

Molnar, P., Anderson, R.S., and Anderson, S.P., 2007, Tectonics, fracturing of rock, and erosion: Journal of Geophysical Research, v. 112, F03014, doi:10.1029/2005JF000433.

Monaghan, M.C., McKean, J., Dietrich, W., and Klein, J., 1992, [10]Be chronology of bedrock-to-soil conversion rates: Earth and Planetary Science Letters, v. 111, p. 483–492.

Monastero, F.C., Katzenstein, A.M., Miller, J.S., Unruh, J.R., Adams, M.C., and Richards-Dinger, K., 2005, The Coso geothermal field; A nascent metamorphic core complex: Geological Society of America Bulletin, v. 117, p. 1534–1553.

Montgomery, D.R., 2001, Slope distributions, threshold hillslopes, and steady-state topography: American Journal of Science, v. 301, p. 432–454.

Montgomery, D.R., 2002, Valley formation by fluvial and glacial erosion: Geology, v. 30, p. 1047–1050.

Montgomery, D.R., and Brandon, M.T., 2002, Topographic controls on erosion rates in tectonically active mountain ranges: Earth and Planetary Science Letters, v. 201, p. 481–489, doi: 10.1016/S0012-821X(02)00725-2.

Montgomery, D.R., and Dietrich W.E., 2002, Runoff generation in a steep, soil-mantled landscape: Water Resources Research, v. 38(9), 1168, doi:10.1029/2001WR000822.

Morewood, N.C., and Roberts, G.P., 1997, The geometry, kinematics and rates of deformation in a normal fault segment boundary, central Greece: Geophysical Research Letters v. 24, p. 3081–3084.

Morewood, N.C., and Roberts, G.P., 1999, Lateral propagation of the surface trace of the South Alkyonides normal fault segment, central Greece: its impact on models of fault growth and displacement-length relationships. Journal of Structural Geology, v. 21, p. 635–652.

Morewood, N.C., and Roberts, G.P., 2000, The geometry, kinematics and rates of deformation within an en echelon normal fault segment boundary, central Italy: Journal of Structural Geology, v. 22, p. 1027–1047.

Morewood, N.C., and Roberts, G.P., 2001, Comparison of surface slip and focal mechanism data along normal faults: an example from the eastern Gulf of Corinth, Greece. Journal of Structural Geology, v. 23, p. 473–487.

Morewood, N.C., and Roberts, G.P., 2002, Surface observations of active normal fault propagation; implications for growth: Journal of the Geological Society, London, v. 159, p. 263–272.

Moon, B.P., 1984, Refinement of a technique for determining rock mass strength for geomorphological purposes: Earth Surface Processes and Landforms, v. 9, p. 189–193.

Mortimer, N., 1993, Geology of the Otago Schist and adjacent rocks, Scale 1:500,000: New Zealand Institute of Geological and Nuclear Sciences Map 7, Lower Hurt, New Zealand.

Morton, D.M., 1975, Seismically triggered landslides in the area above the San Fernando Valley: California Division of Mines and Geology Bulletin 196, p. 145–154.

Müller, L., 1964, The rock slide in the Vaiont Valley: Rock Mechanics and Engineering Geology, v. 2, p. 148–212.

Müller, L., 1987a, The Vaiont catastrophe, A personal review: in, Leonards, G.A. editor, Dam Failures. Engineering Geology, v. 24, p. 423–444.

Mueller, F., Caflisch, T., and Mueller, G., 1976, Firn und Eis der Schweizer Alpen, Gletscherinventar: Geographisches Institut, Eidgenössische Technische Hochschule, Zürich, Publication No. 57, 293 p.

Mueller, K., and Suppe, J., 1997, Growth of Wheeler

Ridge anticline, California: Geomorphic evidence for fault-bend folding behavior during earthquakes: Journal of Structural Geology, v. 19, p. 383–396.

Mueller, K., and Talling, P., 1997, Geomorphic evidence for tear faults accommodating lateral propagation of an active fault-bend fold, Wheeler Ridge, California: Journal of Structural Geology, v. 19, p. 397–411.

Munsell Color, 1990, Munsell Soil Color Charts: Macbeth Division of Kollmorgen Instruments, Baltimore, Maryland.

Myrianthis, M.L., 1982, Geophysical study of the epicentral area of the Alkyonides Islands Earthquakes, central Greece: Geophysical Institute of Hungary Transactions, v. 28, p. 5–17.

Namson, J.S., Davis, T.L., and Lagoe, M.B., 1990, Tectonic history and thrust-fold deformation style of seismically active structures near Coalinga: in, Rymer, M.J., and Ellsworth, W.L., editors, The Coalinga, California, Earthquake of May 2, 1983: U.S. Geological Survey Professional Paper 1487, p. 79–96.

Nicol, A., and Van Dissen, R. V., 2002, Up-dip partitioning of displacement components on the oblique-slip Clarence fault, New Zealand: Journal of Structural Geology, v. 24, p. 1521–1535.

Nilsen, T.H., Taylor, F.A., and Brabb, E., 1976a, Recent landslides in Alameda county, California (1940–71), an estimate of economic losses and correlations with slope, rainfall, and ancient landslide deposits: U.S. Geological Survey Bulletin 1398: 21 p.

Nilsen, T.H., Taylor, F.A., and Dean, R.M., 1976b, Natural conditions that control landsliding in the San Francisco Bay region; an analysis based on data from the 1968–69 and 1972–73 rainy seasons: U.S. Geological Survey Bulletin 1397, 35 p.

Nolan, K.M, Lisle, T.E, Kelsey, H.M., 1987, Bankfull discharge and sediment transport in northwestern California: in, Beschta, R.L, Blinn, T. , Grant, G.E, Ice, G.G, and Swanson, F.J, editors, Proceedings of International Symposium on Erosion and Sedimentation in the Pacific Rim, International Association of Hydrological Sciences Publication 165; p. 439–449.

Nonveiller, E., 1987, The Vaiont reservoir slope failure: in, Leonards, G.A., editor, Dam Failures; Engineering Geology, v. 24, p. 493–512.

Nonveiller, E., 1992. Vaiont slide: influence of frictional heat on slip velocity: in, Semenza, E., Melidoro, G., editors, Proceedings Meeting 1963 Vaiont

Landslide, Ferrara 1986, Univ. of Ferrara, Ferrara, p. 187–197.

Norris, R.J., and Cooper, A.F., 1995, Origin of small-scale segmentation and transpressional thrusting along the Alpine fault, New Zealand: Geological Society of America, Bulletin v. 107, p. 231–240.

Norris, R.J., and Cooper, A.F., 1997, Erosional control on the structural evolution of a transpressional thrust complex on the Alpine fault, New Zealand: Journal of Structural Geology, v. 19, p. 1323–1342.

Norris, R.J., and Cooper, A.F., 2000, Late Quaternary slip rates and slip partitioning on the Alpine Fault, New Zealand: Journal of Structural Geology, v. 23, p. 507–520.

Norris, R.J., and Nicolls, R., 2004, Strain accumulation and episodicity of fault movements in Otago: New Zealand Earthquake Commission Report 01/445, p. 1357.

Norris, R.J., Koons, P.O., and Cooper, A.F., 1990, The obliquely convergent plate boundary in the South Island of New Zealand: Implications for ancient collision zones: Journal of Structural Geology, v. 12, p. 715–725.

Norris, R. J., Koons, P. O., and Landis, C. A. 1994, Seismotectonic evaluation of fault structures in eastern Otago: Report 91/53, New Zealand Earthquake Commission.

Nur, A., Ron, H., and Scotti, O., 1986, Fault mechanics and the kinematics of block rotation: Geology, v. 14, p. 746–749.

Nur, A., Ron, H., and Beroza G., 1993, The nature of the Landers-Mojave earthquake line: Science, v. 261, p. 201–203.

Oberlander, T.M., 1985, Origin of drainage traverse to structures in orogens: in Morisawa, M., and Hack, J.T., editors, Tectonic Geomorphology; The Binghamton Symposia in Geomorphology, International Series, v. 15: London, Allen and Unwin, p. 155–182.

Ochiai, H., Okada, Y., Furuya, G., Okura, Y., Matsui, T., Sammori, T., Terajima, T., and Sassa, K., 2004, A fluidized landslide on a natural slope by artificial rainfall: Landslides, v. 1, p. 211–219.

O'Connor, J.E., Costa, J.E., 1993. Geologic and hydrologic hazards in glacierized basins in North America resulting from 19th and 20th century global warming: Natural Hazards, v. 8, p. 121–140.

O'Connor, J.E., Hardison, J.H., and Costa, J.E., 2001, Debris flows from failure of Neoglacial-Age moraine dams in the Three Sisters and Mount Jefferson Wilderness Areas, Oregon: US Geological Survey Professional Paper 1606, 93 p.

Oldow, J.S., 2003, Active transtensional boundary zone between the western Great Basin and Sierra Nevada block, western U.S. Cordillera: Geology, v. 31, p. 1033–1036. doi: 10.1130/G19838.1

O'Loughlin, C.L., and Pearce, A.J., 1982, Erosion processes in the mountains: in, Soons, J.M., and Selby, M.J., editors, Landforms of New Zealand, Auckland, Longman Paul, p. 67–79.

Oskin, M., and Iriondo, A., 2004, Large-magnitude transient strain accumulation on the Blackwater fault, Eastern California shear zone: Geology, v. 32, p. 313–316, doi: 10.1130/G20223.1

Oskin, M., Stock, J. and Martín-Barajas, A. 2001: Rapid localization of Pacific–North America plate motion in the Gulf of California: Geology, v. 29, p. 459–462.

Ota, Y., Chappell, J., Kelley, R., Yonekura, N., Matsumoto, E., Nishimura, T., Head, J., 1993, Holocene coral reef terraces and coseismic uplift of Huon Peninsula, Papua New Guinea: Quaternary Research, v. 40, p. 177–188.

Ota, Y., Pillans, B., Berryman, K., Beu, A., Fujimori, T. and Miyauchi, T., 1996, Pleistocene marine terraces of the Kaikoura Peninsula and Marlborough coast, South Island, New Zealand: New Zealand Journal of Geology and Geophysics, v. 39, p. 51–73.

Ouchi, S., 1985, Response of alluvial rivers to slow active tectonic movement: American Association of Petroleum Geologists Bulletin, v. 96, p. 504–515.

Owens, I.F., 1992, A note on the Mount Cook rock avalanche of 14 December 1991: New Zealand Geographer, v. 48, p. 74–78

Pantosti D., Collier, R., D'Addezio, G., Masana, E. and Sakellariou, D., 1996, Direct geological evidence for prior earthquakes on the 1981 Corinth fault (central Greece): Geophysical Research Letters, v. 23, p. 3795–3798.

Pavich, M.J., 1989, Regolith residence time and the concept of surface age of the piedmont "peneplain": Geomorphology, v. 2, p. 181–196.

Pavich, M.J., Brown, L., Valette-Silver, J.N., Klein, J., and Middleton, R., 1985, [10]Be analysis of a Quaternary weathering profile in the Virginia Piedmont: Geology, v. 13, p. 39–41.

Pazzaglia, F.J. and Brandon, M.T., 2001, A fluvial record of long-term steady-state uplift and erosion across the Cascadia Forearc High, Western Washington State: American Journal of Science, v. 301, p. 385–431.

Pazzaglia, F.J., and Knuepfer, P.L.K., 2001, Steadystate orogens; Preface: American Journal of Science, v. 301, p. ix–xi.

Pazzaglia, F.J., Braun, D.D., Pavich, M., Bierman, P., Potter, N., Merritts, D., Walter, R., and Germanoski, D., 2006, Rivers, glaciers, landscape evolution, and active tectonics of the central Appalachians, Pennsylvania and Maryland: in, Pazzaglia, F.J., editor, Excursions in Geology and History: Field Trips in the Middle Atlantic States, Geological Society of America Field Guide 8, p. 169–197, doi: 10.1130/2006.fld008(09).

Pearce, A.J., and O'Loughlin, C.L., 1985, Landsliding during a M 7.7 earthquake, influence of geology and topography: Geology, v. 13, p. 855–858.

Pederson, J., Karlstrom, K., Sharp, W., and McIntosh, W., 2002, Differential incision of the Grand Canyon related to Quaternary faulting—Constraints from U series and Ar/Ar dating, Geology, 30, 739–742.

Pederson, J.L., Anders, M.D., Rittenhour, T.M., Sharp, W.D., Gosse, J.C., and Karlstrom, K.E., 2006, Using fill terraces to understand incision rates and evolution of the Colorado River in eastern Grand Canyon, Arizona: Journal of Geophysical Research, v. 111, F02003, doi:10.1029/2004JF000201.

Peltzer, G., Crampé, F., Hensley, S., and Rosen, P., 2001, Transient strain accumulation and fault interaction in the Eastern California shear zone: Geology, v. 29, p. 975–978.

Penck, W. 1953. Morphological Analysis of Landforms: Translated by H. Czech and K. C. Boswell. MacMillan, London.

Penck, W., 1953, Morphological analysis of landforms: MacMillan, London, 429 p.

Perissoratis, C., Mitropoulos, D., and Angelopoulos, I., 1986, Marine geological research at the E. Korinthiakos Gulf: Special Issue of Geology and Geophysical Research, Institute of Geology and Mineral Exploration, Athens, Greece.

Personius, S.F., 1995, Late Quaternary stream incision and uplift in the forearc of the Cascadia subduction zone, western Oregon: Journal of Geophysical Research, v. 100, 20,193–20,210, doi: 10.1029/95JB01684.

Petley, D.N., Bulmer N.H., and Murphy W., 2002, Patterns of movement in rotational and translational landslides. Geology, v. 30, p. 719–722.

Pettinga, J.R., and Wise, D.U., 1994, Paleostress adjacent to the Alpine fault: Broader implications from fault analysis near Nelson, South Island, New Zealand: Journal of Geophysical Research, v. 99, p. 2727–2736.

Pettinga, J.R., Chamberlain, C.G., Yetton, M.D., Van Dissen, R.J., Downes, G., 1998, Earthquake source identification and characterization; Report for the Canterbury Regional Council, Rivers and Coastal Resources, and Hazards Section: Christchurch, Canterbury Regional Council Publication No. U98/10.

Phillips, F.M., 2008, Geological and hydrological history of the paleo Owens River drainage since the late Miocene: in, Reheis, M.C., Hershler, R., and Miller, D.M., editors, Late Cenozoic Drainage History of the Southwestern Great Basin and Lower Colorado River Region, Geologic and Biotic Perspectives; Geological Society of America Special Paper 439, p. 115-150. doi: 10.1130/2008.2439(06).

Phillips, F.M., Zreda, M.G., Benson, L.V., Plummer, M.A., Elmore, D., and Sharma, P., 1996. Chronology for fluctuations in Late Pleistocene Sierra Nevada Glaciers and lakes: Science, v. 274, p. 749–751.

Phillips, F.M., Ayarbe, J.P., Harrison, J.B.J., and Elmore, D., 2003, Dating rupture events on alluvial fault scarps using cosmogenic nuclides and scarp morphology: Earth and Planetary Science Letters, v. 215, p. 203–218.

Phillips, L.F., and Schumm, S.A., 1987, Effects of regional slope on drainage networks: Geology, v. 15, p. 813–816.

Pillans, B.J., 1990, Pleistocene marine terraces in New Zealand, a review: New Zealand Journal of Geology and Geophysics, v. 33, p. 219–232.

Pillans, B.J., 1994, Direct marine-terrestrial correlations, Wanganui Basin, New Zealand: the last 1 million years: Quaternary Science Reviews, v. 13, p. 189–200.

Pinter, N., and Brandon, M.T., 1997, How erosion builds mountains: Scientific American, v. 276, p. 60–65.

Plafker, G. and Ericksen, G.E., 1978, Nevados Huascaran Avalanches, Peru: in, Voight, B. editor, Rockslides and Avalanches, 1 Natural Phenomena; Elsevier, Amsterdam, p. 277–314.

Poage, M.A., and Chamberlain, C.P. , 2002, Stable isotopic evidence for a pre-middle Miocene rain shadow in the western Basin and Range: Implications for the paleotopography of the Sierra Nevada: Tectonics, v. 21, p. 1601–1610.

Poesen, J., 1984, The influence of slope angle on infiltration rate and Hortonian overland flow volume: Zeitschrift für Geomorphologie, v. 49, p. 117–131.

Poland, J.F., 1961, The coefficent of storage in a region of major subsidence caused by compaction of an aquifer system: in, Geological Survey Research 1961, U.S. Geological Survey Professional Paper 424-B, p. B52–B54.

Poland, J.F, 1972, Subsidence and its control: American Association of Petroleum Geologists Memoir 18, p. 50–71.

Poland, J.F., editor, 1984, Guidebook to Studies of Land Subsidence Due to Groundwater Withdrawal: Studies and Reports in Hydrology, 40, UNESCO, Paris.

Poland, J.F., Lofgren, B.E., and Riley, F.S., 1972, Glossary of selected terms useful in studies of the mechanics of aquifer systems and land subsidence due to fluid withdrawal: U.S. Geological Survey Water-Supply Paper 2025, 9 p.

Poland, J.F., Lofgren, B.E., Ireland, R.L., and Pugh, A.G., 1975, Land subsidence in the San Joaquin Valley, California, as of 1972: U.S. Geological Survey Professional Paper 437H, p. H1–H78.

Putirka, K., and Busby C.J., 2007, The tectonic significance of high-K_2O volcanism in the Sierra Nevada, California: Geology, v. 35, p. 923–926. doi: 10.1130/G23914A.1.

Raeside, J.D., 1964, Loess deposits of the South Island, New Zealand, and soils formed on them: New Zealand Journal of Geology and Geophysics, v. 7, p. 811–838.

Ragan, R.M., 1968, An experimental investigation of partial area contribution: International Association of Hydrological Sciences Publication 76, p. 241–249.

Raymo, M.E., and Ruddiman, W.F., 1992, Tectonic forcing of late Cenozoic climate: Nature, v. 359, p. 117–122.

Reheis, M.C., and Dixon, T.H., 1996, Kinematics of the Eastern California shear zone: Evidence for slip transfer from Owens and Saline Valley fault zones to Fish Lake Valley fault zone: Geology, v. 24, p. 339–342.

Reheis, M.C., and Sawyer, T.L., 1997, Late Cenozoic history and slip rates of the Fish Lake Valley, Emigrant Peak, and Deep Springs fault zones, Nevada and California: Geological Society of America Bulletin, v. 109, p. 280– 299, doi: 10.1130/0016-7606 (1997)109<0280:LCHASR>2.3.CO;2

Reheis, M.C., Sawyer, T.L., Slate, J.L., and Gillespie, A.R., 1993, Geologic map of late Cenozoic deposits and faults in the southern part of the Davis Mountain 15' quadrangle, Esmeralda County, Nevada: U.S. Geological Survey Map I-2342, scale 1:24,000.

Reneau, S.L., and Dietrich, W.E 1987, The importance of hollows in debris flow studies; examples from Marin County, California: in, Costa, J.E., and Wieczorek, G.F. editors, Debris Flows/Avalanches; Process, Recognition, and Mitigation, Reviews in Engineering Geology, v. 7, p. 165–180.

Reneau, S.L., and Dietrich, W.E., 1991, Erosion rates in the Southern Oregon Coast Range: Evidence for an equilibrium between hillslope erosion and sediment yield: Earth Surface Processes and Landforms, v. 16, p. 307–322.

Riley, F.S., 1998, Mechanics of aquifer systems – the scientific legacy of Joseph F. Poland: in, Borchers, J.W., editor, Land Subsidence Case Studies and Current Research, Star Publishing Company, Belmont, California, p. 13–27.

Riley, F.S., and McClelland, E.J., 1971, Application of the modified theory of leaky aquifers to a compressible multiple-aquifer system: Mechanics of aquifer systems: Analysis of pumping tests near Pixley, California: U.S. Geological Survey Open-File Report, 96 p.

Rinaldo, G., and Ghirotti, M., 2005, The 1963 Vaiont landslide: Giornale di Geologia Applicata, v. 1, p. 41–52, doi: 10.1474/GGA.2005-01.0-05.0005.

Ritter, D.F., Kochel, R.C., and Miller, J.R., 1995, Process geomorphology, (3rd Edition): William C. Brown, Dubuque, Iowa, 546 p.

Roberts, G.P., 1996a, Variation in fault-slip directions along active and segmented normal fault systems: Journal of Structural Geology, v. 18, p. 835–845.

Roberts, G.P., 1996b, Noncharacteristic normal faulting surface ruptures from the Gulf of Corinth, Greece: Journal of Geophysical Research, v. 101, p. 25255–25267.

Roberts, G.P., and Ganas A., 2000, Fault-slip directions in central and southern Greece measured from striated and corrugated fault planes: Comparison with focal mechanism and geodetic data; Journal of Geophysical Research, v. 105, p. 23443–23462.

Rockwell, T.K., Keller, E.A., and Dembroff, G.R., 1988, Quaternary rate of folding of the Ventura Avenue anticline, western Transverse Ranges, southern California: Geological Society of America Bulletin, v. 100, p. 850–858.

Roe, G.H., Montgomery, D.R., and Hallet, B., 2003, Orographic precipitation and the relief of mountain ranges: Journal of Geophysical Research—Solid Earth, v. 108, no. B6, 2315.

Roering, J.J., Kirchner, J.W., and Dietrich, W.E., 1999, Evidence for nonlinear, diffusive sediment transport on hillslopes and implications for landscape morphology: Water Resources Research, v. 35, p. 853–870.

Roering, J.J., Kirchner, J.W., and Dietrich, W.E., 2001, Hillslope evolution by nonlinear, slope-dependent transport: Steady-state morphology and equilibrium adjustment timescales: Journal of Geophysical Research, v. 106, no. B8, p. 16,499–16,513, doi: 10.1029/2001JB000323.

Roering, J.J., Almond, P., McKean, J., and Tonkin, P., 2002, Soil transport driven by biological processes over millennial timescales: Geology, v. 30, p. 1115–1118.

Roering, J.J., Almond, P., Tonkin, P., and McKean, J., 2004, Constraining climatic controls on hillslope dynamics using a coupled model for the transport of soil and tracers: Application to loess-mantled hillslopes, South Island, New Zealand: Journal of Geophysical Research, v.109, F101010, doi: 10.1029/2003JF000034.

Roering J.J., Kirchner, J.W., and Dietrich, W.E., 2005, Characterizing structural and lithologic controls on deep-seated landsliding: Implications for topographic relief and landscape evolution in the Oregon Coast Range, USA: Geological Society of America Bulletin, v. 117, p. 654–668.

Ross, 1962, Correlation of granitic plutons across faulted Owens Valley, California: U.S. Geological Survey Professional Paper 450-D, p. 86–88.

Rother, H. and Shulmeister, J., 2006, Synoptic climate change as a driver of Late Quaternary glaciations in the mid-latitudes of the Southern Hemisphere: Climate of the Past, v.2, p. 11–19. 1814-9359/cpd/2005-1-1.

Rouse, H., editor, 1950, Engineering Hydraulics: John Wiley and Sons, New York, 1,039 p.

Ruddiman W.F., 2003, The anthropogenic greenhouse era began thousands of years ago: Climate Change, v. 61, p. 261–293.

Ruppert, S., Fliedner, M.M., and Zandt, G., 1998, Thin crust and active upper mantle beneath the southern Sierra Nevada in the western United States: Tectonophysics, v. 286, p. 237–252.

Rymer, M.J., and Ellsworth, W.L., 1990, The Coalinga, California earthquake of May 2, 1983: U.S. Geological Survey Professional Paper 1487, 417 p.

Safran, E.B., Bierman, P.R., Aalto, R., Dunne, T, Whipple, K.X., and Caffee, M., 2005, Erosion rates driven by channel network incision in the Bolivian

Andes: Earth Surface Processes and Landforms, v. 30, p. 1007–1024.

Saleeby, J., and Foster, Z., 2004, Topographic response to mantle lithosphere removal in the southern Sierra Nevada region, California: Geology, v. 32, p. 245–248.

Sarkar, S., Kanungo D.P., and Patra, A.K., 2006, Landslides in the Alaknanda Valley of Garhwal Himalaya, India: Quarterly Journal of Engineering Geology and Hydrogeology, v. 39, p. 79–82.

Sauber, J., Thatcher, W., and Solomon, S.C., 1986, Geodetic measurement of deformation in the central Mojave Desert, California: Journal of Geophysical Research, v. 91, no. B12, p. 12683–12693.

Sauber, J., Thatcher, W., Solomon, S.C., and Lisowski, M., 1994, Geodetic slip rate for the eastern California shear zone and the recurrence time of Mojave desert earthquakes: Nature, v. 367, p. 264–266.

Savage, J.C., Lisowski, M., and Murray, M., 1993, Deformation from 1973 through 1991 in the epicentral area of the 1992 Landers, California, earthquake (Ms=7.5): Journal of Geophysical Research, v. 98, p. 19951–19958.

Scheidegger, A.E., 1961, Theoretical Geomorphology: Springer-Verlag, Berlin, 327 p.

Schlische, R.W., Young S.S., Ackermann, R.V. and Gupta, A. 1996, Geometry and scaling relations of a population of very small rift-related faults: Geology, v. 24, p. 683–686.

Schmidt, K.M., and Montgomery, D.R., 1995, Limits to relief: Science, v. 270, p. 617–620.

Schmidt, K.M., and Montgomery, D.R., 1996, Rock mass strength assessment for bedrock landsliding: Environmental and Engineering Geoscience, v. 2, no. 3, p. 325–338.

Scholz, C. H. 1982, Scaling laws for large earthquakes; consequences for physical models: Seismological Society of America Bulletin, v. 72, p. 1–14.

Schreurs, G., 1994, Experiments on strike-slip faulting and block rotation: Geology, v. 22, p. 567–570.

Schumm, S.A., 1956, The role of creep and rainwash on the retreat of badland slopes: American Journal of Science, v. 254, p. 693–706.

Schumm, S.A., 1963, The disparity between present rates of denudation and orogeny: U.S. Geological Survey Professional Paper 454-H, p. 1–13.

Schumm, S.A., 1985, River channel patterns, braided, meandering, and straight: Annual Review of Earth and Planetary sciences, v. 13, p. 5–27.

Schumm, S.A., 1997, Drainage density, Problems of prediction: in, Stoddart, D.R. editor, Process and Form in Geomorphology, Routledge, London, p. 15–45.

Schumm, S. A., and Chorley, R.J., 1964, The fall of Threatening Rock: American Journal of Science, v. 262, p. 1041–1054.

Schumm, S.A., and Lichty, R.W., 1965, Time, space and causality in geomorphology: American Journal of Science, v. 263, p. 110–119

Schumm, S.A., Dumont, J. F., and Holbrook, J. F., 2000, Active Tectonics and Alluvial Rivers: Cambridge, Cambridge University Press, 290 p.

Scoging, H., 1982, Spatial variations in infiltration, runoff and erosion on hillslopes in semi-arid Spain: in, Bryan, R.B., and Yair, A., editors, Badland Geomorphology and Piping, Geobooks, Norwich, p. 89–112.

Scott, K.M., 1971, Origin and sedimentology of 1969 debris flows near Glendora, California: U.S. Geological Survey Professional Paper 750-C, p. 242–247.

Seaver, D. B., 1986, Quaternary evolution and deformation of the San Emigdio Mountains and their alluvial fans, Transverse Ranges, California: M.S. thesis, Santa Barbara, University of California, 116 p.

Selby, M.J., 1982a, Controls on the stability and inclinations of hillslopes on hard rock: Earth Surface Processes and Landforms, v. 7, p. 449–467.

Selby, M.J., 1982b, Hillslope Materials and Processes, Oxford University Press, New York, 264 p.

Semenza, E., and Ghirotti, M., 2000, History of 1963 Vaiont Slide, The importance of the geological factors to recognise the ancient landslide: Bulletin of Engineering Geological Environments, v. 59, p. 87–97.

Shackleton, N.J., and Opdyke, N.D., 1973, Oxygen isotope and palaeomagnetic stratigraphy of equatorial Pacific core V28-238; Oxygen isotope temperatures and ice volumes on a 10^5 year and 10^6 year scale: Quaternary Research, v. 3, p. 39–55.

Sharpe, C.F.S., 1938, Landslides and Related Phenomena; A Study of Mass Movements of Soil and Rock: Columbia University Press, New York, 137 p.

Shelton, J.S., 1966, Geology Illustrated: Freeman, San Francisco, 434 p.

Shroder, J.F., 1998, Slope failure and denudation in the western Himalaya: Geomorphology, v. 26, p. 81–105.

Sieh, K.E., and Jahns, R.H., 1984, Holocene activity of the San Andreas Fault at Wallace Creek, California: Geological Society of America Bulletin, v. 95, p. 883–896.

Simpson, G., 2004, Role of river incision in enhancing deformation: Geology, v. 32, p. 341–344.

Simpson, G.D.H., Cooper, A.F., and Norris, R.J., 1994, Late Quaternary evolution of the Alpine fault zone at Paringa, South Westland: New Zealand Journal of Geology and Geophysics, v. 37, p. 49–58.

Small, E.E., and Anderson, R.S., 1995, Geomorphically driven late Cenozoic rock uplift in the Sierra Nevada, California: Science, v. 270, p. 277–280.

Small, E.E., Anderson, R.S., Repka, J.L., and Finkel, R., 1997, Erosion rates of alpine bedrock summit surfaces deduced from *in situ* Be-10 and Al-26: Earth and Planetary Science Letters, v. 150, p. 413–425.

Sneed, M., and Galloway, D.L., 2000, Aquifer-system compaction and land subsidence: Measurements, analyses, and simulations-the Holly Site, Edwards Air Force Base, Antelope Valley, California: U.S. Geological Survey Water-Resources Investigations Report 00-4015, 65 p. http://ca.water.usgs.gov/archive/reports/wrir004015/

Snow, J.K., and Prave, A.R., 1994, Covariance of structural and stratigraphic trends; evidence for anticlockwise rotation within the Walker Lane Belt, Death Valley region, California and Nevada: Tectonics, v. 13, p. 712–724, doi: 10.1029/93TC02943.

Snow, J.K., and Wernicke, B.P., 1994, Crustal mass balance in the central Basin and Range: Geological constraints: Eos (Transactions, American Geophysical Union), v. 75, p. 583.

Snyder, N.P., Whipple, K.X., Tucker, G.E., and Merritts, D.J., 2000, Landscape response to tectonic forcing; digital elevation model analysis of stream profiles in the Mendocino triple junction region, northern California: Geological Society of America Bulletin, v. 112, p. 1250–1263.

Stark, C.P., and Stark, G.J., 2001, A channelization model of landscape evolution: American Journal of Science, v. 301, p. 455–485.

Starkel, L, 1979, The role of extreme meteorological events on the shaping of mountain relief: Geographic Polonica, v. 41, p. 13–20.

Starkel, L., 2003, Climatically controlled terraces in uplifting mountain areas: Quaternary Science Reviews, v. 22, p. 2189–2198.

Stegmann, S., Strasser, M., Anselmetti, F.S., and Kopf, A., 2007, Geotechnical *in situ* characterisation of subaquatic slopes: The role of pore pressure transients versus frictional strength in landslide initiation: Geophysical Research Letters v. 34,

doi:10.1029/2006GL029122,

Stein, R.S., and Ekstrom, G., 1992, Seismicity and geometry of a 110-km-long blind thrust fault: 2. Synthesis of the 1982–1985 [Coalinga] California earthquake sequence: Journal of Geophysical Research, v. 97, p. 4865–4883.

Stein, R.S., and Thatcher, W., 1981, Seismic and aseismic deformation associated with the 1952 Kern County, California, earthquake and relationship to the Quaternary history of the White Wolf fault: Journal of Geophysical Research, v. 86, p. 4913–4928.

Stein, R.S., and Yeats, R.S., 1989, Hidden earthquakes: Scientific American, v. 260, p. 48–57. http://quake.wr.usgs.gov/research/deformation/modeling/papers/scientam/scientam.html

Stein, R.S., King, G.C.P., and Rundle, J.B., 1988, The growth of geological structure by repeated earthquakes 2; field examples of continental dip-slip faults: Journal of Geophysical Research, v. 93, p. 13,319–13,331.

Stenner, H.D., Lund, W.R., Pearthree, P.A., and Everitt, B.L., 1999, Paleoseismologic investigations of the Hurricane fault in northwestern Arizona and southwestern Utah: Arizona Geological Survey Open–File Report 99-8, 137 p.

Stewart, J.H., 1983, Extensional tectonics in the Death Valley area, California, transport of the Panamint Range structural block 80 km northwestward: Geology, v. 11, p. 153–157.

Stirling, C., Esat, T.M., Lambeck, K., and McCulloch, M.T., 1998. Timing and duration of the Last Interglacial: implications for a restricted interval of widespread coral growth. Earth and Planetary Science Letters: v. 160, p. 745–762.

Stock, G.M., Anderson, R.S., and Finkel, R.C., 2004, Pace of landscape evolution in the Sierra Nevada, California, revealed by cosmogenic dating of cave sediments: Geology, v. 32, p. 193–196.

Stock, G.M., Anderson, R.S., and Finkel, R.C., 2005, Rates of erosion and topographic evolution of the Sierra Nevada, California, inferred from cosmogenic 26Al and ^{10}Be concentrations: Earth Surface Processes and Landforms, v. 30, p. 985–1006.

Stock, J., and Dietrich, W. E., 2003, Valley incision by debris flows; Evidence of a topographic signature: Water Resources Research, v. 39, 1089, doi:10.1029/2001WR001057.

Stockli, D.F., Dumitru, T.A., McWilliams, M.O., and Farley, K.A., 2003, Cenozoic tectonic evolution of the White Mountains, California and Nevada: Geological Society of America Bulletin, v. 115, p. 788–816.

Stolar, D., Roe G., and Willett S., 2007, Controls on the patterns of topography and erosion rate in a critical orogen: Journal of Geophysical Research, v. 112, F04002, doi:10.1029/2006JF000713.

Stone, P., Dunne G.C., Moore, J.G., and Smith, G.I., 2000, Geologic Map of the Lone Pine 15' Quadrangle, Inyo County, California: U.S. Geological Survey Geologic Investigations Series Map I–2617.

Stout, M.L., 1977, Radiocarbon dating of landslides in southern California: California Geology, v. 30, p. 99–105.

Stout, M.L., 1992, Mega-landslides in Southern California: in, Pipkin, B.W., and Proctor, R.J., editors, Engineering geology practice in Southern California, Association of Engineering Geologists Special Publication 4, p. 575–578.

Strahler, A.N., 1950, Equilibrium theory of erosional slopes approached by frequency distribution analysis: American Journal of Science, v. 248, p. 673–696; 800–814.

Strahler, A.N., 1952, Dynamic basis of geomorphology: Geological Society of America Bulletin, v. 63, p. 923–938.

Strahler, A.N., 1957, Quantitative analysis of watershed Geomorphology: American Geophysical Union Transactions, v. 38, p. 913–920.

Strahler, A.N., 1958, Dimensional analysis applied to fluvially eroded landforms: Geological Society of America Bulletin, v. 69, p. 279–300.

Strahler, A.N., 1964, Quantitative geomorphology of drainage basins and channel networks: in, Chow, V.T., editor, Handbook of Applied Hydrology, McGraw-Hill, New York, p. 4-40–4-74.

Stuiver, M., Reimer, P.J., Bard, E., Beck, J.W., Burr, G.S., Hughen, K.A., McCormac, G., van der Plicht, J., and Spurk, M., 1998, INTCAL98 radiocarbon age calibration, 24,000-0 cal BP: Radiocarbon, v. 40, p. 1041–1083.

Sturman A.P., and Wanner, H., 2001, A comparative review of the weather and climate of the New Zealand Southern Alps and the European Alps: Mountain Research and Development, v. 21, p. 359–369.

Stüwe, K., White L., and Brown, R., 1994, The influence of eroding topography on steady-state isotherms; Application to fission track analysis: Earth and Planetary Science Letters, v. 124, p. 63–74.

Suggate, R.P., 1963, The Alpine fault: Transactions of the Royal Society of New Zealand, v. 2, p. 105–129.

Suggate, R.P., 1968, The Paringa Formation, Westland, New Zealand: New Zealand Journal of Geology and Geophysics, v. 11, p. 345–355.

Summerfield, M.A., 1996, Understanding landscape development: the evolving interface between geomorphology and other earth sciences: Area, v. 28, p. 211–220.

Summerfield, M.A., 2005, The changing landscape of geomorphology: Earth Surface Processes and Landforms, v. 30, p. 779–781. doi 10.1002/esp.1250.

Summerfield, M.A., and Hulton, N.J., 1994, Natural controls of fluvial denudation rates in major world drainage basins: Journal of Geophysical Research, v. 99, p. 13,871–13,883.

Sutherland, R. 1994, Displacement since the Pliocene along the southern section of the Alpine Fault, New Zealand: Geology, v. 22, p. 327–330.

Sutherland, R., 1999, Cenozoic bending of New Zealand basement terranes and Alpine Fault displacement: a brief review: New Zealand Journal of Geology and Geophysics, v. 42, p. 295–301.

Surpless, B.E., Stockli, D.F., Dumitru, T.A., and Miller, E.L., 2002, Two-phase westward encroachment of Basin and Range extension into the northern Sierra Nevada: Tectonics, v. 21, 1002, doi: 10.1029/2000TC001257.

Sylvester, A.G., and Smith, R.R., 1976, Tectonic transpression and basement-controlled deformation in the San Andreas fault zone, Salton trough, California: American Association of Petroleum Geologists Bulletin, v. 60, p. 74–96.

Talling, P.J., and Sowter, M.J., 1999, Drainage density on progressively tilted surfaces with different gradients, Wheeler Ridge, California: Earth Surface Processes and Landforms, v. 24, p. 809–824.

Terzaghi, K., 1925, Principles of Soil Mechanics: Addison-Wesley, London, 550 p.

Terzaghi, K., 1925, Erdbaumechanik und bodenphysikalische Grundlage: Deutike, Leipzig, 399 p.

Terzaghi, K., and Peck, R.B., 1948, Soil Mechanics in Engineering Practice: John Wiley and Sons, New York, 566 p.

Terzaghi, K., and Peck , R.B., 1967, Soil Mechanics in Engineering Practice: John Wiley, Hoboken, New Jersey, 729 p.

Thiede, R. C., Bookhagen, B., Arrowsmith, J. R., Sobel, E. R., and Strecker, M. R., 2004, Climatic control on rapid exhumation along the Southern Himalayan Front: Earth and Planetary Science Letters, v. 222, p. 791–806.

Thompson, H. D., 1949, Drainage evolution in the

Appalachians of Pennsylvania: Annals of the New York Academy of Science, v. 52, p. 31–63.

Thornes, J.B., and Brunsden, D., 1977, Geomorphology and time: John Wiley, New York, 208 p.

Tippett, J.M., and Kamp, P.J.J., 1993, Fission track analysis of the late Cenozoic vertical kinematics of continental Pacific crust, South Island, New Zealand: Journal of Geophysical Research, v. 98, p. 16,119–16,148.

Tolman, C.F., and Poland, J.F., 1940, Ground-water, salt-water intrusion, and ground-surface recession in Santa Clara Valley, Santa Clara County, California: American Geophysical Union Transactions, part 1, p. 23–34.

Tonkin, P.J., Runge, E.C.A., and Ives, D.W., 1974, A study of Late Pleistocene loess deposits, South Canterbury, New Zealand, Part 2—Paleosols and their stratigraphic significance: Quaternary Research, v. 4, p. 217–231.

Troeh, F.R., 1965, Landform equations fitted to contour maps: American Journal of Science, v. 263, p. 616–627.

Tucker, G.E., and Bras, R.L., 1998, Hillslope processes, drainage density, and landscape morphology: Water Resources Research, v. 34, p. 2751–2764.

U.S. Department of the Interior, 1946, The Colorado River: "A Natural Menace Becomes a National Treasure": A comprehensive departmental report on the development of the water resources of the Colorado River basin for review prior to submission to the Congress (March, 1946), 293 p.

U.S. Geological Survey, 2004, United States Geological Survey Quaternary Fault and Fold Database for the United States; Nevada, http://earthquakes.usgs.gov/qfaults/nv/index.html.

Unruh, J.R., 1991, The uplift of the Sierra Nevada and implications for late Cenozoic epeirogeny in the western Cordillera: Geological Society of America Bulletin, v. 103, p. 1395–1404.

Unruh, J.R., and Sawyer, T.L., 1997, Assessment of blind seismogenic sources, Livermore Valley, eastern San Francisco Bay region: U.S. Geological Survey National Earthquake Hazards Reduction Program Report for Grant 1434-95-G-2611, 88 p.

Unruh, J.R., Twiss, R.J., and Hauksson, Egill, 1996, Seismogenic deformation field in the Mojave block and implication for tectonics of the eastern California shear zone: Journal of Geophysical Research, v. 101, p. 8335–8361.

Van der Beek, P., Champel, B., and Mugnier, J-L., 2002, Control of detachment dip on drainage development in regions of active fault-propagation folding: Geology, v. 30, p. 471–474.

Van Dissen, R.J., 1989, Late Quaternary faulting in the Kaikoura region, southeastern Marlborough, New Zealand: M.S. Geology thesis, Oregon State University, 72 p.

Van Dissen, R.J., and Yeats, R.S., 1991, Hope fault, Jordan thrust, and uplift of the Seaward Kaikoura Range: Geology, v. 19, p. 393–396.

Veveakis, E., Vardoulakis, I., and Di Toro, G., 2007, Thermoporomechanics of creeping landslides; The 1963 Vaiont slide, northern Italy: Journal of Geophysical Research, v. 112, F03026, doi:10.1029/2006JF000702.

Vincent, K.R., 1995, Implications for models of fault behavior from earthquake surface-displacement along adjacent segments of the Lost River Fault, Idaho: Ph.D. Dissertation, University of Arizona Tucson, AZ, 152 p.

Vine, F.J., and Matthews, D.H., 1963, Magnetic anomalies over oceanic ridges: Nature, v. 199, p. 947–949.

Vita-Finzi, C., 1993, Evaluating Late Quaternary uplift in Greece and Cyprus: in, Pritchard, H.M., Alabaster, T., Harris, N.B.W., and Neary, C.R., editors, Magmatic Processes and Plate Tectonics; Special Publication of the Geological Society of London, v. 76, p. 417–424.

Wadia, D.N., 1931, The syntaxis of the northwest Himalaya: its rocks, tectonics and orogeny: Records of the Geological Survey of India, v. 65, p. 189–220.

Wagner, G., and Van den Haute, P., 1992, Fission Track Dating: Kluwer Academic Publishing, Dordrecht, Netherlands, 285 p.

Wahrhaftig, C., 1965, Stepped topography of the southern Sierra Nevada: Geological Society of America Bulletin, v. 76, p. 1165–1190.

Wakabayashi, J., and Sawyer, T.L., 2000, Neotectonics of the Sierra Nevada and the Sierra Nevada-Basin and Range Transition, California, with field trip stop descriptions for the northeastern Sierra Nevada: in Brooks, E.R., and Dida, L.T., editors, Field Guide to the Geology and Tectonics of the Northern Sierra Nevada, California Division of Mines and Geology Special Publication 122, p. 173–212.

Wakabayashi, J., and Sawyer, T.L., 2001, Stream incision, tectonics, uplift, and evolution of topography of the Sierra Nevada, California: Journal of Geology, v. 109, p. 539–562.

Wakabayashi, J., and Smith, D.L., 1994, Evaluation of recurrence intervals, characteristic earthquakes and slip rates associated with thrusting along the Coast Range–Central Valley geomorphic boundary, California: Bulletin of the Seismological Society of America, v. 84, p. 1960–1970.

Wakabayashi, J., Hengesh, J.V., and Sawyer, T.L., 2004, Four-dimensional transform fault processes: progressive evolution of step-overs and bends: Tectonophysics v. 392, p. 279– 301.

Walcott, R.I., 1978, Present tectonics and late Cenozoic evolution of New Zealand: Geophysical Journal of the Royal Astronomical Society, v. 52, p. 137–164.

Walcott, R.I., 1998, Modes of oblique compression: Late Cenozoic tectonics of the South Island, New Zealand: Reviews in Geophysics, v. 36, p. 1–26.

Wallace, R.E., 1977, Profiles and ages of young fault scarps, north-central Nevada: Bulletin of the Geological Society of America, v. 88, p. 1267–1281.

Wallace, R.E., 1984, Patterns and timing of Late Quaternary earthquakes in the Great Basin Province and relation to some regional tectonic features: Journal of Geophysical Research, v. 89, p. 5,763–5,769.

Wang, Q., Zhang, P., Freymueller, J., Bilham, R., Larson, K., Lai, X., You, X., Niu, Z., Wu, J., Li, Y., Liu, J., Yang, Z., and Chen, Q., 2001, Present-day crustal deformation in China constrained by Global Positioning System measurements: Science, v. 294, p. 574–577, doi:10.1126/ Science.1063647.

Wang, Y.J., Cheng, H., Edwards, R.L., An, Z.S., Wu, J.Y., Shen, C.C., and Dorale, J.A., 2001, A high resolution absolute dated Late Pleistocene monsoon record from Hulu Cave, China: Science, v. 294, p. 2345–2348. doi: 10.1126/science.1064618.

Ward, C.M., 1988, New Zealand marine terraces: uplift rates: Science v. 240, p. 803.

Watterson, J., 1986, Fault dimensions, displacements and growth: Pure and Applied Geophysics, v. 124, p. 365–373.

Weissmann, G. S., Mount, J. F., and Fogg, G. E., 2002, Glacially driven cycles in accumulation space and sequence stratigraphy of a stream-dominated alluvial fan, San Joaquin Valley, California, U.S.A.: Journal of Sedimentary Research, v. 72, p. 270–281.

Wellman, H.W., 1953, Data for the study of Recent and Late Pleistocene faulting in the South Island of New Zealand: New Zealand Journal of Science and Technology, v. 34B, p. 270–288.

Wellman, H.W., 1979, An uplift map for the South Island of New Zealand, and a model for uplift of the Southern Alps: in, Walcott, R.I., and Cresswell, M.M., editors, The Origin of the Southern Alps; Bulletin 18 of the Royal Society of New Zealand, p. 13–20.

Wellman, H.W., and Willett, R.W., 1942. The geology of the West Coast from Abut Head to Milford Sound: Transactions of the Royal Society of New Zealand v. 71, p. 282–306.

Wentworth, C.M., 1986, Maps of debris-flow features evident after the storms of December 1955 and January 1982, Montara Mountain area, California: U.S. Geological Survey Open-File Report, 2 sheets.

Wentworth, C.M., and Zoback, M.D., 1989, The style of Late Cenozoic deformation at the eastern front of the California Coast Ranges: Tectonics, v. 8, p. 237–246.

Wentworth, C.M., and Zoback, M.D., 1990, Structure of the Coalinga area and thrust origin of the earthquake, in Rymer, M. J. and W. L. Ellsworth, editors, The Coalinga, California, Earthquake of May 2, 1983: U.S. Geological Survey Professional Paper 1487, p. 41–68.

Wentworth, C.M., Zoback, M.D., and Bartow, J.A., 1983, Thrust and reverse faults beneath the Kettleman Hills anticlinal trend, Coalinga earthquake region, California, inferred from deep seismic-reflection data: Eos, Transactions, American Geophysical Union, v. 64, p. 747.

Wernicke, B.P., Axen, G.J., and Snow, J.K., 1988, Basin and Range extensional tectonics at the latitude of Las Vegas, Nevada: Geological Society of America Bulletin, v. 100, p. 1738–1757.

Wernicke, B., Clayton, R., Ducea, M., Jones, C.H., Park, S., Ruppert, S., Saleeby, J., Snow, J.K., Squires, L., Fliedner, M., Jiracek, G., Keller, R., Klemperer, S., Luetgart, J., Malin, P., Miller, K., Mooney, W., Oliver, H., Phinney, R., 1996, Origin of high mountains in the continents: The southern Sierra Nevada: Science, v. 271, p. 190–193.

Wernicke, B., Davis, J.L., Bennett, R.A., Normandeau, J.E., Friedrich, A.M., and Niemi, N.A., 2004, Tectonic implications of a dense continuous GPS velocity field at Yucca Mountain, Nevada: Tectonics, v. 109, doi: 10.1029/2003JB002832.

Wesnousky, S.G., 1986, Earthquakes, Quaternary faults, and seismic hazard in California: Journal of Geophysical Research, v. 91, p. 12,587–12,631.

Wesnousky, S.G., Barron, A.D., Briggs, R.W., Caskey, J.S., Kumar, S., and Owen, L., 2005, Paleoseismic transect across the northern Great Basin: Journal of Geophysical Research, v. 110, B05408, doi: 10.1029/2004JB003283.

Whipple, K.X., 2001, Fluvial landscape response time,

How plausible is steady-state denudation?: American Journal of Science, v. 301, p. 313–325.

Whipple K.X, and Meade B.J., 2004, Controls on the strength of coupling among climate, erosion, and deformation in two-sided, frictional orogenic wedges at steady-state: Journal of Geophysical Research–Earth Surface, v. 109, doi: 10.1029/2003JF000019

Whipple, K.X., and Trayler, C.R. 1996, Tectonic control of fan size, the importance of spatially variable subsidence rates: Basin Research, v. 8, p. 351–366. doi:10.1046/j.1365-2117.1996.00129.

Whipple, K.X., Kirby, E., and Brocklehurst, S.H., 1999, Geomorphic limits to climate-induced increases in topographic relief: Nature, v. 401, p. 39–43, doi: 10.1038/43375.

Wieczorek, G.F., and Snyder, J.B., 1999, Rock falls from Glacier Point above Camp Curry, Yosemite National Park, California: U.S. Geological Survey Open-file Report 99-385, http://pubs.usgs.gov/of/1999/ofr-99-0385/.

Wieczorek, G.F., Morrissey, M.M., Iovine, G., and Godt, J., 1998, Rock-fall hazards in the Yosemite Valley: U.S. Geological Survey Open-file Report 98-467, 1:12,000, 7 p. http://pubs.usgs.gov/of/1998/ofr-98-0467/

Wieczorek, G.F., Eaton, S.L., Yanosky, T.M., and Turner E. J., 2006, Hurricane-induced landslide activity on an alluvial fan along Meadow Run, Shenandoah Valley, Virginia (eastern USA): Landslides v. 3, p. 95–106. doi: 10.1007/s10346-005-0029-5.

Willett, S.D., 1999, Orogeny and orography: the effects of erosion on the structure of mountain belts: Journal of Geophysical Research, v. 104, p. 28,957–28,981.

Willett, S.D., and Brandon, M.T., 2002, On steady states in mountain belts: Geology, v. 30, p. 175–178.

Willett, S.D., Slingerland, R. and Hovius, N., 2001, Uplift, shortening, and steady state topography in active mountain belts: American Journal of Science, v. 301, p. 455–485.

Willgoose, G.R., Bras, R.L., and Rodriguez-Iturbe, I., 1991, A physically based coupled network growth and hillslope evolution model: Water Resources Research, v. 27, p. 1671–1684.

Wills, C.J., and Borchardt, G., 1993, Holocene slip rate and earthquake recurrence on the Honey Lake fault zone, northeastern California: Geology, v. 21, p. 853–856.

Wilson, J.T., 1965, A new class of faults and their bearing on continental drift: Nature, v. 207, no. 4995, p. 343–347.

Wilson, L., 1968, Morphogenetic classification, in, Fairbridge, R.W. editor, Encyclopedia of Geomorphology: Reinhold Book Corp., New York, p. 717–728.

Wilson, R.C., 1997, Normalizing rainfall/debris-flow thresholds along the U.S. Pacific coast for long-term variations in precipitation climate: in, Chen, C-L., editor, Proceedings, First International Conference on Debris-Flow Hazards Mitigation, Hydraulics Division, American Society of Civil Engineers, August 7–9, 1997, San Francisco, California, p. 32–43.

Wilson, R.C., and Jayko, A.S., 1997, Preliminary maps showing rainfall thresholds for debris-flow activity, San Francisco Bay Region, California: U.S. Geological Survey Open-File Report 97-745F.

Winograd, I.J., Szabo, B.J., Coplen, T.B., and Riggs, A.C., 1988, A 250,000 year climatic record from Great Basin vein calcite—Implications for Milankovitch theory: Science, v. 242, p. 1275–1280.

Winograd, I.J., Coplen, T.B., Landwehr, , J.M., Riggs, A.C., Ludwig, K.R., Szabo, B.J., and Ravesz, K.M., 1992, Continuous 500,000-year climate record from vein calcite in Devils Hole, Nevada: Science, v. 258, p. 255–260.

Wolkowinsky, A.J., and Granger, D.E., 2004, Early Pleistocene incision of the San Juan River, Utah, dated with ^{26}Al and ^{10}Be: Geology, v. 32, p. 749–752, doi: 10.1130/G20541.1.

Wolman, M.G., and Miller, J.P., 1960, Magnitude and frequency of forces in geomorphic processes: Journal of Geology, v. 68, p. 54–74.

Wright, J.D., 1998, Global climate change in marine isotope records: in, Sowers, J.M., Noller, J.S., and Lettis, W.R., editors, Dating and Earthquakes, Review of Quaternary Geochronology and its Applications to Paleoseismology, U. S. Nuclear Regulatory Commission Report, NUREG/CR 5562, p. 2-671–682.

Wright, L.A., Otton, J. K., and Troxel, B.W., 1974, Turtleback surfaces of Death Valley viewed as phenomena of extensional tectonics: Geology, v. 2, p. 53–54.

Yamanoi, T., 1979, The development of the Yokkamachi landslide area, Koide-Machi, Niigata Prefecture: Journal of the Japan Society of Landslides, v. 16, p. 29–36.

Yeats, R., Sieh, K., and Allen, C., 1997, The Geology of Earthquakes: Oxford University Press, New York, 568 p.

Yerkes, R.F., 1990, Tectonic setting: in, Rymer, M.J., and Ellsworth, W.L., editors, The Coalinga, California Earthquake of May 2, 1983; U.S. Geological Survey Professional Paper 1487, p. 13–22.

Yetton, M.D., 1998, Progress in understanding the paleoseismicity of the central and northern Alpine fault, Westland, New Zealand: New Zealand Journal of Geology and Geophysics, v. 41, p. 475–483.

Yetton, M.D., and Nobes, D.C., 1998, Recent vertical offset and near-surface structure of the Alpine Fault in Westland, New Zealand, from ground penetrating radar profiling: New Zealand Journal of Geology and Geophysics, v. 41, p. 485–492.

Yokoyama, Y., Lambeck, K., DeDeckker, P., Johnston, P., and Fifield, K., 2000, Timing of the Last Glacial Maximum from observed sea-level minima: Nature, v. 406, p. 713–716.

Young, A., 1972, Slopes: Oliver and Boyd, Edinburgh/London, 288 p.

Young, D.J., 1964, Stratigraphy and petrography of north-east Otago loess: New Zealand Journal of Geology and Geophysics, v. 7, p. 839–863.

Youngson, J.H., Bennett, E.R., Jackson, J.A., Norris, R.J., Raisbeck, G.M., and Yiou, F., 2005, "Sarsen Stones" at German Hill, Central Otago, New Zealand, and their potential for in situ cosmogenic isotope dating of landscape evolution: Journal of Geology, v. 113, p. 341–354.

Zandt, G., 2003, The southern Sierra Nevada drip and the mantle wind direction beneath the southwestern United States: International Geological Review, v. 45, p. 213–223.

Zandt, G., and Furlong K.P., 1982, Evolution and thickness of the lithosphere beneath coastal California: Geology, v. 10, p. 376–381.

Zeitler, P.K., Meltzer, A.S., Koons, P.O., Craw, D., Hallet, B., Chamberlain, C.P., Kidd, W.S.F., Park, S.K., and Seeber, L., 2001, Erosion, Himalayan geodynamics, and the geomorphology of metamorphism: GSA Today, v. 11, no. 1, p. 4–9.

Zellmer, J.T., 1980, Recent deformation in the Saline Valley region, Inyo County, California: Ph.D thesis, University of Nevada, Reno, 224 p.

Zepeda, R.L., 1993, Active tectonics and soil chronology of Wheeler Ridge, southern San Joaquin Valley, California, Ph.D. dissertation, University of California, Santa Barbara, 180 p.

Zepeda, R.L., Keller, E.A., and Rockwell, T.K., 1986, Rates of active tectonics at Wheeler Ridge, southern San Joaquin Valley, California: Geological Society of America Abstracts with Programs, v. 18, p. 202.

Zhang, P., Ellis, M., Slemmons, D.B., and Mao, F., 1990, Right-lateral displacements and the Holocene slip rate associated with prehistoric earthquakes along the Southern Panamint Valley fault zone: Implications for Southern Basin and Range tectonics and coastal California deformation: Journal of Geophysical Research, v. 95, no. B4, p. 4857–4872.

Zhang, P.Z., Shen, Z., Wang, M., Gan, W., Bürgmann, R., Molnar, P., Wang, Qi, Niu, Z., Sun, J., Wu, J., Hanrong, S., Xinzhao, Y., 2004, Continuous deformation of the Tibetan Plateau from global positioning system data: Geology, v. 32, p. 809–812.

Index

Where feasible, this index organizes multiple entries under topics that organize subjects and themes of this book. Examples include "base level", "concepts", "earthquakes", "models", "plate tectonics", "process-response models" and "reaches".

Semibold italics page numbers (261) refer to illustrations. **Bold** page numbers (236) refer to tables. Use of – (such as 105–11) flags a continuous discussion of a topic